GENERAL AQUACULTURE, FISHERIES & FISH SCIENCES

GENERAL AQUACULTURE, FISHERIES & FISH SCIENCES

Editor

Sharlene Albinson

scitus
academics

General Aquaculture, Fisheries & Fish Sciences
Edited by **Sharlene Albinson**

This Book is Printed on Acid-Free Paper. ♾

ISBN: 9781681179544
Library of Congress Control Number: 2017949099

SCITUS Academics LLC,
616, Corporate Way, Suite 2, 4766,
Valley Cottage, NY 10989

www.scitusacademics.com

Notice

Printed in United States of America.

Preface

Fish is an excellent source of animal protein and a wide range of essential nutrients and contributes significantly to food security. Aquaculture is currently responsible for one third of the fish consumed globally. When managed responsibly, aquaculture has the potential to help take some of the pressure off of capture fisheries and to provide employment and revenue for coastal communities. However, as aquaculture production continues to grow, so do concerns over aquaculture's impacts on the environment and wild fish species. In view of the large number and diversity of species cultivated, it is perhaps not surprising that aquaculture has emerged into the world's fastest growing food production system. Fisheries and Aquaculture are at the crux of development in many developing countries having a sea coast. Fisheries and aquaculture, directly or indirectly, play an essential role in the livelihoods of millions of people around the world, from the small-scale inland fishers who harvest fish from lakes and rivers to the men and women who work in large processing plants. Taking family members and other dependents into account, almost 540 million people, or nearly 8 percent of the world population, rely on the sector for their livelihoods.

This book 'General Aquaculture, Fisheries & Fish Sciences' covers research studies and reviews within the whole field of fisheries and Aquaculture Sciences, and it will continue to provide information on the latest trends and developments in this ever-expanding subject. It covers descriptions of aquatic plants and animals and feeding to in-depth coverage of economics, marketing, management and diseases of aquatic animals and plants. It covers basic and applied research articles in a broad range of subject areas relevant to fisheries science in such important topics as fisheries, biology, aquaculture, environment, chemistry and biochemistry, and food science and technology.

Table of Contents

CHAPTER 1

Development of Freshwater Native Species with Aquacultural Potential

Fernando Garcia-Trejo[1], Silvia Hurtado-Gonzalez[1], Genaro M. Soto-Zarazua[1] and P. J Gutierrez-Yurrita[2]

[1] División de Estudios de Posgrado Facultad de Ingeniería, Universidad Autónoma de Querétaro, C.U. Cerro de las Campanas, Qro, México

[2] Centro Interdisciplinario de Investigaciones y Estudios sobre Medio Ambiente y Desarrollo (CIIEMAD), Instituto Politécnico Nacional, México

1. INTRODUCTION

Aquaculture is an activity practiced by Chinese people since 2800 B.C. The first culture species were the carp (*Cyprinus carpio*) and mullet (*Mugil cephalus*), the first documented knowledge about fish culture in captivity belong to the carp. This activity has been supporting human demands for fish products for centuries and now is an important worldwide industry. Over the years this practice has become more technical with the objective to make the work easily but mainly in order to increase production. Nowadays this activity has grown to an entire industry that handles both supplements as the product itself. Global production from aquaculture now supplies one third of seafood consumed worldwide. With this massive increase in world production the current aquaculture industry is one of the fastest growing sectors in food production [1]. However, this rapid growth in the aquaculture industry has generated ecological damage due to a huge extractive use of water, land, and feeds. Besides, is important to consider that these requirements are associated to another impacts such as; polluting, salinization of soils, nutrient-loading, clearing of natural habitat, overexploit of ground water reserves, introduction and transmission of diseases [2]. At this respect Bailey [3] establish a new term "blue revolution", which describes the expansion of fish-farming in tropical regions, according to this idea aquaculture must provide huge quantities of fish and help to solving problems of world food security and alleviating poverty. However, production increase in aquaculture demands feeds, energy for the cultured species and almost always is obtained through catch, so the fastest growing sectors in food production threatens its ability to continue to provide increasing yields in a sustainable manner, and concerns with the resulting from fish-farming have led to calls for the "greening of the blue revolution" [4].

The situation in México is not an exception; the expansion of aquaculture has been accompanied by degradation of the natural environment, especially on marine aquaculture [5]. As it happens in the world directly impacts of fisheries and aquaculture are:

- Introduction of nonnative fish species:
- Introduction of exogenous parasites:
- Nutrient pollution:
- Habitat modification:
- Overcollection of wild seed stock:
- Changes of food webs
- Increase of interspecific competition:

It is clear that current food productions techniques in aquaculture are good just under financial point of view but always leave aside the environment aspects. The relationship between aquaculture and environment is complex specifically the biodiversity topic. Many examples of positive and negative impacts have been documented, however until now there is no solution which allows the development of a relationship between food production and the environment. This solution must be adopting a new paradigm based in ecological concepts of extreme resource efficiency and the closing of nutrient and waste cycles, resource-use optimization [6]. As can be seen, this is not an easy task since it requires the creation of multidisciplinary teams which can see the problem holistically and try to give a solution that benefits all parties involved in the process. The efforts and the perception of the environment are different between countries, but ultimately the problem to be addressed holistically. However, studies on fish typically focus on species that currently have commercial value, causing species that lack such market value to be ignored. This is the case of several freshwater native species, which can be founded in central and South America. Some attempts to cultivate native species have occurred mainly in areas or rural communities, where in addition to enhance the conservation of species protein contributes to the diet of the community [7]. One of the most interesting case studies in Mexico is growing "white fish" (*Chirostoma estor*) with the aim repopulate some areas where the introduction of their populations has been declined [8]. Most documented is that of the native Central American cichlid (*Cichlasoma urophthalmus*), of which there have been many studies to support its culture [9]. In southern region some attempts to grow some silversides in Argentina (*Odontestes Basilichthys*), some Characidae family members have been grown in Brazil and more recently three species of carnivorous cichlids aquarium purposes. In Peru it has favored the cultivation of called piracucu (*Arapaima gigas*), one of the largest fish of fresh water. Possibly the Cichlidae family members are those that show the greatest potential for cultivation [10, 11].

The main purpose of this chapter is to show the experience of three studies with native species; one refers to a small native species located in the state of Querétaro and with a great ecological importance, *Girardinichthys multiradiatus* [12]. The study of this fish focused on the description of its habitat throughout a hydrologic cycle in which ecophysiological responses were determined in order to stablish guidelines for its management and to preserve its population. In this work population structure and dynamics were getting and trophic and ecophysiological responses to fluctuations in environmental factors were also identified in order to have the possibility of laboratory reproduction and growth. On the other hand, native mojarra *Herinchthys cyanogutattus* founded on the basin Pánuco river. In this case the purpose was to evaluate its useful in fishery and later in the aquaculture. The work consisted of two stages: First, the characterization of their environment in order to locate stable populations of the mojarra and to characterize ecologically its habitat. Second, the mojarra was moved to the laboratory to try different forms of acclimatization for its future use giving them tried food. Once acclimated, the stock was use to carried out density studies of individuals for culture (capacity of load), as well as of ideal thermal for its production. Finally, *Procambarus digueti* which faces severe ecological problems (over fishing, no control of heights neither of sexes, there is not articulated extraction methods neither fishing seasons and restrains), since they are captured as food and as curative remedies from pre-Hispanic eras. To this situation the strong environmental pressure is added by the disturbance of their habitat, what has carried a decrease in numerical abundance. The objective of this work was determined the optimum cultivation conditions with respect to the load capacity and diet in the growth of *P. digueti* in intensive production systems. The controlled production of this specie will reduce the fishing pressure and it will be able to serve to repopulate the sites where may have been decline the natural population.

To successfully achieve the cultivation of a native species, compared to the technological advantages offered by more exotic trading requires knowledge of the biology, ecology and aquaculture potential (ability to live at high densities, accept food encapsulation, and withstand high environmental variations) of each species. If aquaculture potential studies are performed with ecophysiological and bioenergetics approach may be developed predictive models of how to develop a population under different environmental factors, and even develop experimentally testable hypotheses [13].

2. THE EXPERIENCE WITH GIRARDINICHTHYS MULTIRADIATUS

Studies on the biological aspects of fish typically focus on species that currently have commercial value, causing species that lack such market value to be ignored. This is the case of several freshwater fish, specifically of several members of the Goodeidae family. This is a diversified and small family of cyprinodontoid fish, confined to the central plateau of Mexico where its dispersion center lies in the well-isolated Lerma basin. Four species of the Godeidae family have been reported in Querétaro: three species are distributed widely in the Lerma basin (*Goodea atripinis, Xenotoca variata, Goodea gracilis*), and one species (*Girardinichthys multiradiatus*) can only be found in one body of water in the municipality of Amealco [14]. Scientific knowledge about it focuses on sexual dimorphism, peculiar courtship rituals, and viviparity [15], taxonomic aspects [16, 17], ethology [18], biology [19], and trophic ecology [20].

2.1. Environmental conditions

G. multiradiatus was founded on San Martín Dam, located at 60 km south of Queretaro city, near to Amealco municipality (100° 09’ 43” W; 20° 15’ 02” N), at 2600 meters above sea level. The climate is subhumid with summer rains (Cw1) with an average temperature of 15.1°C, the months of May and June are those with the highest temperatures. The average annual rainfall is 659.5 mm, occurring mainly during the summer [21]. The main contribution to the dam water is from rain. Sampling was made over a full hydrological cycle (one year, beginning in February) in which the *G. multiradiatus* population was monitored once every two months, at the same time physical factors were measure (T°, pH, dissolved O_2, turbidity, depth). The physicochemical parameters of water showed stable behavior during the studied hydrological cycle (pH = 7 to 9; dissolved oxygen = 6.5 to 7 ppm). On the other hand, the temperature showed significant variation, with the highest temperatures (20-25°C) recorded between April and August, with the lowest (10-18°C) recorded between October to February. Also, due to seasonal differences in water usage, the water level of the dam was low from April to August and high from October to March (Fig. 1).

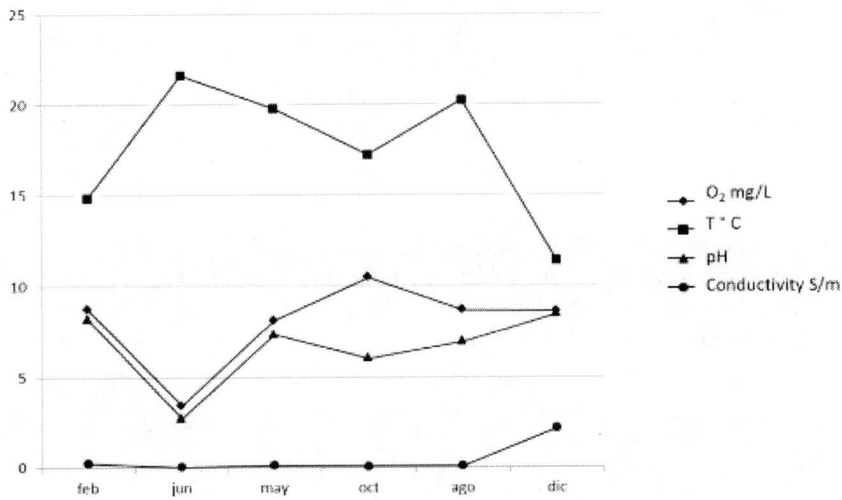

Figure 1. Environmental parameters during a hydrological cycle in San Martín Dam.

2.2. Population ecology

According to the Cassie method, the population of *G. multiradiatus* consists of 12 classes, ranging from 8 to 48 mm standard length. Figure 2 shows the general structure of the mexcalpique population of San Martín. The numbers in parentheses indicate the percentage of each size class of the total population obtained through a year. Two of these size classes were found only in laboratory studies due to their small sizes; these sizes were smaller than what the nets in situ could catch.

Figure 3 shows the bimonthly structure of the population of *G. multiradiatus* in San Martín and the seasonal pattern of population growth, demonstrated in Von Bertalanffy equation with $L\infty = 47$; $K = 0.8870$ and, $t0 = -0.2103$. The variations throughout the year are present in both, in the structure of the population and its growth rate, with the shorter pattern cohorts having a higher growth rate and no increase in length during the period of February to June.

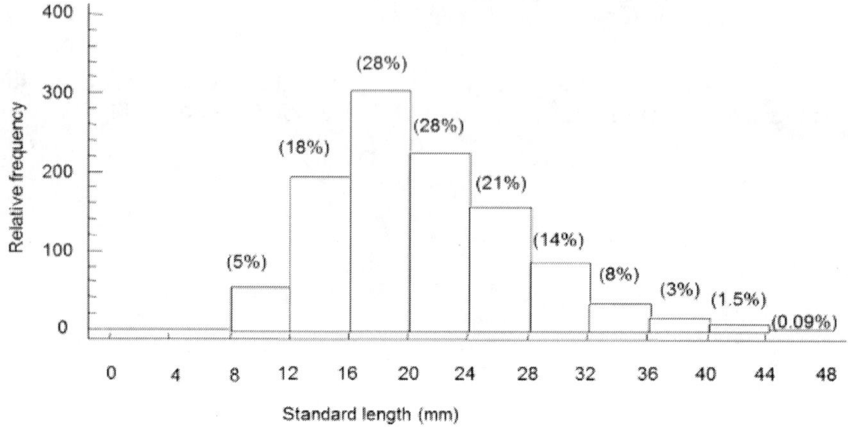

Figure 2. General structure of the population of San Martín mexcalpique. The numbers in parentheses indicate the percentage of each size class of the total population obtained through a year.

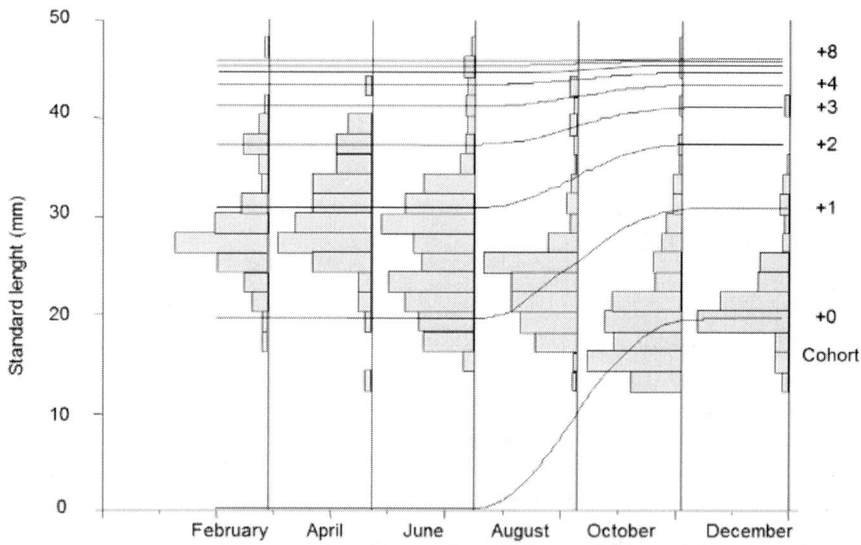

Figure 3. Bimonthly structure of the population of G. multiradiatus in San Martin and the curves of growth with the von Bertalanffy model for highly seasonal cycles. Upon reaching the asymptotic curve determines the final class of each age cohort.

2.3. Feeding habits

To assess the daily feeding activity of the *G. multiraditus* in San Martín Dam samples were collected with spoon nets every four hours during a period of 24 hours (10:00, 14:00, 18:00, 22:00, 02:00 and, 06:00 hours). These catches allowed determining the feeding ecology of the species (feeding time, type of diet at different times of day, food components). The relative density in activity was measured using the catch per unit effort method (CPUE) based on the number of individuals caught by dragging. The fish were fixed in 70% alcohol and then transported to the laboratory (Nielsen et al., 1983). From 1022 stomachs analyzed, (1022), 18 food components were identified. The most abundant component found were insects (47%), especially the Diptera order, followed by detritus (24.0%) and Cladocera (17.5%), with the remaining components accounting for 10.7%. Unusual food components (less than 10% of the total), were only found at a specific times of the hydrological cycle, Table 1. The benthic review showed 20 components and trophic index was calculated indicating that *G. multiradiatus* is a polytrophic specie (H ≠ 0). The maximum numbers of categories found in their stomachs were eight while minimum was two. Circadian sampling showed that this specie is polytrophic with two daily feeding periods (5:00 to 8:00 and 14:00 to 18:00 hours).

Table 1. Occurrence of food components (N = number sampled, n = number of organisms that have the category, % occurrence rate) by sex.

Category	Youth N = 654		Males N = 195		Females N = 173	
	n	%	n	%	n	%
Detritus	230	16.08	141	37.03	127	38.49
Diptera	517	55.31	146	34.32	142	34.41
Cladocera	231	21.5	51	13.73	38	13.73
Copepoda	53	3.78	25	2.26	31	2.26
Animal remains	41	1.27	29	2.56	47	2.56
Amphipoda	20	0.67	27	3.1	34	3.01
Hemiptera	16	0.52	27	2.96	38	2.96
Odonata	13	0.38	21	3.21	28	3.21
Himenoptera	1	0.06	4	0.11	12	0.66
Coleoptera	0	0	2	0.04	6	0.6
Crustacea	1	0.004	4	0.41	4	0.14
Ephemeroptera	3	0.06	0	0	5	0.38
Gasteropoda	3	0.05	1	0.02	6	0.17
Plecoptera	3	0.03	2	0.11	0	0
Thrichoptera	0	0	1	0.1	0	0
Homoptera	0	0	1	0.05	0	0

2.4. Bioenergetic

To quantify the aerobic metabolism and nitrogen excretion, animals were placed in a semi closed system with (0.5L) respirometric chambers, using a method which assumes that the reduction of oxygen and the increase of nitrogen in the chamber depend on the weight of the animal, the volume of water, the period of time in which no water circulated in the chamber and the ambient temperature [22]. After each cycle, sample was oven-dried in an electric oven between 70-80°C until the samples had constant weight. From each composite sample 2 g were measured and taken as analytical sample; the samples were digested with concentrated nitric acid. The determination of the percentage proximate composition was chemically analyzed according to the method of analysis described by the Association of Official Analytical Chemist [23]. While other 2 g sample tissue were combusted in a Parr bomb calorimeter to obtain oxycalorific measurements. The flow of energy that was used to determine the energy efficiency and assimilation is show in the next equation:

$$C = P + G + ER + EU + F$$

Where; C is the total energy content of food consumed, P and G are the energy equivalents of somatic and gonadal growth respectively, ER is the energy utilized in respiration, EU is the energy lost as nitrogenous and other waste compounds excreted in the urine, and F is the unabsorbed energy voided with the faeces (Bolduc et al., 2002; Bradshaw, 2003). All variables expressed in calories per gram of dry weight (cal/g).The ratio used to transform measured aerobic metabolism into calories was the standard oxycaloric coefficient for fish which mainly excrete $N\text{-}NH_4$ (Qox = 3.20 cal mg-1 O_2). Nitrogen excretion was estimated using literary references, taking into consideration the type of fish, size, feeding habits, and physiological status [24].

Physiological experiments showed higher energy expenditures in August with values of 2500 cal/g and minimal values for December with 200 cal/g. The increased energy expenditure was found in the early hours of the day (daylight hours) and then declined, reaching minimum values at night, with the exception of October, which displayed an inverse pattern. Calorimetric analysis did not show statistically significant difference between the energy provided by sex ($p > 0.05$). Main food energy intake was 4.8 ± 0.3 Kcal g-1 of dry weight, with the total weight of the mexcalpique consisting of, on average, 85.49 ± 2.49% organic matter, and 14.50 ± 2.49% mineral matter. By replacing the caloric values in the energy balance equation, was determined that *G. multiradiatus* uses approximately 81% of the energy consumed in the production of tissue and gametes (P and G), respiration process spent 5.7% (ER) and the rest 13.3% is invested in maintenance (EU and F). Multivariate analysis of environmental factors on the metabolism, showed no significant differences, however the temperature showed the lowest value of significance ($p = 0.08$).

3. HERINCHTHYS CYANOGUTATTUS

Texas Cichlids were formerly given the scientific name of *Cichlasoma cyanoguttatum*, but are now known by the name *Herichthys cyanoguttatus*. The genus *Herichthys* has been through several changes, and currently consists of nine species, native to lakes and rivers in south Texas and northern Mexico, making them the most northern naturally occurring species of cichlid in the world. It's the only native cichlid in the US and amongst the first cichlids imported to Europe, having first been imported in 1912. This species has also been introduced into areas they are not indigenous to, sometimes on purpose, but often by aquarium owners desperate to divest themselves of a fish the can no longer take care of. The areas of non-indigenous populations range from northern Texas to Florida, where it has become a popular game fish. This is due to having a tasty flavor similar to that of their distant relative, Tilapia.

In Mexico this species is called "Mojarra del norte" and It could be distinguished by a coupled of dark spots and a tiny blue circles on its sides. Adults show a olivaceus iridescent spots when viewed in the sun, there are also lines of the same color on the head, body and fins. During reproduction it is possible to see white region in the front part and a dark in the back especially in females, while males develop a prominent hump.

3.1. Biology

This fish could live in a wide range of temperature, between 5° and 30°C., [25, 26]. Trophic spectrum show variations between each population according to the region but in general is consider an omnivorous fish [27]. Many studies have described aspects of reproduction; the most relevant aspect is the monogamous behavior, when a male select a coupled it becomes aggressive and territorial [28].

3.2. Environmental conditions

The first step in this research was to looking-for a population in order to do the ecological description of its habitat. H. *cyanoguttatus* was found in several places around Queretaro State, but only one was selected due to accessibility. This place is called Taxhido river and is located at 70km E, from the Queretaro capital 20° 35' 18"O and 99° 40' 47"N, the climate is subhumid with summer rains (Cw1) with an average temperature of 15.1°C, the months of May and June are those with the highest temperatures. The average annual rainfall is 659.5 mm, occurring mainly during the summer and the main contribution to the dam water is from a spring. Sampling was made over a full hydrological cycle (one year, beginning in February) in which the *H. cyanoguttatus* population was monitored once every two months, at the same time physical factors were measure (T°, pH, dissolved O_2, turbidity, depth). The physicochemical parameters of water showed stable behavior during the studied hydrological cycle (pH = 7.4 to 8.2; dissolved oxygen = 4.5 to 8.3 ppm). On the other hand, the temperature was constant between 29 and 31°C., Fig. 4.

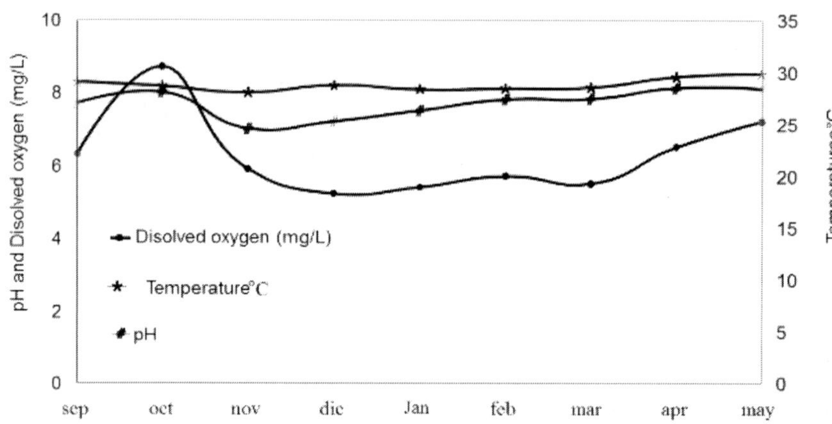

Figure 4. Environmental parameters tendency during an hydrologic cycle in Taxhido river.

3.3. Laboratory studies

Organisms were collected in Taxhido and then transport to laboratory and maintained for acclimation during a month. After this time a group was used to optimal temperature determination and other group for an optimal density experiment.

3.3.1. Optimal temperature

The fish were placed in 15 rectangular glass tanks distributed using a Latin square scheme in order to avoid spatial effects. The tanks´ dimensions were of 0.4.5 m depth, 0.8 m wide and 0.3 m long, with a water storage capacity of 100 L. Five triplicated treatments with 15 organisms were applied using environmental temperature as a medium value, so the treatments were; 24, 26, 28, 30, and 32°C. The handling of tanks involves, the feces removal and partial water change (30%) weekly. The fish were feeding with a commercial diet for Tilapia (Api-Tilapia 1, maltaCleyton® with 50% protein, 12% lipid, 13% ash, 3% fiber, 12 moisture) throughout the experiment. Feeding frequency was adjusted to three provisions offered three times daily starting at 8 AM, 1 PM and 6 PM. The results show that 28°C., is the best temperature for *H. cyanoguttatus*, fig. 5.

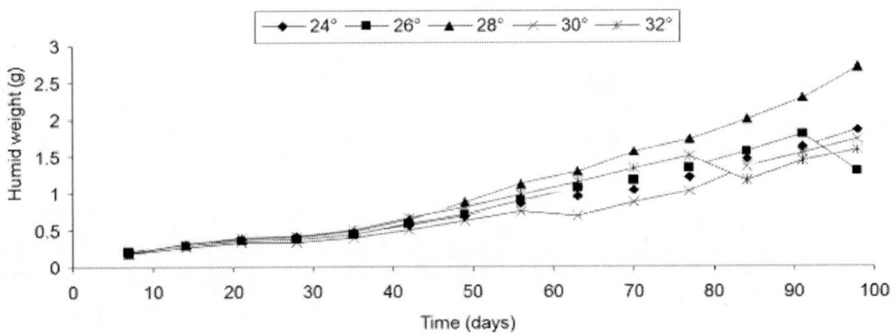

Figure 5. Humid weight behavior for the different temperature treatments

3.3.2. Optimal density

Once the temperature was determined a similar experiment was carry out, but in this case the variable was the density. Five densities were probed, T1= 5, T2=10, T3=15, T4=20 ind/per aquarium. It is important to consider that control temperature was implemented in each aquarium in order to avoid an effect for spatial distribution. The results show that 15 individuals is the best density for *H. cyanoguttatus*, table 2.

3.3.3. Bioenergetics

Aerobic metabolism were determine in natural conditions with a semi closed system with (0.5L) respirometric chambers, using a method which assumes that the reduction of oxygen and the increase of nitrogen in the chamber depend on the weight of the animal, the volume of water, the period of time in which no water circulated in the chamber and the ambient temperature [22]. After each cycle, sample was oven-dried in an electric oven between 70-80°C until the samples had constant weight. From each composite sample 2 g were measured and taken as analytical sample; the samples were digested with concentrated nitric acid. The determination of the percentage proximate composition was chemically analyzed according to the method of analysis described by the Association of Official Analytical Chemist [23].

Table 2. Performance parameters

Performance parameters	T1= 5 ind	T2=10 ind	T3=15 ind	T4=20 ind
Initial number (n)	5	10	14	19
Final average number (n)	1	6	9	7
Survival rate (%)	26.66	56.66	61.7	38.88
Initial Total weight (g)	2.16	2.84	4.28	16.92
Initial individual average weight (g)	0.43	0.28	0.29	0.91
Final Total weight (g)	2.36	9.26	14.05	23.81
Final individual average weight (g)	0.59	1.09	1.49	2.05
Weight gain (%)	95.16	293.44	400.59	94.86

Physiological experiments showed higher oxygen consume at 15:00 hrs, while a minimum consumption was founded during the morning, Fig. 6.

After three weeks in laboratory condition (acclimatization) oxygen consumption was measured to know if a metabolism was changed. The results show that higher consumption was found at 15:00 and minimal during the morning so the fish did not show modification in oxygen consumption.

Finally, oxygen consumption was measured for each of the temperature treatments in order to know in which an alteration occurs. The results show that maximum values for oxygen were founded at T3= 30° minimal at T2 =26° treatments.

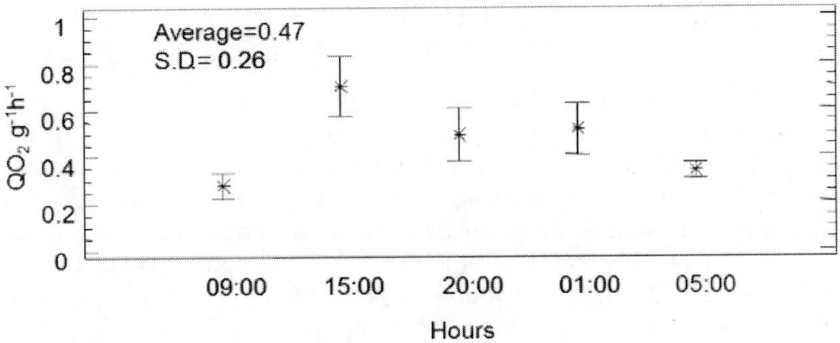

Figure 6. Oxygen consumption in natural condition for H. cyanogutattus

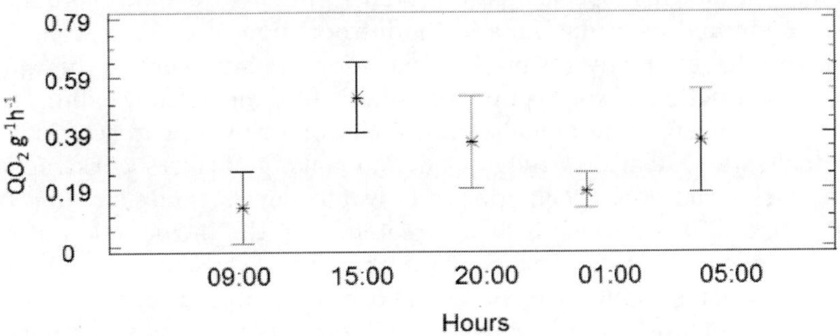

FIGURE 7. Oxygen consumption under laboratory condition for H. cyanogutattus

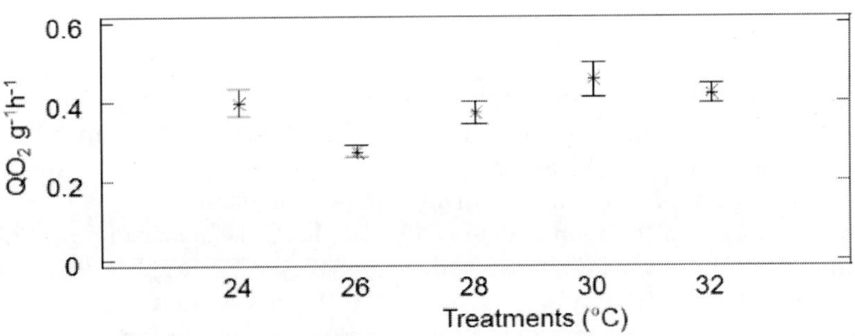

Figure 8. Oxygen consumption for each temperature treatment at the end of the experiment.

4. PROCAMBARUS DIGUETTI

The cambarids, known in Mexico as acociles (in Náhuatl), makaxil (in Mayan), chapos (in Purépecha), or freshwater crayfish, according to the region, are well known, and were regularly consumed by the Aztecs and other ethnic groups based around the Valle de Mexico already during prehispanic times [29]. *Procambarus digueti* is endemic to the Mexican Central-Occidental Plateau and is found only in certain parts of the Duero and Camécuaro rivers. The Duero River is 10 km long, whist the river Camécuaro is just 2 km long. The two rivers meet at Las Adjuntas. Furthermore, this species is known to occur in Lake Chapala, though samples were collected 40 to 50 years, so is believed to have naturally disappeared from some locations, though these local extinctions may in fact be unnatural This species is threatened by over-exploitation, habitat destruction and degradation, and the introduction of exotic species.These processes have already resulted in reduction in abundance of this species, and the extinction of some populations. Much of the natural habitat of this species has been altered by chemical pollution or by human activities such as canalization, clearing, dredging and embanking of rivers, construction of reservoirs, and the regulation of water levels and stream flows. Furthermore, this species is also threatened by the introduction of exotic crayfish such as *Procambarus clarkii* and *Cherax* species from Australia, which competes for food resources and refuges, and also alters the total production of the native ecosystems (Gutiérrez-Yurrita and Latournerié-Cervera 1999). In addition to over explosion of population, a cultural pressure exists, purepecha people attributes curative properties at *Procambarus* [30].

4.1. Study area

The first part of the experiment was carry out in the national park of Camecuaro lake, which is found in Michoacan State over the municipality called Tangancícuaro, 19°54′10″N;102°12′20″O, at 1,700 meters above sea level. The national park has a spring called Camecuaro lake which is the main contribution for Camecuaro river, the depth is between 1.5 a 1.8 m; temperature 17.7 y 21°C; dissolved oxygen 7.3 y 7.5 mg/ L; visibility of 100%; hardness 138.8 y 145.5 mg/L $CaCO_3$. A handled extraction was practice in order to have a desirable sample size for the laboratory work. The organisms were transported in plastic bags with a supplemented oxygen and ice to avoid the over heat.

4.2. Laboratory studies

Acclimatization of organisms was carrying out during a month in glass aquariums considering environmental conditions. The principal problems for the maintenance of the organism were the feed and the density, so in this case these two experiments were carry out.

4.2.1. Establishing the diet

The organism were placed in 15 rectangular glass tanks distributed using a Latin square scheme in order to avoid spatial effects. The tanks´ dimensions were of 0.45 m depth, 0.8 m wide and 0.3 m long, with a water storage capacity of 100 L. Three commercial diets were probed T1= Trucha initial; T2= Camaronina, and T3=Tilapia initical, table 3. The handling of tanks involves, the feces removal and partial water change (30%) weekly. Feeding frequency was adjusted to two provisions offered three times daily starting at 8 AM and 6 PM.

Table 3. Proximal Chemical Composition of the diets tested for growth

Diet Compounds	**Trucha Iniciador (50:15)**	**Camaronina (35:8)**	**Tilapia Iniciador (32:4)**
Protein (%)	50	35	32
Humidity (%)	12	12	12
Grass (%)	15	8	4
Crude fiber (%)	4	5	10
Ash (%)	12	10	10
Calcium (%)	2	1.4	
Phosphorous (%)	1.2	0.9	
E. L. N.	7	30	

Multiple condition factor (K) was calculate for each treatment, with this factor is possible to know the relative health for organisms [31].

$$K=(10^2*W)/L^b$$

Where;

K= Multiple condition factor

W= Weight

L= Length

b= exponent from $W = KL^b$

The type of growth was determine for the Ricker equation [32];

$W=aL^b$

Where;

W=weight

a = intercept

b= slope

Aerobic metabolism were determine in natural conditions with a semi closed system with (0.5L) respirometric chambers, using a method which assumes that the reduction of oxygen and the increase of nitrogen in the chamber depend on the weight of the animal, the volume of water, the period of time in which no water circulated in the chamber and the ambient temperature [22]. After each cycle, sample was oven-dried in an electric oven between 70-80°C until the samples had constant weight. After three weeks in laboratory under the diets treatment, oxygen consumption was measured to know if a metabolism was changed. The results show that higher consumption was found at 15:00 and minimal during the morning so the fish did not show modification in oxygen consumption, Fig. 10.

At the end production of total biomass was estimated, the results show that T1 is the best feed for *Procambarus digueti*, and T3 is the diet with a minor biomass production.

4.2.2. Establish an optimal density

The organism were placed in 15 rectangular glass tanks distributed using a Latin square scheme in order to avoid spatial effects. The tanks´ dimensions were of 0.4.5 m depth, 0.8 m wide and 0.3 m long, with a water storage capacity of 100 L. Five densities were probed, T1= 5, T2=10, T3=15, T4=20 ind/per aquarium. The handling of tanks involves, the feces removal and partial water change (30%) weekly.

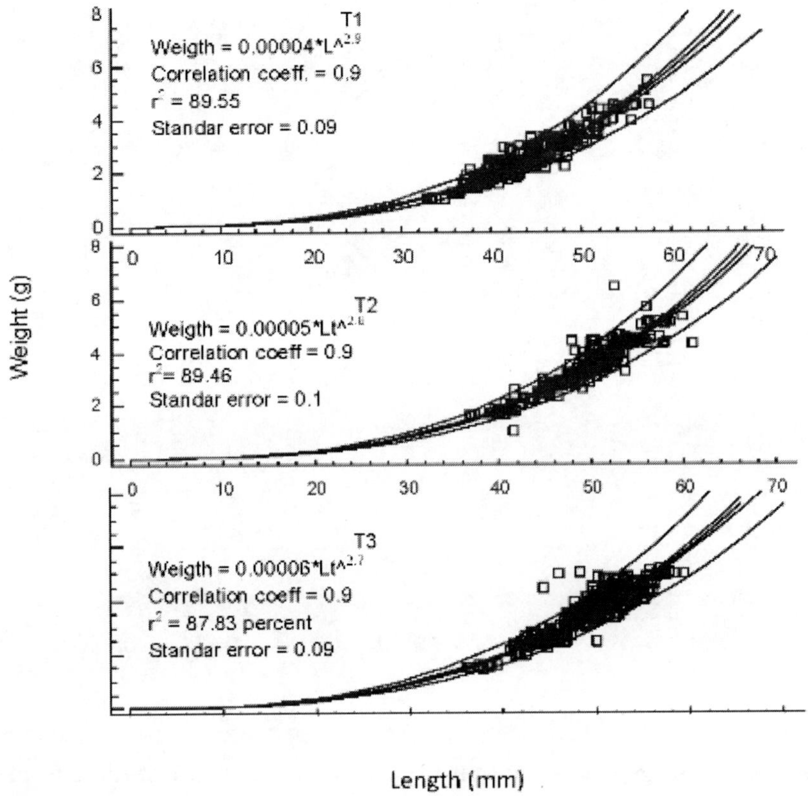

Figure 9. Regression for each one of treatments, values for the equation $W=KL^b$

The fish were feeding with a commercial diet for trucha (50% protein, 10% humidity, 12% lipid, 12% ash, 4% fiber, 12% moisture) throughout the experiment. Feeding frequency was adjusted to three provisions offered two times daily at 8 AM, and 6 PM. The results show that a density of 15 organisms is optimal for *Procambarus* growth, table 4.

4.2.3. Bioenergetics

The flow of energy that was used to determine the energy efficiency and assimilation is show in the next equation:

$$C = P + G + ER + EU + F$$

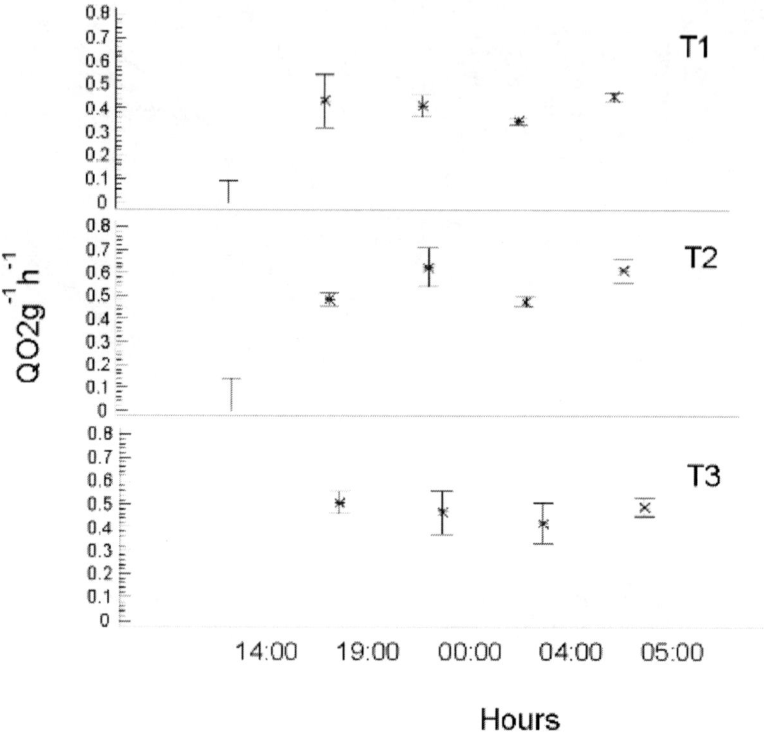

Figure 10. Oxygen consumption for each diet treatment at the end of the experiment

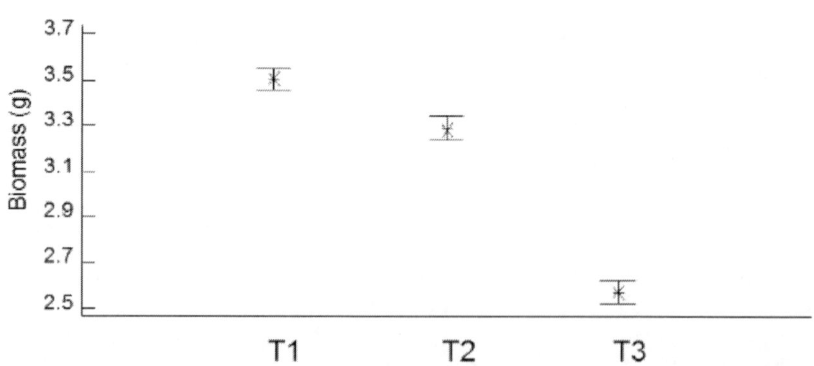

Figure 11. Biomass production at the end of experiment for each of the treatments

Table 4. Performance growth for the densities treatment in Procambarus digueti

Treatment	5	10	15	20
Initial number (n)	5	10	15	20
Final average number (n)	5	8	9	6
Survival rate (%)	100	80	60	30
Initial individual average weight (g)	4.8	5.23	5.53	5.18
Final individual average weight (g)	7.12	5.67	6.36	2.82
SGR (%/dia)	1.94	1.71	1.83	1.018
Weight gain (%)	612.9	467.2	536	182
Production	0.302	0.158	0.091	0.063

Table 5. Energetic balance for *Procambarus digueti*

Diet	C	P	R	U	F
	Consume cal/mg	Biomass cal/mg	Oxygen cal/mg	Nitrogen cal/mg	Feces g
Tilapia (32:4)	5423.44	3.664	1.309	0.450	0.014
Camaronina (34:8)	5545.03	3.845	1.175	0.525	0.009
Trucha (50:15)	5706.273	3.873	1.264	0.569	0.006

Where; C is the total energy content of food consumed, P and G are the energy equivalents of somatic and gonadal growth respectively, ER is the energy utilized in respiration, EU is the energy lost as nitrogenous and other waste compounds excreted (50% protein) is the best food for *Procambarus digueti.* The energy balance is show in the table 5.

5. CONCLUSION

Aquaculture has been supporting human demands for fish products for centuries and is an important industry worldwide. Global production from aquaculture has been increasing steadily, having more than doubled in the last decade; aquaculture now supplies one third of seafood consumed worldwide. With the massive increase in world aquaculture production in 1990s, the current aquaculture industry is one of the fastest growing sectors in world food production. However, the expansion of aquaculture has been accompanied by degradation of the natural environment, especially on marine aquaculture. Directly impacts of fisheries and aquaculture are habitat modification, collection of wild seedstock, changes of food webs, introduction of nonnative fish species and diseases that harm wild fish populations, and nutrient pollution. According to the FAO, major issues that need to be addressed are problems with access to proper technology and financial resources, together with environmental impacts and diseases. Another argues that further increases in aquaculture production will come mainly from further investment in biotechnology. The development of new strategies or technologies does not imply that the old one disappears; to the contrary the main idea is to use the experience and improve existing technology.

These three experiences and review of similar cases that have been developed in Mexico allow us to establish a general methodology in order to know the aquaculture potential for native species.

1) Knowledge.- The first step consist in to know and get the information above the specie. In the case of *G. multiradiatus* the knowledge practically doesn't exists so the research was oriented to the basic biology and ecology aspects. For example, with *G. multiradiatus* the main objective was to obtain reproduction but to reach these aspects it is required the maintenance under laboratory conditions and for this the knowing of food habits are essential. In the case of *H. cyanogutattus* the main problem was its taxonomic status, so the principal problem to obtain the basic knowledge is that some aspect were publish under the scientific name *Cichlasoma cyanoguttatum* and others aspects with the actual name *Herinchthys cyanogutattus.*

2) Environmental prospections.- It is necessary to know the basic physicochemical parameters (T°, pH, DO_2) in order to establish the strategy for transportation and laboratory maintenance. Field observations and ecological description is necessary in order to know the feeding habits, interspecific competence and the disposition of resources. With these parameters is possible to start the research.

3) Feed.- Under laboratory conditions feeding is the main problem in order to continue so the proofs needed is the acceptance of commercial food. The three species show an acceptance for commercial food but the problem here is the metabolism aspects such as assimilation and performance growth. This kind of problems could be solved with a bioenergetics approach.

4) Bioenergetics.- Ecophysiological basis of species should take into consideration the physiological characteristics and ecological role of the organism in question. Physiological analysis will reflect the conditions which affect population characteristics, such as population growth, intraspecific competition, and functional and numerical responses. These studies can do more than being simply descriptive, since they enable the development of scenarios that can be tested either through strictly controlled laboratory experiments or field experiments.

The new aquaculture research must be consider to add native species, this work is an effort to get the basic information in order to development of biotechnology and a link between the basic and applied science.

ACKNOWLEDGEMENTS

This work was partially finance by CONACYT.

REFERENCES

1. FAO, editor. Review of the State of the World Marine Fishery Resources2011; Rome: FAO.
2. Pillay TV. Aquaculture and the environment. second ed. Iowa: Blacwell publishing; 2004.
3. Bailey C. Blue revolution: the impact of technological innovation on Third World fisheries. The Rural Sociologist. 1985;5(4): 259-66.
4. Costa-Pierce B. Ecological Aquaculture: the evolution of the Blue Revolution second ed. Iowa: Blackwell Science; 2002. p. 383.
5. Saborio Coze A, Flores Nava A. Review of environmental impact assessment and monitoring of aquaculture in Latin America Environmental impact assessment and monitoring of aquaculture. Rome: FAO; 2009. p. 395–454.
6. García-Trejo F, Soto- Zarazúa GM. La nueva ecología, una perspectiva integradora hacia un paradigma de sustentabilidad. CIENCIA@UAQ. 2012;5(1).

7. Ross LG, Beveridge MC. Is a better strategy necessary for development of native species for aquaculture? A mexican case of study. Aquaculture research. 1995[29]:539-47.
8. Martínez-Palacios CA, Barriga-Tovar E, Taylor JF, Ríos-Durán G, Ross LG. Effect of temperature on growth and survival of Chirostoma estor estor, Jordan 1879, monitored using a simple video technique for remote measurement of length and mass of larval and juvenile fishes. Aquaculture. 2001(1-4):369-77.
9. Ross LG, Martinez-Palacios CA, Morales E. Developing native fish species for aquaculture: the interacting demands of biodiversity, sustainable aquaculture and livelihoods. Aquaculture Research. 2008[39]:675–83.
10. Bardach JE, Ryther JH, Mclarney WO. Acuacultura, crianza y cultivo de organismos marinos y de agua dulce. México: AGT 1986.
11. Martinez-Palacios CA, Ross LG. The reproductive biology and growth of the Central American cichlid Cichlasoma urophthalmus (Günther) Journal of Applied Ichthyology. 1992;8:99–109.
12. Garcia-Trejo F, Hurtado Gonzalez S, Soto Zarazua G, Alatorre-Jácome O, Rico-García E, Gutiérrez-Yurrita PJ. Ecophysiological responses to the effect of annual management on an endemic viviparous fish in central plateau of México Neotropical ichthyology 2013;11(1):117-23.
13. Gutiérrez-Yurrita PJ, Montes C. Environmental factors controlling the crayfish Procambarus clarkii activity in Doñana National Park temporary freshwater marsh, SW. Spain Comparative Biochemistry and Physiology, Part A 1998;120:713-21.
14. De la Vega-Salazar MY. Estado de conservación de los peces de la familia Goodeidae (Cyprinodontiformes) en la mesa central de México. Revista Biología Tropical. 2006;54(1):163-77.
15. Macias GC. Social behavior and operational sex ratios in the viviparous fish Girardinichthys multiradiatus. Copeia. 1994;4:919-25.
16. Álvarez del Villar J. Peces Mexicanos (Claves). In: Pesca. CNCd, editor. Instituto Nacional de Investigaciones en Biología Pesquera Mexico 1970. p. 166.
17. Miller RR, Fitzsimmons JM. Ameca splendens, a new genus and species of goodeid fish from Western Mexico, with remarks on the classification of the Goodeidae. Copeia. 1971;1:1–13.
18. Uyeno T, Miller R, Fitzimons J. Karyology of cyprinodontoid fishes of the Mexican family Goodeidae Copeia. 1983;2:497–510.
19. Macias-Garcia C, Burt de Perea T. Ultraviole-based female preferences in a viviparous fish Behavioral Ecology and Sociobiology 2002;52(1):1-6.
20. Trujillo-Jiménez P, Viveros EE. La ecología alimentaria del pez endémico Girardinichthys multiradiatus (Cyprinidontiformes: Goodeidae), en el Parque Nacional Lagunas de Zempoala, México. Revista de Biología Tropical 2006;54 1247-55.
21. García E. Modificaciones al Sistema de Clasificación Climática de Koppen. México: Enriqueta García; 1988.
22. Steffensen JF. Some errors in respirometry of aquatic breathers: how to avoid and correct for them Fish Physiology and Biochemistry. 1989;6:49-59.
23. AOAC. Official Methods of Analysis of AOAC International,. sixteen ed. Gaithersburg, MD, USA.: AOAC; 1998.
24. Klekowsky RZ, Duncan A. Physiological approach to ecological energetics. In: Grodzinzky H, Groves TDD, editors. Methods for ecological bioenergetics. Oxford: Blacwell Scientific Publication; 1975. p. 227-61.
25. Hubbs CT, Lucier GP, Garrett RJ, Edwards SM, Dean E, Marsh E, et al. Survival and abundance of introduced fishes near San Antonio, Texas. Texas Journal of Science. 1978;4[30]:369-76.
26. Shafland PL, Pestrak JM. Lower lethal temperatures for fourteen non-native fishes in Florida. Environmental Biology of Fishes 1982(7):149–56.
27. Lee DS, Gilbert CR, Hocutt CH, Jen RE. Atlas of North American freshwater fishes. Raleigh, NC: North Carolina State Museum of Natural History; 1980.
28. Itzkowitz M, Nyby J. Field Observations of Parental Behavior of the Texas Cichlid Cichlasoma cyanoguttatum. American Midland Naturalist. 1982;2[108]:364-8.

29. Gutierrez-Yurrita PJ, Morales-ortiz A, Oviedo A, Ramirez-Perez C. Distribution, spread, habitat characterisation and conservation of the crayfish species (Cambaridae) in Querétaro (central México). Freshwater Crayfish. 2001;13:349-58.
30. Gutiérrez-Yurrita PJ, R LCJ, editors. Preliminary studies on exploitation of two native crayfish species of the genus Procambarus (CRUSTACEA: DECAPODA) by purépecha people in Michoacán, México. Proceedings of the III International Congress of Ethnobiology; 1992.
31. Kuri-Nivon E. Manual para la producción de trucha y salmón.. Idaho: Universidad de Idaho; 1976. Manual para la producción de trucha y salmón. Universidad de Idaho.
32. Richter JB. Anfangsgründen der Stöchiometrie oder Messkunst chemischer Elemente (Foundations of stoichiometry or surveying chemical elements). Breslau1792.

CHAPTER 2

Molecular markers and their applications in fisheries and aquaculture

Tanya Chauhan[1], Kumar Rajiv[2]

[1]*National Institute of Criminology and Forensic Science, Rohini, New Delhi, India*

[2]*Department of Chemistry, (SC), University of Delhi, Delhi, India.*

ABSTRACT

Genetic variation in a species enhances the capability of organism to adapt to changing environment and is necessary for survival of the species. Genetic variation arises between individuals leading to differentiation at the level of population, species and higher order taxonomic groups. The genetic diversity data has varied application in research on evolution, conservation and management of natural resources and genetic improvement programmes, etc. Development of Molecular genetic markers has powerful ability to detect genetic studies of individuals, populations or species. These molecular markers combined with new statistical developments have revolutionized the analytical power, necessary to explore the genetic diversity. Molecular markers and their statistical analysis revolutionized the analytical power, necessary to explore the genetic diversity. Various molecular markers, protein or DNA (mt-DNA or nuclear DNA such as microsatellites, SNP or RAPD) are now being used in fisheries and aquaculture. These markers provide various scientific observations which have importance in aquaculture practice recently such as: 1) Species Identification 2) Genetic variation and population structure study in natural populations 3) Comparison between wild and hatchery populations 4) Assessment of demographic bottleneck in natural population 5) Propagation assisted rehabilitation programmes. In this review article, we have concentrated on the basics of molecular genetics, overview of commonly used markers and their application along with their limitations (major classes of markers) in fisheries and aquaculture studies.

Keywords: Genetic Diversity; Molecular Markers; Microsatellite; Aquaculture

1. INTRODUCTION

All organisms are subject to mutations because of normal cellular operations or interactions with the environment, leading to genetic variation (polymorphism). Genetic variation in a species enhances the capability of organism to adapt to changing environment and is necessary for survival of the species [1]. In conjunction with other evolutionary forces like selection and genetic drift, genetic variation arises between individuals leading to differentiation at the level of population, species and higher order taxonomic groups. Molecular genetic markers are powerful tools to detect genetic uniqueness of individuals, populations or species [2,3]. These markers have revolutionized the analytical power, necessary to explore the genetic diversity [4]. The conclusion from genetic diversity data has varied application in research on evolution, conservation and management of natural resources and genetic improvement programmes, etc [5-10]

In addition to protein markers, application of DNA markers is finding wide acceptance in population genetics. With DNA markers, it is theoretically possible to observe and exploit genetic variation in the entire genome. Both genomic and mitochondrial DNA is used for varied applications. The commonly used technique are allozyme analysis, types of restriction fragment length polymorphism (RFLP), randomly amplified polymorphic DNA (RAPD), amplified fragment length polymorphism (AFLP), microsatellite typing, single nucleotide polymorphism (SNP), and expressed sequence tag (EST) markers, etc.

Molecular markers can be classified into type I and type II markers. Type I markers are associated with genes of known function, while type II markers are associated with anonymous genomic regions [11]. Under this classification, allozyme markers are type I markers because the protein they encode has known function. RAPD markers are type II markers because RAPD bands are amplified from anonymous genomic regions via the polymerase chain reaction (PCR). Microsatellite markers are also type II markers unless they are associated with genes of known function. The significance of type I markers is becoming extremely important for aquaculture genetics. Type I markers serve as a bridge for comparison and transfer of genomic information from a map-rich species into a relatively map-poor species. In general, type II markers such as RAPDs, microsatellites, and AFLPs are considered non-coding and therefore selectively neutral. Such markers have found widespread use in population genetic studies to characterize genetic divergence within and among the populations or species [12]

2. ALLOZYME MARKERS

Analysis of allozyme loci remained one of the most popular approaches in examining population genetics and stock structure questions in fishes [13]. The technique is rapid, relatively inexpensive and provides an independent estimate of level of variation within a population without an extensive morphological and quantitative survey [14]. Isohyets are structurally different molecular forms of an enzyme system with qualitatively the same catalytic function encoded by one or more loci [15]. Isohyets, which are encoded by different alleles of the same gene locus, are designated as "allozymes" or "alloenzymes" [16]. Amino acid differences in the polypeptide chain of the different allelic forms of an enzyme reflect changes in the underlying DNA sequence. Depending on the nature of the amino acid changes, the resulting protein products may migrate at different rates (due to charge and size differences) when run through a gel subjected to an electrical field. Differences in the relative frequencies of alleles are used to quantify genetic variation and distinguish among genetic units at the levels of populations, species, and higher taxonomic designations. Disadvantages associated with allozymes include occasional heterozygote deficiencies due to null (enzymatically inactive) alleles and sensitive to the amount as well as quality of tissue samples. In addition, some changes in DNA sequence are masked at the protein level, reducing the level of detectable variation. Some changes in nucleotide sequence do not change the encoded polypeptide (silent substitutions), and some polypeptide changes do not alter the mobility of the protein in an electrophoretic gel (synonymous substitutions). At present 75 isozyme systems representing several hundred genetic loci are known [17]. With the strength as codominant marker, ease of use, and low cost, the allozyme markers are popular in population structure and phylogenetic studies, though has limited role in aquaculture genetics.

3. MITOCHONDRIAL DNA MARKERS

Mitochondrial DNA (mtDNA) analysis is being increasingly used in recent population and phylogenetic surveys of organisms. Studies of vertebrate species generally have shown that sequence divergence accumulates more rapidly in mitochondrial than in nuclear DNA [18]. This has been attributed to a faster mutation rate in mtDNA that may result from a lack of repair mechanisms during replication [19] and smaller effective population size due to the strict maternal inheritance of the haploid mitochondrial genome [20]. Due to its rapid rate of evolution, mtDNA analysis has proven useful in clarifying relationships among closely related species. Different parts of the mitochondrial genome are known to evolve at different rates [21]. Almost the entire mtDNA molecule is transcribed except for the approximately 1-kb control region (D-loop), where replication and transcription of the molecule is initiated. In general, non-coding segments like the D-loop exhibit elevated levels of variation relative to coding sequences such as the cytochrome b gene [22], presumably due to reduced functional constraints and relaxed selection pressure. The 16S rRNA gene in the mitochondrial genome is one of the slowest evolving genes [21] whereas rapidly evolving regions are control regions [23,24]. Due to non-Mendelian mode of inheritance, the mtDNA molecule is considered as a single locus [2]. In addition, because mtDNA is maternally inherited, the phylogenies and population structures derived from mtDNA data may not reflect complete picture of the nuclear genome if gender-biased migration or selection [20] or introgression [25] exists.

Analyses of mtDNA markers have been used extensively to investigate stock structure in a variety of vertebrates including fishes [26-30], birds [31-34], mammals [35] and reptiles [36-39].

4. RANDOM AMPLIFIED POLYMORPHIC DNA (RAPD) MARKERS

RAPD markers are the amplified products of less functional part of the genome that do not strongly respond to selection on the phenotypic level. Such DNA regions may accumulate more nucleotide mutations with potential to assess inter-population genetic differentiation [40]. The amplification of genomic DNA by PCR with arbitrary nucleotide sequence primers, RAPD can detect high levels of DNA polymorphisms [41,42]. The technique detects coding as well as non-coding DNA sequences, and many of the most informative polymorphic sequences are those derived from repetitive (non-coding) DNA sequences in the genome [43]. Because 90% of the vertebrate nuclear genome is non-coding, it is presumed that most of the amplified loci will be selectively neutral. RAPD loci are inherited as Mendelian markers in a dominant fashion and scored as present/absent. RAPDs have all the advantages of a PCR-based marker, with the added benefit that primers are commercially available and do not require prior knowledge of the target DNA sequence or genome organization. Other advantages of RAPDs include the ease with which a large number of loci and individuals can be screened simultaneously. Shortcomings of this type of marker include the difficulty of demonstrating Mendelian inheritance of the loci and the inability to distinguish between homozygotes and heterozygotes. Analysis follows the assumption that populations under study follow Hardy-Weinberg expectations. In addition, the presence of paralogous PCR product (different DNA regions which have the same lengths and thus appear to be a single locus), low reproducibility due to the low annealing temperature used in the PCR amplification, have limited the application of this marker in fisheries science [44].

5. SINGLE NUCLEOTIDE POLYMORPHISM (SNP)

Single nucleotide polymorphism (SNP) describes polymorphisms caused by point mutations that give rise to different alleles containing alternative bases at a given nucleotide position within a locus. SNPs are becoming a focal point in molecular marker development since they represent the most abundant polymorphism in any organism's genome (coding and non-coding regions), adaptable to automation, and reveal hidden polymorphism not detected with other markers and methods [9, 10]. Theoretically, a SNP within a locus can produce as many as two alleles, each containing one of two possible base pairs at the SNP site. Therefore, SNPs have been regarded as bi-allelic. SNP markers are inherited as co-dominant markers. Several approaches have been used for SNP discovery including SSCP analysis [45], heteroduplex analysis, and direct DNA sequencing. DNA sequencing has been the most accurate and most used approach for SNP discovery. SNPs are not without their limitations, however, might provide marginal additional, or even less, utility in some applications (e.g. relatedness) [9].

6. MICROSATELLITE MARKERS

Microsatellites consist of multiple copies of tandemly arranged simple sequence repeats (SSRs) that range in size from 1 to 6 base pairs [e.g., ACA or GATA; 46,47]. Abundant in all species studied to date, microsatellite motifs have been estimated to occur as often as once every 10 kb in fishes

[48]. Microsatellites tend to be evenly distributed in the genome on all chromosomes and all regions of the chromosome. However, data from whole genome sequencing has somewhat contradicted this statement. They have been found inside gene coding regions [49], introns, and in the non-gene sequences. Most microsatellite loci are relatively small, ranging from a few to a few hundred repeats. Regardless of specific mechanisms, changes in numbers of repeat units can result in a large number of alleles at each microsatellite locus in a population. Microsatellites have been inherited in a Mendelian fashion as codominant markers. Microsatellites were found to be informative in several species, which showed almost no variation at other markers [50]. However, use of microsatellite markers involves a large amount of up-front investment and effort. Each microsatellite locus has to be identified and its flanking region sequenced to design of PCR primers. Due to polymerase slippage during replication, small size differences between alleles of a given microsatellite locus (as little as 2 bp in a locus comprised of di-nucleotide repeats) are possible. Microsatellites recently have become an extremely popular marker type in a wide variety of genetic investigations.

7. NEW DEVELOPING MARKERS IN FISHERIES AND AQUACULTURE

Various type of DNA markers have been developed, including Allozymes, microsatellites, RAPDs, mt-DNA and SNPs. These markers in fish populations have revealed high levels of genetic variation distributed throughout the fish genome. A recent initiative has been made to accelerate efforts of DNA marker development, genome mapping and species identification. Major progress has been made toward Expressed Sequence Tags (EST) and DNA barcode development in several aquaculture species.

8. EXPRESSED SEQUENCE TAGS (ESTS)

Expressed sequence tags (ESTs) are single-pass sequences generated from random sequencing of cDNA clones [51]. The EST is use to identify genes and analyze their expression by means of expression profiling. It helps for rapid and valuable analysis of genes expressed in specific tissue types, under specific physiological conditions, or during specific developmental stages. ESTs offer the development of cDNA microarrays that allow analysis of differentially expressed genes to be determined in a systematic way [52], in addition to their great value in genome mapping [53]. For genome mapping, ESTs are most useful for linkage mapping and physical mapping in animal genomics such as those of cattle and swine, where radiation hybrid panels are available for mapping non-polymorphic DNA markers [54]. A radiation panel is composed of lines of hybrid cells, with each hybrid cell containing small fragments of irradiated chromosomes of the species of interest. Typically, the cells from species of interest are radiated to break chromosomes into small fragments. The radiated cells are unable to survive by themselves. However, the radiated cells can be fused with recipient cells to form hybrid cells retaining a short segment of the radiated chromosome. Characterization of the chromosomal break points within many hybrid cell lines would allow linkage and physical mapping of markers and genes. In spite of its popularity in mammalian genome mapping [55, 56], radiation hybrid panels are not yet available for any aquaculture species. Development of radiation hybrid panels from aquaculture species is not expected in the near future, given the fact that physical mapping using BAC libraries can provide even higher resolution and the fact that BAC libraries are already available from several aquaculture species. Therefore, ESTs are useful for mapping in aquaculture species only if polymorphic ESTs are identified [57]. The value of EST resources and applications of bioinformatics in aquaculture genetics/genomics is inevitable, and it is expected that various EST databases will serve as rich sources of genomic information not only for aquaculture geneticists, but also for aquaculture physiologists, immunologists and biotechnologists.

9. DNA BARCODING

The principle of conservation biology is the preservation and management of biodiversity. The two major problems to such an endeavor are the difficulty of developing an assessment of this diversity for prioritization of hotspots of species richness [58] and the identification of lineages particularly worthy, or in need, of preservation [59-64]. Understudied taxa are greatly susceptible to extinction [65], suggesting there is a conservation penalty for our ignorance. Even there are millions of unidentified and unknown species [66]. DNA barcodes, segments of approximately 600 base pairs of the mitochondrial gene cytochrome oxidase I (COI), have been proposed as a fast, efficient, and inexpensive technique to catalogue all biodiversity [67-70]. Barcoding is the use of universal polymerase chain reaction (PCR) primers to amplify and sequence an approximately 600-basepair fragment of the COI gene. That portion of sequence is then compared using distance-based algorithms with an existing database of "known" sequences from specimens previously identified by taxonomists. DNA barcodes from a small portion of the mitochondrial genome might seem like an effective and rapid way to assess at least some, perhaps minimal, level of biodiversity. And for groups that are already relatively well known, especially birds and mammals, molecular studies based on barcode sized sequences have revealed cryptic DNA lineages and may be helpful [70].

10. APPLICATION OF MOLECULAR MARKERS SPECIES IDENTIFICATION

The inter-specific genetic divergence established through species specific diagnostic molecular markers provides precise knowledge on phylogenetic relationships and also resolve taxonomic ambiguities [71-74]. These markers can be used to detect hybrid and introgressed or backcrossed individuals [75], distinguish early life history stage of morphologically close species [76] both in hatchery and in natural populations.

Species-specific allozyme markers have been identified in many fishes [Tilapia: 72,77,78; Sciaenid: 73; Anguilla sp: 79; Mugilidae: 80] Specific diagnostic allozyme loci were used for different species: apache trout (Oncorhynchus apache), cutthroat (Oncorhynchus clarki) and rainbow trout (Oncorhynchus mykiss) [81] and Gambusia affinis and G. holbrooki [82]. Allozyme markers have also been used for individual classification in cyprinid species Zacco pachycephalus and Z. platypus [83], in cyprinodontid species V. letourneuxi and V. hispanica [84], in mullets Mullus barbatus and M. surmuletus [85] and hake species Merluccius australis and M. hubbsi [86].

Species-specific diagnostic RAPD fingerprints were generated in several fish species and their taxonomic relationship has been analyzed. The RAPD-PCR technique was employed to identify three endemic morphologically similar Spanish species of Barbus: Barbus bocagei, B. graellsii and B. sclateri that have similar morphologies [87]. RAPD markers were characterized to identify five species of family Cyprinidae: Chondrostoma lemmingii, Leuciscus pyrenaicus, Barbus bocagei, Barbus comizo, all endemic in the Iberian Peninsula, and introduced Alburnus alburnus [88], for studying genetic relationship and diversities in four species of Indian Major carps (family Cyprinidae): rohu (Labeo rohita), kalbasu (L. calbasu), catla (Catla catla) and mrigal (Cirrhinus mrigala) [89], for identification of three eel species, A. japonica, A. australis and A. bicolor [90] and to estimate the population structure and phylogenetic relationships among the eight species of the genus Barbus [88].

Large variation in mtDNA sequences among species can be utilized to produce species-specific markers. Since the structures of mitochondrial

RNA genes (tRNA and rRNA) and the functional molecule of the 16S rRNA are highly conserved among the animal taxa that are related even distantly [21], change of even few nucleotides in such a gene between closely related taxa might indicate a substantial degree of genetic divergence [2]. Mt-DNA sequences have been used as useful marker for species-specific identification in many fishes [Tuna: 91; Billfish: 92 Snappers: 29, 93; Myctophidae: 94; Grey mullets: 95]. Comparable levels of divergence based on 12S rRNA and 16S rRNA sequences have been reported for several recently diverged fish species [genus Sternoptyx: 96; Cyclothone sp: 97]. Sequence variation in the control region (D-loop) of the mitochondrial DNA (mtDNA) was examined to assess the genetic distinctiveness of the short-jaw cisco, Coregonus zenithicus [98] and revealed high similarity of C. zenithicus and the related species C. artedi, C. hoyi, C. kiyi, and C. clupeaformis Identification of Astyanax altiparanae (Teleostei, Characidae) in the Iguacu River, Brazil, was done on the basis of mitochondrial DNA and RAPD markers [99]. Two species, Acipenser baeri, and A. stellatus, was studied using mitochondrial DNA (D-loop, cytochrome b (cyt-b) and ND5/6 genes) sequencing to determine whether traditionally defined subspecies correspond to taxonomic entities and conservation management units [100].

11. GENETIC VARIATION AND POPULATION STRUCTURE STUDY IN NATURAL POPULATIONS

Molecular markers provide direct assessment of pattern and distribution of genetic variation [5] thus helping in answering, "if the population is single unit or composed of subunits". Several evolutionary forces affect the amount and distribution of genetic variation among populations and thereby population differentiation [101]. Geographic distance and physical barriers enhance reproductive isolation by limiting the migration and increase genetic differentiation between populations [102]. Impact of migration and gene flow on genetic differentiation also depends upon effective size of receiving population and number of migrants. Increased computational power and mathematical models have enhanced the scope of conclusions that can be drawn out of genotype data generated through molecular markers. Some of the possibilities are assignment of migrants [103], determination of genetic bottlenecks [104], effective breeding population estimates [105] besides genetic variation and differentiation estimations [106-108]. These markers have been extensively employed across various taxonomic groups [mosquito: 109; turtle: 39; amphibians: 7; panda: 110; five vertebrate classes including fish, amphibian, reptiles, birds and mammals: 6]. Experiments on fish populations have significantly contributed towards development of science of population genetics, models and analytical softwares.

Population genetic structure has been investigated using allozyme markers in many fish species [Oncorhynchus gorbuscha: 111; Tenualosa ilisha: 112 and Lal et al., 113; Pagrus auratus: 114].

Fifteen random primers were used to analyze the genome DNA of Jian carp (Cyprinus carpio var jian) by the RAPD technique [115]. Study on cold tolerant traits for common carp Cyprinus carpio was conducted by Chang et al. [116] and nine RAPD-PCR markers associated with cold tolerance of common carp were identified. The genetic diversity has been studied using RAPD markers in Carassius auratus [117], Epinephelus merra population [118] and Solea solea [119].

Genetic variation have been assessed with Allozyme and RAPD markers on Mullus surmuletus L., [120] and three species of Pimelodidae catfish [121].

Population structure has been examined using microsatellite markers of sockeye salmon [122], Chinook salmon [123] and Arctic charr populations [124]. Genetic variation have been assessed using microsatellite genetic markers to identify the population structure of brook charr, Salvelinus fontinalis [125] and 14 populations of northern pike (Esox lucius) in the North Central United States and in six populations from Quebec, Alaska, Siberia, and Finland [126].

Based on five microsatellite loci, the genetic structure of endangered fish species Anaecypris hispanica was studied in eight distinct populations in the Portuguese Guadiana drainage to determine levels of genetic variation within and among populations and suggested implications for conservation of the species [127].

Combination of allozyme and microsatellites was used to investigate genetic divergence in Salmo trutta [128] and Salmo salar [129].

Alarcon et al. [130] represents population genetic analysis of gilthead sea bream (Sparus aurata) and Kanda [131], Kanda and Allendorf [132] examine population genetic structure of bull trout Salvelinus confluentus using a combination of allozyme, microsatellite and mtDNA variation.

Genetic variability of Salmo trutta [133] and Sparus aurata [130] was evaluated on the basis of Allozyme, Microsatellites and RAPD markers.

Patterns of population subdivision and the relationship between gene flow and geographical distance in the tropical estuarine fish Lates calcarifer (Centropomidae) were investigated using mtDNA control region sequences [134].

Allozymes and mtDNA sequences were assessed to evaluate the genetic variability in small marine fish Pomatoschistus microps [135], brown trout [136] and Macquaria novemaculeata [137].

12. COMPARISON OF GENETIC VARIATION BETWEEN WILD AND HATCHERY POPULATIONS

Molecular markers also find application in aquaculture to assess loss of genetic variation in hatcheries through, comparison of variation estimates between hatchery stocks and wild counterparts. The information is useful obtained in monitoring farmed stocks against inbreeding loss and to plan genetic up gradation programmes. A major aspect such studies address is concerned with the assessment of farm escapes into the natural population and introgression of wild genome.

Brook trout Salvelinus fontinalis from ustocked waters, naturalized lakes, and hatcheries in New York and Pennsylvania were analyzed electrophoretically for allozyme expression [138]. All wild-unstocked samples were highly differentiated populations and significantly different from each other and from hatchery samples.

Genetic diversity was investigated using microsatellites between farmed and wild populations of Atlantic salmon [139]. Farmed salmon showed less genetic variability than natural source population in terms of allelic diversity.

Variation in allozymes and three microsatellite loci was assessed in populations of wild and cultured stocks of Sparus aurata [140] and Sparius auratus [130]. The microsatellite heterozygosity values were high in wild, but lower in the cultured samples.

13. ASSESSMENT OF DEMOGRAPHIC BOTTLENECK IN NATURAL POPULATION

Demographic bottlenecks occur when populations experience severe, temporary reduction in size. Because bottlenecks may influence the distribution of genetic variation within and among populations, the genetic effects of reductions in population size have been studied extensively by evolutionary biologist [141,142].

It may often be necessary to perform genetic analyses of temporal replicates to estimate the significance of spatial variation independently from that of temporal variation in order to ensure the reliability of estimates of a defined population structure. Such estimates provide understanding about changes in genetic variation, effective population size and other historical bottlenecks and can be extrapolated to define evolutionary trends of species. Today various models are available that can resolve bottlenecks or effective population size changes through use of heterozygosity excess, linkage disequilibrium etc. However, estimates through temporal changes are considered more accurate. Analysis of temporal changes is limited due to lack of historical data as well as samples. Therefore, such studies are limited and mostly use archived samples, wherever available. In vertebrates, a limited number of studies have specifically assessed the temporal changes in genetic variation for more than one generation.

Microsatellite DNA markers have been used to assess bottlenecks in many fish species. A microsatellite analysis of DNA was performed, from archived scales to compare the population structure among four sympatric landlocked populations of Atlantic salmon [143], Atlantic salmon [144], European hake [145] and steelhead from [146].

Larson et al. [147] recommended close monitoring of negative effects on sea otter population based on the conclusion from mtDNA, D-loop, microsatellite variability comparison between prefur trade and present population. Prefur trade DNA samples were obtained from excavated bones.

14. PROPOGATION ASSISTED REHABILITATION PROGRAMMES

Habitat alterations and over harvesting have contributed to the decline or disappearance of numerous natural populations. In addition, reinforcement programs of wild populations based on releases of hatchery reared fish of non-native origin compromise the conservation of remnant native trout resources. Effect of these programmes through releases in natural populations has been studies in many fishes through molecular markers.

Beaudou et al. [148] found through allozyme polymorphism that brown trout (Salmo trutta L.) in the Abatesco river basin on the eastern coast of Corsica restoration was mainly due to the populations of the tributaries, which had been less disturbed by the spate. This study has shown that the wild population was primarily restored by the surviving individuals, particularly those from the tributaries that escaped the spate.

To assess the levels of gene introgression from cultured to wild brown trout populations, four officially stocked locations and four non-stocked locations were sampled for one to three consecutive years and compared to the hatchery strain used for stocking. Allozyme analysis for 25 loci included providing allelic markers distinguishing hatchery stocks and native populations [133]. Different levels of hybridization and introgression with hatchery individuals were detected in stocked drainages as well as in protected locations.

Tanya Chauhan, Kumar Rajiv The foregoing review incorporates the wide spectrum of information that the molecular markers provide. The literature indicates that different markers have been employed depending upon the question to be answered. The importance of the research on molecular markers improved due to enhanced computational power, large data available that has enabled researchers to derive various mathematical estimators. Such innovations provide insight concerning the population bottleneck, migration patterns besides the genetic structure in natural populations.

REFERENCES

1. Fisher, R.A. (1930) The Genetical Theory of Natural Selection. Oxford University Press, UK.
2. Avise, J.C. (1994) Molecular Markers, Natural History and Evolution. Chapman and Hall, New York, London.
3. Linda, K.P. and Paul, M. (1995) Developments in molecular genetic techniques in fisheries. In: G.R. Carvalho and T.J. Pitcher, Eds., Molecular Genetics in Fisheries, Chapman and hall, London, 1-28.
4. Hillis, D.M., Mable, B.K. and Moritz, C. (1996) Applications of molecular systematics: The state of the field and a look to the future. In: Hillis, D.M., Moritz, C. and Mable, B.K. Eds., Molecular systematics, Sinauer Associates, Massachusetts, 515-543.
5. Ferguson, A., Taggart, J.B., Prodohl, P.A., McMeel, O., Thompson, C., Stone, C., McGinnity, P. and Hynes, R.A. (1995) The application of molecular markers to the study and conservation of fish populations with special reference to Salmo. Journal of Fish Biology, 47(A), 103-126.
6. Neff, B.D. and Gross, M.R. (2001) Microsatellite evolution in vertebrates: Inference from AC dinucleotide repeats. Evolution, 55(9), 1717-1733.
7. Jehle, R. and Arntzen, J.W. (2002) Microsatellite markers in amphibian conservation genetics. Herpetological Jour nal, 12, 1-9.
8. Wasko, A.P., Martins, C., Oliveira, C. and Foresti, F. (2003) Non-destructive genetic sampling in fish. An improved method for DNA extraction from fish fins and scales. Hereditas, 138(3), 161-165.
9. Morin, P.A., Luikart, G., Wayne, R.K. and the SNP working group, SNPs in ecology, evolution and conservation. Trends in Ecology and Evolution, 19(4), 208-216.
10. Liu, Z.J. and Cordes, J.F. (2004) DNA marker technologies and their applications in aquaculture genetics. Aquaculture, 238, 1-37.
11. O'Brien, S.J. (1991) Molecular genome mapping: lessons and prospects. Current Opinion in Genetic Development, 1(1), 105-111.
12. Brown, B. and Epifanio, J. (2003) Nuclear DNA. In: Hallermann, E.M. Ed., Population Genetics: Principles and Applications for Fisheries Scientists, American Fisheries Society, Bethesda, 458-472.

13. Suneetha, B.K. (2000) Interspecific and inter specific genetic variation in selected mesopelagic fishes with emphasis on microgeographic variation and species characterization. Dr. Scient. Dissertation, Department of Fisheries and Marine Biology, University of Bergen, Bergen, Norway.
14. Menezes, M.R., Naik, S. and Martins, M. (1993) Genetic characterization in four sciaenid species from the Arabian Sea. Journal of Fish Biology, 43(1), 61-67.
15. Markert, C.L. and Moller, F. (1959) Multiple forms of enzymes: Tissue, ontogenetic and species-specific patterns. Proceedings of the Naionall Academy of Science (USA), 45(5), 753-763.
16. Starck, M.G. (1998) Isozymes in Molecular tools for screening biodiversity. In: Angela, K., Peter, G.I. and. David, S.I. Eds., Chapmann and Hall, London, 75-80.
17. Murphy, R.W., Sites, J.J.W., Buth, D.G. and Haufler, C.H. (1996) Proteins I: Isozyme electrophoresis. In: Hillis, D.M., Moritz, C. and Mable, B.K. Eds., Molecular Systematics, Sinauer Associates, Sunderland, 51-132.
18. Brown, W.M. (1985) The mitochondrial genome of animals. In: Maclntyre, R.J. Ed., Molecular Evolutionary Genetics, Plenum, New York, 95-130.
19. Wilson, A.C., Cann, R.L., Carr, S.M., George, M., Gyllensten, U.B., Helm-Bychowski, K.M., Higuchi, R.G., Palumbi, S.R. and Prager, E.M. (1985) Mitochondrial DNA and two perspectives on evolutionary genetics. Biological Journal of Linnean Society, 26(4), 375-400.
20. Birky, C.W., Fuerst, P. and Maruyama, T. (1989) Organelle gene diversity under migration, mutation, and drift: equilibrium expectations, approach to equilibrium, effect of heteroplasmic cells, and comparison to nuclear genes. Genetics, 121(3), 613-627.
21. Meyer, A. (1993) Evolution of mitochondrial DNA in fishes. In: Mochachka, P.W. and Mommsen, T.P. Eds., Biochemistry and molecular biology of fishes, Elsevier Press Amsterdam, New York, 1-38.
22. Brown, J.R., Bechenbach, A.T. and Smith, M.J. (1993) Intraspecific DNA sequence variation of the mitochondrial control region of white sturgeon (Acipenser transmontanus). Molecular Biology Evolution, 10(2), 326-341.
23. Chow, S., Okamoto, H., Uozumi, Y., Takeuchi, Y. and Takeyama, H. (1997) Genetic stock structure of the swordfish (Xiphias gladius) inferred by PCR-RFLP analysis of the mitochondrial DNA control region. Marine Biology, 127(3), 359-367.
24. Gold, J.R. Sun, F. and Richardson, L.R. (1997) Population structure of red snapper from the Gulf of Mexico as inferred from analysis of mitochondrial DNA. Transaction of American Fisheries Society, 126(3), 386-396.
25. Chow, S. and Kishino, H. (1995) Phylogenetic relationships between tuna species of the genus Thunnus (Scombriidae: Teleosrei): Inconsistent implications from morphology, nuclear and mitochondrial genomes. Journal of Molecular Evolution, 41, 741-748.
26. Avise, J.C., Helfman, G.S., Saunders, N.C. and Hales, L.S. (1986) Mitochondrial DNA differentiation in North Atlantic eels: Population genetic consequences of an unusual life history pattern. Proceeding of the National Academy Science (USA), 83(12), 4350-4354.
27. Graves, J.E., McDowell, J.R. and Jones, M.L. (1992) A genetic analysis of weakfish Cynoscion regalis stock structure along the mid-Atlantic coast. Fisheries Bulletin, 90, 469-475.
28. Gold, J.R., Richardson, L.R., Furman, C. and King, T.L. (1993) Mitochondrial DNA differentiation and population structure in red drum (Sciaenops ocellatus) from the Gulf of Mexico and Atlantic Ocean. Marine Biology, 116(2), 175-185.
29. Chow, S., Clarke, M.E. and Walsh, P.J. (1993) PCR-RFLP analysis of thirteen western Atlantic snappers (subfamily Lutjaninae): A simple method for species and stock identification. Fisheries Bulletin (US), 91, 619-627.
30. Heist, E.J. and Gold, J.R. (1999) Microsatellite DNA variation in sandbar sharks (Carcharhinus plumbeus) from the Gulf of Mexico and mid-Atlantic bight. Copeia, 1, 182-186.
31. Baker, A.J. and Marshall, H.D. (1997) Molecular evolution of the mitochondrial genome. In: Mindell, D.P. Ed., Avian Molecular Evolution and Systematics, Academic Press, San Diego, 51-82.

32. Greenberg, R., Cordero, P.J., Droege, S. and Fleischer, R.C. (1998) Morphological adaptation with no mitochondrial DNA differentiation in the coastal plain swamp sparrow. Journal of American Ornithologists' Union, 115(3), 706-712.
33. Mila, B., Girman, D.J., Kimura, M. and Smith, T.B. (2000) Genetic evidence for the effect of a postglacial population expansion on the phylogeography of a North American songbird. Proceedings of the Royal Society London Series B, 267(1447), 1033-1040
34. Zink, R.M., Barrowclough, G.F., Atwood, J.L. and Blackwell-Rago, R.C. (2000) Genetics, taxonomy, and conservation of the threatened California gnatcatcher. Conservation Biology, 14(5), 1394-1405.
35. Menotti-Raymond, M. and O'Brien, S.J. (1993) Dating the genetic bottleneck of the African cheetah. Proceedings of the National Academy of Science (USA.), 90(8), 3172-3176.
36. Avise, J.C., Walker, D. and Johns, G.C. (1998) Speciation durations and Pleistocene effects on vertebrate phylogeography. Proceedings of the Royal Society of London Series B, 265(1407), 1707-1712.
37. Serb, J.M., Phillips, C.A. and Iverson, J.B. (2001) Molecular phylogeny and biogeography of Kinosternon flavenscens based on complete mitochondrial control region sequences. Molecular Phylogenetics and Evolution, 18(1), 149-162.
38. Riberon, A., Sotiriou, E., Miaud, C., Andreone, F. and Taberlet, P. (2002) Lack of genetic diversity in Salamandra lanzai revealed by cytochrome b gene sequences. Copeia, 2002, 229-232.
39. Shanker, K., Ramadevi, J., Choudhaury, B.C., Singh, L. and Aggarawal, R.K. (2004) Phylogeny of olive ridley turtles (Lepidochelys olivacea) on the east coast of India: implications for conservation theory. Moelcular Ecology, 13(7), 1899-1909.
40. Mamuris, Z., Sfougaris, A.I., Stamatis, C. and Suchentrunk, F. (2002) Assessment of genetic structure of Greek Brown Hare (Lepus europeaus) populations based on variation in Random Amplified Polymorphic DNA (RAPD). Biochemical Genetics, 40(9-10), 323- 338.
41. Williams, J.G.K., Kubelik, A.R., Livak, K.J., Rafalski, J.A. and Tingey, S.V. (1990) DNA polymorphisms amplified by arbitrary primers are useful as genetic markers. Nucleic Acids Research, 18(22), 6531-6535.
42. Welsh, J. and McClelland, M. (1990) Fingerprinting genomes using PCR with arbitrary primers. Nucleic Acids Research, 18(24), 7213-7218.
43. Haymer, D.S. (1994) Random amplified polymorphic DNAs and microsatellites: What are they, and can they tell us anything we don't already know? Annals of Entomological Society of American, 87, 717-722.
44. Wirgin, I.I. and Waldman, J.R. (1994) What DNA can do for you? Fisheries, 19, 16-27.
45. Hecker, K.H., Taylor, P.D. and Gjerde, D.T. (1999) Mutation detection by denaturing DNA chromatography using fluorescently labeled polymerase chain reaction products. Analytical Biochemistry, 272(2), 156-164.
46. Tautz, D. (1989) Hypervariability of simple sequences as a general source for polymorphic DNA markers. Nucleic Acids Research, 17(16), 6463-6471.
47. Litt, M. and Luty, J.A. (1989) A hypervariable microsatellite revealed by in-vitro amplification of dinucleotide repeat within the cardiac muscle actin gene. American of Journal of Human Genetics, 44(3), 397-401.
48. Wright, J.M. (1993) DNA fingerprinting in fishes. In: W. Hochachka, P. and Mommsen, T. Eds., Biochemistry and Molecular Biology of Fishes, Elsevier, Amsterdam, 58-91.
49. Liu, Z.J., Li, P., Kocabas, A., Ju, Z., Karsi, A., Cao, D. and Patterson, A. (2001) Microsatellite-containing genes from the channel catfish brain: evidence of trinucleotide repeat expansion in the coding region of nucleotide excision repair gene RAD23B. Biochemical Biophysical Research Communication, 289(2), 317-324.
50. Taylor, A.C., Sherwin, W.B. and Wayne, R.K. (1994) Genetic variation of microsatellite loci in a bottlenecked species: The northern hairy-nosed wombat Lasiorhinus krefftii. Molecular Ecology, 3(4), 277-290.
51. Adams, M.D., Kelley, J.M., Gocayne, J.D., Dubnick, M., Polymeropoulos, M.H., Xiao, H., Merril, C.R., Wu, A., Olde, B., Moreno, R.F., Kerlavage, A.R., McCombie, W.R. and Venter, J.C. (1991) Complementary DNA sequencing: Expressed sequence tags and human genome project. Science, 252(5013), 1651-1656.

52. Wang, K., Gan, L., Jeffry, E., Gayle, M., Gown, A.M., Skelly, M., Nelson, P.S., Ng, W.V., Schummer, M., Hood, L. and Mulligan, J. (1999) Monotoring gene expression profile changes in ovarian carcinomas using cDNA microarray. Gene, 229(1-2), 101-108.
53. Boguski, M.S. and Schuler, G.D. (1995) Establishing a human transcript map. Nature Genetics, 10(4), 369-371.
54. Cox, D.R., Burmeister, M., Price, E., Kim, S. and Myers, R.M. (1990) Radiation hybrid mapping: a somatic cell genetic method for constructing high-resolution map of mammalian chromosomes. Science, 250(4978), 245-250.
55. Korwin-Kossakowska, A., Reed, K.M., Pelak, C., Krause, E., Morrison, L. and Alexander, L.J. (2002) Radiation hybrid mapping of 118 new porcine microsatellites. Animal Genetics, 33(3), 224-227.
56. McCoard, S.A., Fahrenkrug, S.C., Alexander, L.J., Freking, B.A., Rohrer, G.A., Wise, T.H. and Ford, J.J. (2002) An integrated comparative map of the porcine X chromosome. Animal Genetics, 33(3), 178-185.
57. Liu, Z.J., Karsi, A. and Dunham, R.A. (1999) Development of polymorphic EST markers suitable for geneticlinkage mapping of catfish. Marine Biotechnology, 1(5), 437-447.
58. Dobson, A.P., Rodriguez, J.P., Roberts, W.M. and Wilcove, D.S. (1997) Geographic distribution of endangered species in the United States. Science, 275(5299), 550- 555.
59. Daugherty, C.H., Cree, A., Hay, J.M. and Thompson, M.B. (1990) Neglected taxonomy and continuing extinctions of tuatara (Sphenodon). Nature, 347(6289), 177- 179.
60. Faith, D.P. (1994) Genetic diversity and taxonomic priorities for conservation. Biological Conservation, 68(1), 69-74.
61. Crozier, R.H. (1997) Preserving the information content of species: Genetic diversity, phylogeny, and conservation worth. Annual Review of Ecology and Systematics, 28(1), 243-268.
62. Haig, S.M. (1998) Molecular contributions to conservation. Ecology, 79(2), 413-425.
63. Soltis, P.S. and Gitzendanner, M.A. (1999) Molecular systematics and the conservation of rare species. Conservation Biology, 13(3), 471-483.
64. Moritz, C. (2002) Strategies to protect diversity and the evolutionary processes that sustain it. Systematic Biology, 51(2), 238-254.
65. McKinney, M.L. (1999) High rates of extinction and threat in poorly studied taxa. Conservation Biology, 13(6), 1273-1281.
66. Novotny, V., Basset, Y., Miller, S.E., Weiblen, G.D., Bremer, B., Cizek, L. and Drozd, P. (2002) Low host specificity of herbivorous insects in a tropical forest. Nature, 416, 841-844.
67. Hebert, P.D.N., Ratnasingham, S. and deWaard, J.R. (2003) Barcoding animal life: Cytochrome c oxidase subunit 1 divergences among closely related species. Royal Society London, 270 (Suppl 1), S96-S99.
68. Stoeckle, M. (2003) Taxonomy, DNA, and the bar code of life. BioScience, 53(9), 2-3.
69. Stoeckle, M., Janzen, D., Hallwachs, W., Hanken, J. and Baker, J. (2003) Taxonomy, DNA, and the barcode of life," Draft conference report. Rubinoff DNA Barcodes and Conservation Barcode Conference, The Rockefeller University, New York, http://phe.rockefeller.edu/Barcode Confeence/docs/B2summary.doc
70. Hebert, P.D.N., Stoeckle, M.Y., Zemlak, T.S. and Francis, C.M., (2004) Identification of birds through DNA barcodes. PLoS Biology, 2(10), 312-316.
71. Rocha-Olivares, A., Moser, H.G. and Stannard, J. (2000) Molecular identification and description of pelagic young of the rockfishes Sebastes constellatus and Sebastes ensifer. Fisheries Bulletin, 98, 353-363.
72. Backer, J., Bentzen, P. and Moran, P. (2002) Molecular markers distinguish coastal cutthroat trout from coastal rainbow trout/steelhead and their hybrids. Transaction of American Fisheries Society, 131(3), 404-417.
73. Asensio, L., Gonzalez, I., Fernandez, A., Rodriguez, M.A., Lobo, E., Hernandez, P.E., Garcia, T. and Martin, R. (2002) Application of random amplified polymorphic DNA (RAPD) analysis for identification of grouper (Epinephelus guaza), wreckfish (Polyprion americanus), and nile perch (Lates niloticus) fillets. Journal of Food Product, 65(2), 432-435.

74. Rasmussen, C., Ostberg, C.O., Clifton, D.R., Holloway, J.L. and Rodriguez, R.J. (2003) Identification of a genetic marker that discriminates ocean-type and stream-type Chinook salmon in the Columbia River Basin. Transaction American Fisheries Society, 132(1), 131-142.
75. Compton, D. and Utter, F.M. (1985) Natural hybridization between steelhead trout (Salmo gairdneri) and coastal cutthroat trout (Salmo clarki clarki) in two Puget sound streams. Canadian Journal of Fisheries and Aquatic Science, 42, 110-119.
76. Olivar, M.P., Moser, H.G. and Beckley, L.E. (1999) Lantern fish larvae from the Agulhas current (SW Indian Ocean). Science of Marine, 63, 101-120.
77. McAndrew, B.J. and Majumdar, K.C. (1983) Tilapia stock identification using electrophoretic markers. Aquaculture, 30(1-4), 249-261.
78. Bartly, D.M. and Gall, G.A.E. (1991) Genetic identification of native cutthroat trout (Onchorynchus clarki) and introgressive hybridization with introduced rainbow trout (O. mykiss) in streams associated with the Alvord basin, Oregon and Nevada. Copeia, 3, 854-859.
79. Lee, S.C., Tosi, S.C.M., Cheng, H.L. and Chang, J.T. (1997) Identification of Anguilla Japonica and A. marmorata elvers by allozyme electrophoresis. Journal of Fish Biology, 51(1), 208-210.
80. Rossi, A.R., Capula, M., Crosetti, D., Sola, L. and Campton, D.E. (1998) Allozyme variation in global populations of striped mullet, Mugil cephalus (Pisces: Mugilidae). Marine Biololgy, 131(2), 203-212.
81. Carmichael, G.J., Hanson, J.N., Schmidt, M.E. and Morizot, D.C. (1993) Introgression among apache, cutthroat and rainbow trout in Ariozona. Transaction of American Fisheries Society, 122, 121-130
82. Wooten, M.C. and Lydeard, C. (1990) Allozyme variation in a natural contact zone between Gambusia affinis and Gambusia holbrooki. Biochemical Systematics and Ecology, 18(2-3), 169-173.
83. Wang-Hurng, Y., Lee, S.C. and Yu, M.J. (1997) Genetic evidence to clarify the systematic status of the genra Zacco and Candidia (Cypriniformes: Cprinidae). Zoological Studies, 36(3), 170-177.
84. Perdices, A., McChordom, A. and Doadrio, I. (1996) Allozyme variation and relationships of the endangered cyprinodontid genus Valencia and its implications for conservation. Journal of Fish Biology, 49(6), 1112-1127.
85. Mamuris, Z., Apostolidis, A.P. and Triantaphyllidis, C. (1998) Genetic protein variation in red mullet (Mullus barbatus) and striped red mullet (M. surmuletus) populations from the Mediterranean sea. Marine Biology, 130(3), 353-360.
86. Roldán, M.I. and Pla, C. (2001) Species identification of two sympatric hakes by allozyme markers. Science of Marine, 65, 81-84.
87. Callejas, C. and Ochando, M.D. (1998) Identification of Spanish barbel species using the RAPD technique. Journal of Fish Biology, 53(1), 208-215.
88. Callejas, C. and Ochando, M.D. (2002) Phylogenetic relationships among Spanish Barbus species (Pisces, Cyprinidae) shown by RAPD markers. Heredity, 89(1), 36- 43.
89. Barman, H.K., Barat, A., Yadav, B.M., Banerjee, S., Meher, P.K., Reddy, P.V.G.K. and Jana, R.K. (2003) Genetic variation between four species of Indian major carps as revealed by random amplified polymorphic DNA assay. Aquaculture, 217(1-4), 115-123.
90. Takagi, M. and Taniguchi, N. (1995) Random amplified polymorphic DNA (RAPD) for identification of three species of Anguilla, A. japonica, A. australis and A. bicor. Fisheries Science, 61, 884-885.
91. Chow, S. and Inoue, S. (1993) Intra-and interspecific restriction fragment length polymorphism in mitochondrial genes of Thunnus tuna species. Bulletin of National Research Institute of Far Seas Fisheries, 30, 229-248.
92. Finnerty, J.R. and Block, B.A. (1992) Direct sequencing of mitochondrial DNA detects highly divergent haplotypes in blue marlin (Makaira nigricans). Molecular Marine Biology and Biotechnology, 1(3), 206-214.

93. Hare, J.A., Crown, R.K., Zehr, J.P., Juanes, F. and Day, K.H. (1998) A correction to: biological and oceanographic insights from larval labrid (Pisces: Labridae) identification using mtDNA sequences. Marine Biology, 130(4), 589-592.
94. Suneetha, B.K. and Dahle, G. (2000) Analysis of mitochondrial DNA sequences from Benthosema glaciale and two other myctophids (Pisces: Myctophiformes): Intraand interspecific genetic variation. Interspecific and inter specific genetic variation in selected mesopelagic fishes with emphasis on microgeographic variation and species characterization, University of Bergen, Bergen, Norway.
95. Murgia, R., Tola, G., Archer, S.N., Vallerga, S. and Hirano, J. (2002) Genetic identification of grey mullet species (Mugilidae) by analysis of mitochondrial DNA sequence: application to identify the origin of processed ovary products (bottarga). Marine Biotechnology, 4(2), 119-126.
96. Miya, M. and Nishida, M. (1998) Molecular phylogeny and evolution of the deep-sea fish genus Sternoptyx. Molecular Phylogenetics and Evolution, 10(1), 11-22.
97. Miya, M.and Nishida, M. (1996) Molecular phylogenetic perspective on the evolution of the deep-sea fish genus Cyclothone (Stomiiformes: Gonostomatidae). Ichthyolgical Research, 43(4), 375-398.
98. Reed, K.M., Dorschner, M.O., Todd, T.N. and Phillips, R.B. (1998) Sequence analysis of the mitochondrial DNA control region of ciscoes (genus Coregonus): Taxonomic implications for the Great Lake species flock. Molecular Ecology, 7(9), 1091-1096.
99. Prioli, S.M.A.P., Prioli, A.J. and Julio, H.F.J. (2002) Identification of Astyanax altiparanae (Teleostei, Characidae) in the Iguacu River, Brazil, based on mitochondrial DNA and RAPD markers. Genetics and Molecular Biology, 25(4), 421-430.
100. Doukakis, P., Birstein, V.J., Ruban, G.I. and Desalle, R. (1999) Molecular genetic analysis among subspecies of two Eurasian sturgeon species, Acipenser baerii and A. stellatus. Molecular Ecology, 8(12 Suppl 1), 117-127.
101. Felsenstein, J. (1985) Confidence limits on phylogenies: An approach using the bootstrap. Evolution, 39, 783-791.
102. Ryman, N., (2002) Population genetic structure. NOAA Technical Memoranda, Northwest Fisheries Science Centre Publication Page,http://www.nwfsc.noaa.gov/ publications/ tecmemos/index.cfm
103. Piry, S., Alapetite, A., Cornuet, J.M., Paetkau, D., Baudoiin, L. and Estoup, A. (2004) Geneclass2: A software for genetic assignment and first generation migrant detection. Journal of Heredity, 95(6), 536-539.
104. Luikart, G. and Cornuet, J.M. (1998) Empirical evaluation of a test for identifying recently bottlenecked populations from allele frequency data. Conservation Biology, 12(1), 228-237.
105. Luikart, G. and Cornuet, J.M. (1999) Estimating the effective number of breeders from heterozygote excess in progeny. Genetics, 151(3), 1211-1216.
106. Weir, B.S. and Cockerham, C.C. (1984) Estimating F-statistics for the analysis of population structure. Evolution, 38(6), 1358-1370.
107. Nei, M. (1983) Genetic Polymorphism and the role of mutation in evolution. In: Nei, M. and Kohen, R.K. Eds., Evolution of Gene and Proteins, Sinaver Associates, Sunderlans, 165-190.
108. Raymond, M. and Rouseet, F. (1995) An exact test for population differentiation. Evolution, 49, 1280-1283.
109. Fonseca, D.M., Dennis, A., Pointe, L. and Fleischer, C. (2000) Bottlenecks and multiple introductions: Population genetics of the vector of avian malaria in Hawaii. Molecular Ecology, 9(11), 1803-1814.
110. Zhang, Y.P., Wang, X.X., Ryder, O.A., Li, M.P., Zhang, Y. Yong, H.M. and Wang, P.Y. (2002) Genetic diversity and conservation of endangered animal species. Pure Applied Chemestry, 74(4), 575-584.
111. Efremov, V.V., (2002) Allozyme Variation in Pink Salmon Oncorhynchus gorbuscha from Sakhalin Island. Journal of Ichthyology, 42, 339-347.

112. Salini, J.P., Milton, D.A., Rahaman, M.J. and Hussein, M.G. (2004) Allozyme and morphological variation throughout the geographic range of the tropical shad, hilsa Tenualosa ilisha. Fisheries Research, 66(1), 53-69.
113. Lal, K.K., Kumar, D., Srivastava, S.K., Mukherjee, A., Mohindra, V., Prakash, S., Sinha, M. and Ponniah, A.G. (2004) Genetic variation in Hilsa shad (Tenualosa ilisha) population in River Ganges. Indian Journal of Fisheries, 51, 33-42.
114. Meggs, L.B., Austin, C.M. and Coutin, P.C. (2003) Low allozyme variation in snapper, Pagrus auratus, in Victoria, Australia. Fisheries Management and Ecology, 10(3), 155-162.
115. Dong, Z., Zhu, J., Yuan, X. and Wang, J. (2002) RAPD analysis of the genome DNA of Jian carp. Journal of Zhanjiang Ocean University, 22, 3-6.
116. Chang, Y., Sun, Y. and Liang, A. (2003) Study on cold tolerant traits for common carp Cyprinus carpio. Journal of Shanghi Fisheries University, 12, 102-105.
117. Feng, J., Zhang, X., Zhou, X., Chen, J. and Wang, L. (2003) RAPD markers and genetic diversity of Carassius auratus in the Qihe River. Transaction of Oceanology and Limnology, 4, 90-94.
118. Zheng, L. and Liu, C. (2002) Studies on random amplified polymorphic DNA (RAPD) of Epinephelus merra Bloch. Journal of Zhanjiang Ocean University, 22, 14- 18.
119. Exadactylos, A., Geffen, A.J., Panagiotaki, P. and Thorpe, J.P. (2003) Population structure of Dover sole Solea solea: RAPD and allozyme data indicate divergence in European stocks. Marine Ecollogy Progress Series, 246, 253-264.
120. Mamuris, Z., Stamatis, C. and Triantaphyllidis, C. (1999) Intraspecific genetic variation of striped red mullet (Mullus surmuletus L.) in the Mediterranean Sea assessed by allozyme and random amplified polymorphic DNA (RAPD) analysis. Heredity, 83(pt1), 30-38.
121. Almeida, F.S., Fungaro, M.H.P. and Sodré, L.M.K. RAPD and isoenzyme analysis of genetic variability in three allied species of catfishes (Siluriformes: Pimelodidae) from the Tibagi river, Brazil. Journal of Zoology, 253, 113-120.
122. Nelson, R.J., Wood, C.C., Cooper, G., Smith, C. and Koop, B. (2003) Population structure of sockeye salmon of the central coast of British Columbia: Implications for recovery planning. North American Journal of Fisheries Management, 23, 703-720.
123. Beacham, T.D., Supernault, K.J., Wetklo, M., Deagle, B., Labaree, K., Irvine, J.R., Candy, J.R., Miller, K.M., Nelson, R.J. and Withler, R.E. (2003) The geographic basis for population structure in Fraser River chinook salmon (Oncorhynchus tshawytscha). Fisheries Bulletin, 101, 229-242.
124. Brunner, P.C., Douglas, M.R. and Bernatchez, L. (1998) Microsatellite and mitochondrial DNA assessment of population structure and stocking effects in Arctic charr Salvelinus alpinus (Teleostei: Salmonidae) from central Alpine lakes. Molecular Ecology, 7, 209-223.
125. Adams, B.K. and Hutchings, J.A. (2003) Microgeographic population structure of brook charr: A comparison of microsatellite and mark-recapture. Journal of Fish Biology, 62(3), 517-533.
126. Senanan, W. and Kapuscinski, A.R. (2000) Genetic relationships among populations of northern pike (Esox lucius). Canadian Journal Fisheries Aquatic Science, 57, 391-404.
127. Salgueiro, P., Carvalho, G., Collares-Pereira, M.J. and Coelho, M.M. (2003) Microsatellite analysis of genetic population structure of the endangered cyprinid Anaecypris hispanica in Portugal: Implications for conservation. Biological Conservation, 109(1), 47-56.
128. Palm, S., Dannewitz, J., Jaervi, T., Petersson, E., Prestegaard, T. and Ryman, N. (2003) Lack of molecular genetic divergence between sea-ranched and wild sea trout (Salmo trutta). Molecular Ecology, 12(8), 2057- 2071.
129. Elliott, N.G. and Reilly, A. (2003) Likelihood of bottleneck event in the history of the Australian population of Atlantic salmon (Salmo salar L.). Aquaculture, 215(1-4), 31-44.
130. Alarcon, J.A., Magoulas, A., Georgakopoulos, T., Zouros, E. and Alvarez, M.C. (2004) Genetic comparison of wild and cultivated European populations of the Gilthead Sea bream (Sparus aurata). Aquaculture, 230(1-4), 65-80.
131. Kanda, N. (1998) Genetics and conservation of Bull trout: Comparison of population genetic structure among different genetic markers and hybridization with brook trout. Dissertation Abstracts International Part B: Science and Engineering, University of Montana, Missoula.

132. Kanda, N. and Allendorf, F.W. (2001) Genetic population structure of Bull trout from the Flathead River basin as shown by microsatellite and mitochondrial DNA marker. Transaction of American Fisheries Society, 130, 92-106.
133. Cagigas, M.E., Vazquez, E., Blanco, G. and Sanchez, J.A. (1999) Genetic effects of introduced hatchery stocks on indigenous brown trout (Salmo trutta L.) populations in Spain. Ecological of Freshwater Fish, 8(3), 141-150.
134. Chenoweth, S.F., Hughes, J.M., Keenan, C.P. and Shane, L. (1998) Concordance between dispersal and mitochondrial gene flow: Isolation by distance in a tropical teleost, Lates calcarifer (Australian barramundi). Heredity, 80, 1897-1907.
135. Gysels, E.S., Hellemans, B., Pampoulie, C. and Volckaert, F.A. (2004) Phylogeography of the common goby, Pomatoschistus microps, with particular emphasis on the colonization of the Mediterranean and the North Sea. Molecular Ecololgy, 13(2), 403-417.
136. Marzano, F.N., Corradi, N., Papa, R., Tagliavini, J. and Gandolfi, G. (2003) Molecular evidence for introgression and loss of genetic variability in Salmo (trutta) macrostigma as a result of massive restocking of Apennine populations (Northern and Central Italy). Environmental Biology of Fishes, 68(4), 349-356.
137. Jerry, D.R. (1997) Population genetic structure of the catadromous Australian bass from through out its range. Journal of Fish Biology, 51(5), 909-920.
138. Perkins, D.L. and Krueger, C.C. (1993) Heritage brook trout in northeastern USA: Genetic variability within and among populations. Transaction of American Fisheries Society, 122, 1515-1532.
139. Norris, A.T., Bradley, D.G. and Cunningham, E.P. (1999) Microsatellite genetic variation between and within farmed and wild Atlantic salmon (Salmo salar) populations. Aquaculture, 180(3-4), 247-264.
140. Palma, J., Alarcon, J.A., Alvarez, C., Zouros, E., Magoulas, A. and Andrade, J.P. (2001) Developmental stability and genetic heterozygosity in wild and cultured stocks of gilthead sea bream. Journal of Marine Biological Association of United Kingdom, 81(2), 283-288.
141. Wright, S. (1931) Evolution in Mendelian populations. Genetics, 16(2), 97-159.
142. Nei, M., Maruyama, T. and Chakraborty, R. (1975) The bottleneck effect and genetic variability in populations. Evolution, 29(1), 1-10.
143. Tessier, N. and Bernatchez, L. (1999) Stability of population structure and genetic diversity across generations assessed by microsatellites among sympatric populations of landlocked Atlantic salmon (Salmo salar L.). Molecular Ecology, 8, 169-179.
144. Nielsen, E.E., Hansen, M.M. and Loeschcke, V. (1999) Genetic variation in time and space: Microsatellite analysis of extinct and extant populations of Atlantic salmon. Evolution, 53, 261-268.
145. Lundy, J.C., Rico, C. and Hewitt, M.G. (2000) Temporal and spatial genetic variation in spawning grounds of European hake (Merluccius merluccius) in the Bay of Biscay. Molecular Ecology, 9(12), 2067-2079.
146. Heath, D.D., Bryden, C.A., Shrimpton, J.M., Iwama, G.K., Kelly, J. and Heath, J.W. (2002) Relationships between heterozygosity, allelic distance (^{d2}), and reproductive traits in Chinook salmon, Oncorhynchus tshawytscha. Canadian Journal of Fisheries and Aquatic Science, 59, 77-84.
147. Larson, S., Jameson, R., Etnier, M., Flemings, M. and Bentzen, P. (2002) Loss of genetic diversity in sea otters (Enhydra lutris) associated with the fur trade of the 18th and 19th centuries. Molecular Ecology, 11(10), 1899- 1903.
148. Beaudou, D., Baril, D., Roche, Â.B. and Le. Baron, M. (1995) Recolonization in a devastated Corsican river: Respective contribution of wild and domestic brown trout. Bulletin Francais de la Peche et de la Pisciculture, 337, 259-266.

CHAPTER 3

Identification and Characterization of Métiers in Multi-Species Artisanal Fisheries. A Case Study in Northwest Spain

Javier González-Álvarez[1], Laura García-de-la-Fuente[1*], Lucía García-Flórez[2], Mª del Pino Fernández-Rueda[2], Jorge Luis Alcázar-Álvarez[2]

[1]*INDUROT (Instituto de Recursos Naturales y Ordenación del Territorio), University of Oviedo, Mieres, Spain*

[2]*CEP (Centro de Experimentación Pesquera), Regional Ministry of Rural Development and Natural Resources of the Principality of Asturias, Gijón, Spain*

ABSTRACT

Artisanal fisheries are characterized by using a vast array of different fishing gears and target species. Such heterogeneity has traditionally complicated their management and the availability of accurate data on fishing effort and incomes. This study identifies and characterizes the métiers currently used by an artisanal fishing fleet based in a coastal Atlantic region in South-Europe (Asturias, Northwest Spain), by applying an integrated method of analysis which combines administrative data on fishing licenses, daily sales statistics and information from surveys. As a result, 21 artisanal fishing métiers have been identified, with special relevance of the trammel net (targeting Lophius spp.), the Merluccius longline and the Merluccius gillnet métiers. Spatio-temporal patterns of fishing effort have been also described, and outputs and inputs demands per métier have been quantified. It has been also found a high degree of alternation between different métiers during the fishing season and significant differences in terms of technical aspects and workforce. Finally, fishermen's short- and long-term strategies (métier choice) are discussed, different comparisons are made on the basis of other South-European studies and main findings are linked to artisanal fisheries management.

Keywords: Métier, Artisanal Fisheries, Multi-Species Fleet Management, Bioeconomic Sustainability, North West Spain

1. INTRODUCTION

Despite their socioeconomic importance in many coastal communities, European artisanal fisheries have traditionally received less interest than industrial fisheries in marine and fisheries sciences. In fact, they, play a fundamental role in terms of human and territorial development at a local-regional scale [1] - [4] . In Spain, many studies have outlined that artisanal fisheries are suffering from a heavy decline in recent times [5] [6] . Although the high heterogeneity of these activities across different South-European countries and regions has impeded so far a standardized definition of these fisheries[1] [2] [7] - [9] , most of them are characterized by a high diversification in gears and fishing techniques and changing spatio-temporal patterns of their use. These features make difficult artisanal fishing management [2] [10] - [12] .

In such a complex context, the concept of métier has been traditionally used to describe and characterize fisheries, as it is able to capture the major part of their heterogeneity [13] - [15] . Métier has been defined by European Decision 2008/949/CE[2] as "a group of fishing operations targeting a similar (assemblage of) species, using similar gear, during the same period of the year and/or the same area and which are characterised by a similar exploitation pattern". Therefore, the concept of "métier" is closely linked to the fishing activities, patterns, traditions and gears employed by artisanal fishermen: each métier gathers a group of fishing operations defined by the combination of fishing gear, target species, area and season [11] [16] , making up homogeneous units that summarize in a single variable the main characteristics of a large number of fishing trips [17] .

As a consequence, a métier-based approach is very useful for understanding the spatio-temporal patterns of fishing effort allocation and resulting catches [10] [11] . Moreover, technical and bioeconomic characterization of fishing métiers is a basic step to accurately assess the real fishing effort and pressure exerted by artisanal and small-scale multi-specific fisheries. This understanding constitutes a key step for the improvement of economic and ecosystem fisheries management, as it helps to select appropriate management strategies and measures [12] [18] [19] , and provides basic information to address marine spatial planning concerning artisanal fleets [20] .

As Forcada et al. [10] have already acknowledged, identification of métiers has been widely based on the analysis of large historical datasets on species composition or catches data [17] [21] - [27] . The problem arises when such information is not available for artisanal fisheries. Recent research projects addressing métiers identification or applying a métier-based approach in the South-European area constitute illustrative examples of these limitations. On the one hand, the IBERMIX Project [28] was aimed at redefining and providing métier-disag- gregated fishing data for the Spanish fleets operating in the Atlantic Iberian Peninsula waters oriented to ICES assessments and STECF mixed-fisheries management; as the segmentation was based on logbooks data, small- scale fleets were excluded from the analysis[3]. On the other hand, the TECTAC Project [29] has been aimed at addressing and supplying fisheries managers with a modelling tool to evaluate the impact of regulations on the dynamics of fleets and fishing mortality, on the basis of fishing tactics and strategies identification by métier. To do so, 30 demersal fleets from Denmark, the United Kingdom (England), the Netherlands, France and Spain were investigated, but in almost all the cases, vessels chosen for the project registered total length ranges above 10 - 12 metres, as well as a weak representativeness of so-called small-scale fisheries[4].

In many cases identification and characterization of artisanal métiers have been carried out based on alternative approaches such as onboard samples or surveying methods [10] [11] [30] - [32] , although researchers have also pointed out reliability problems that may arise when data elicited from resource users are employed [33] . Nevertheless, there is no precedent in exploring the potential of other sources of information to identify and characterize poor-data artisanal fisheries, as for example the administrative registers of fishing licences used by artisanal vessels: in many South-European coastal regions, artisanal vessels are required to communicate to fishing Administration each daily change in the fishing gears they used (equivalent to different artisanal fishing licences), so that it supplies an interesting source to be applied in monitoring their activity.

All these things considered, the main objective of this study has been to carry out a detailed and updated characterization of the main technical, biological and economic issues concerning the métiers used nowadays by the artisanal fishing fleet of Asturias, an Atlantic region located in Northwest Spain. To reach this goal, a novel approach using administrative data on fishing licenses combined with daily sales registers and information collected through a surveying process has been applied in order to: i) identify and classify the artisanal fishing métiers performed by the aforementioned fleet, ii) characterize their main technical, bioeconomic and temporal patterns, iii) provide basic information to improve decision-making processes and iv) discuss the relevance of the findings for the design of specific sustainable management measures in the context of South-European artisanal fisheries.

The present study has been part of a wider project about artisanal fishing fleets of the European Atlantic Arc ("PRESPO: Sustainable Development of Atlantic Arc Artisanal Fisheries"), developed between 2009 and 2012 within the Atlantic Area Transnational Cooperation Programme 2007-2013.

The structure of this article is as follows. Section 2 describes the case study and summarizes the data collection process, explaining the main characteristics of data sources used and the methodology applied, in order to identify and characterize the métiers. In Section 3 results are presented addressing the following topics: métier identification and classification, level of activity/use, alternation and seasonality, target species and species composition, weight and incomes per daily sale bill and, finally, technical and workforce demands associated to each métier. Last part of the article includes a chapter of discussion (Section 4) about the main results achieved linked to artisanal fisheries management and remaining constraints that need further research and a section. To conclude, Section 5 shows the main conclusions.

2. CASE STUDY DESCRIPTION, MATERIAL AND METHODS

2.1. Case Study

This study was focused on the Principality of Asturias, an Atlantic region in Northwest Spain (Figure 1) that has 335 km of shoreline along the Cantabrian Sea and about one million inhabitants. Regional fishing grounds, take up 984,938 hectares[5]. A detailed description of this area was included in Garcia-de-la-Fuente et al. [12] . Asturian artisanal vessels are allowed to use a large range of traditional fishing gears such as trammel nets, longlines, traps, gill nets, etc., targeting many valuable species sold in local markets and changing frequently the fishing gears and techniques along the year [12] . These issues allow them to develop a highly adaptive and flexible activity in order to maximize profitability. On the other hand, the high variety and alternation between different métiers during the fishing season makes difficult to estimate with confidence fishing effort, real pressure over marine resources and coastal ecosystems, the performance of each fishery and any accurate economic indicator about the activity's return.

Regional experts and public administration agree to consider as artisanal at least those vessels registered as "artes menores" in the National Census of the Operating Fishing Fleet. This group comprises 233 boats[6] distributed along 19 ports from where they exploit areas that can be reached in a few hours. The average total length registered by this fleet is 8.8 m, as well as a capacity of 4.5 GT (gross tonnage) and an engine power of 51 kW. Furthermore, vessels from other census categories are allowed to operate with a temporary "artes menores" license upon previous request to the regional fishing administration.

2.2. Data Collection

Official statistics, administrative data and information from a specific fishers' survey were used in this study. Administrative data on artisanal fishing modalities (licenses) used per day and vessel and daily sales bills[7] of the artisanal fleet during 2009 have been exploited together to identify and characterize artisanal fishing métiers. Detailed information about total catch, including discards and unsold catches, was not available for this study.

The regional ministry with powers in marine fishing affairs establishes the fishing modalities (licenses) that can be used by the artisanal fleet. They present a high level of disaggregation, considering as different modalities those similar gears with minor technical differences (e.g. mesh size) and targeting different species. Although Asturian artisanal vessels are allowed to alternate between fishing modalities, they are requested to use just one per fishing day and to communicate to the fishing administration any change in the modality used.

The analysis of both information sources resulted in an integrated database for 205[8] vessels. A matrix was built from this database, whose rows represented daily sales bills per vessel (a total of 14,254 fishing operations or cases), while columns contained information about the vessel, fishing modality, landings weight (per species) and income obtained. Taking into account the activity patterns of the Asturian artisanal fleet [12], it was assumed that daily sales bills correspond to the landings and retained captures for commercial purposes of a single fishing day for the analyzed vessels.

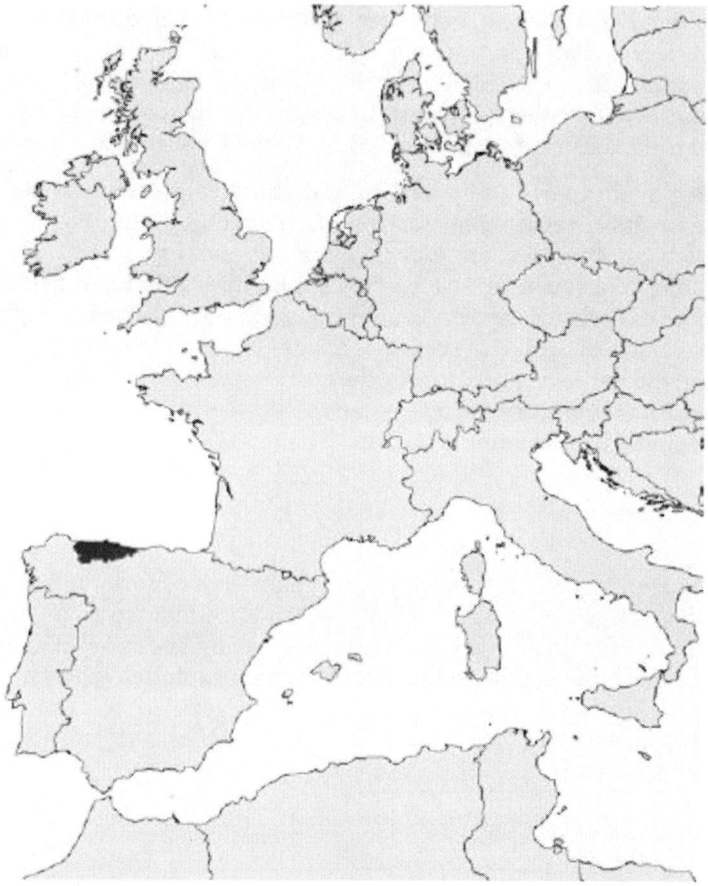

Figure 1. Map of the study area.

As complementary source, a survey was carried out through face-to-face interviews with ship-owners and ship's masters, and information regarding the use of different métiers was applied for this research. Studies elsewhere using interview data show that this approach can produce highly valuable information for fisheries assessment [10] [12] [34] . The interviews were carried out between May 2009 and March 2010 following a stratified random sampling methodology based on the number of boats per fishing port in order to avoid overestimating the importance of highly located métiers [26] . As result, a total of 146 vessels from 18 Asturian fishing ports were interviewed, which implies coverage of 71.2% of the artisanal vessels with landings in 2009. This sample was considered also representative by port: in 13 fishing ports, a sample covering between 40% - 64% of the local artisanal vessels was surveyed, while in other 2 fishing ports this coverage raised up to 73% and 88% respectively. Finally, only in 3 cases the sample was under the 30% of vessels based in the local fishing port. Among other issues[9], in the questionnaire fishermen had to declare all the gears they used, specifying (per fishing gear) the month/s of the year, main target species, frequency of hauls (daily, other), percentage of total annual landings caught, fishing ground/s and distance from the coast to the fishing areas.

2.3. Methodology for Métiers Identification and Characterization

Métiers can be identified at different levels of aggregation, and depending on this, results obtained can vary (mainly in number). Literature on métiers identification outlines the importance of linking them to the target species; this question is especially relevant in artisanal métiers, as many of them target one specific species. On this basis, some limitations arise when a classification framework as the one included in Decision 2008/949/ EC[10] [35] is taken as reference to identify artisanal fishing métier, as in many cases its maximum level of disaggregation (level 6, specifying gear type, species target assemblage and mesh size and other devices) does not allow reflecting all the rich and vast array (amount and variety) of artisanal métier. Thus, this research proposes the need of setting out alternative identification/classification frameworks with higher resolution and level of disaggregation. At the same time, it is desirable to put them in correspondence with European DCF.

To do so, this article proposes as novelty that administrative data on fishing licenses are used to get a highly detailed level of métiers practiced by artisanal fleets, taking into account that some fishing authorities (above all, at a local-regional scale) have developed management instruments well-fitted to particularities of the métier practiced by the artisanal fleets under their competence. Thus, the existing artisanal fishing modalities (administrative licenses) were used to identify most of artisanal métiers in a first stage for the region studied. As a result, all those operations using the same fishing modality and targeting the same species were a priori considered as constituting a distinct métier. On the other hand, during the interviewing process artisanal fishers highlighted that they used some of the existing fishing modalities to develop different fishing operations in terms of seasonality and objective species. Hence, in a second stage further métiers were identified thanks to the surveying process[11]. This information was later confirmed by the analysis of daily sales data.

The final characterization of métiers was based on a) the administrative license used, b) species declared as targeted by the fishers during the interviews, c) the weight, income and species composition according to the sales register and d) the months of the year when most of the operations were carried out. To assess significant differences in landings and profitability, average weight and income per daily sales among the identified métiers were tested by one-way analysis of variance (ANOVA). However, given the large size sample (24,254 cases) and number of factors included in the analysis (16 different métiers), none of the analyses met the prerequisite of variance homogeneity (Levene's test for equality of variances) which is needed to address the ANOVA. Alternatively, a non-parametric Kruskal-Wallis test was applied[12] using the software SPSS [36] : this test is well suited for testing whether sample populations come from the same distribution, does not assume normal distributions of the residuals for such samples and is suitable when outliers in data are suspected.

2.4. Typology of Vessels per Métier

Describing the different métiers may not be sufficient to understand the mechanisms involved in the dynamics of artisanal fishing activity [17] . Due to this, it is necessary to analyze not only the main fishing métiers at trip level, but to identify the scale and characteristics of vessels carrying out each different métiers and how those characteristics may influence fishing decisions and strategies. To do so, information from 205 vessels about boat total length, gross tonnage, engine power and crewmembers was collected from the National Census of the Operating Fishing Fleet and the Social Marine Institute respectively, which allowed estimating the average technical characteristics demanded per métier.

Finally, an ANOVA was carried out to test significant differences in such variables between the different métiers. When the normality and homogeneity of variance requisites were not met, log-transformation $x' = \log(x + 1)$ was applied (tonnage and crew members).

3. RESULTS

3.1. Métier Identification and Technical Classification

In order to organize results according to European Data Collection Framework (2008/949/EC) [35] , the identified métier are shown in correspondence with level 4 (gear type) of this regulation (Table 1). A total of 21 artisanal fishing métiers were identified[13], using 8 types of fishing gears according to the aforementioned DCF. Of these métiers, 18 correspond directly to fishing modalities as established by the regional fishing administration, while 3 new métiers were identified (Conger longline, Nephrops trammel net and Homarus pot). Although these métiers are practiced under some of the existing fishing licenses, they target different species and present slight technical variations, which justifies considering them as a distinct métier. In order to identify and characterize these new three métiers within the daily sales bills database, all of those fishing days using longlines, trammel nets or pots in which the declared target species for these métiers (Conger conger, Nephrops norvegicus and Homarus gammarus) represented more than 40% of the total sales were considered to belong to the respective new identified métier.

The main métiers performed by the Asturian artisanal fleet in 2009 were Lophius spp. trammel net[14] (31% of the fishing days), Merluccius longline (19%) and Merluccius gillnet (12%). Other important métier were Octopus pot, Maja pot and Mullus trammel net.

Table 1. Métiers identified in the Asturian artisanal fleet, according to gear type categories established by the European DCF.

Gear type (DCF)	Identified métiers		Target species	No of fishing days in 2009
	Métiers	Local name		
Set longlines (LLS)	*Merluccius* longline	Palangrillo	*Merluccius merluccius*	2672
	Conger longline[a]	Palangrillo congrio[a]	*Conger conger*	562
	Dicentrarchus longline	Palangrillo lubina	*Dicentrarchus labrax*	847
Hand and Pole lines (LHP)/ (LHM)	Rod-hook[b]	Pincho-caña[b]	*Merluccius merluccius*	0
	Squid jig-hook	Potera	*Loligo vulgaris*	731
	Rod-hook[c]	Anzuelo para caballa	*Scomber scombrus*	278
Trammel net (GTR)	*Mullus* trammel net	Trasmallo	*Mullus surmuletus*	902
	Nephrops trammel net[a]	Trasmallo cigala[a]	*Nephrops norvegicus*	5
	Lophius tramelnet	Miño	*Lophius* spp.	3012
Set gillnet (GNS)	*Mullus* gillnet	Beta salmonetera	*Mullus surmuletus*	629
	Merluccius gillnet	Beta	*Merluccius merluccius*	1658
	Maja gillnet[b]	Volanta marisquera[b]	*Maja squinado*	0
Driftnet (GND)	Driftnet[b]	Abareque[b]	*Sardina pilchardus*	0
Purse seine (PS)	Purse seine[b]	Cerco chicharro[b]	*Trachurus trachurus*	0
Pots and traps (FPO)	*Octopus* pot	Nasa pulpo	*Octopus vulgaris*	880
	Maja pot	Nasa marisco	*Maja squinado*	1203
	Homarus pot[a]	Nasa bugre[a]	*Homarus gammarus*	163
	Anguilla pot	Nasa butrón	*Anguilla anguilla*	3
Miscellaneous	Glass eel	Cedazo	*Anguilla anguilla*	689
	Seaweed	Algas	*Gelidium* spp	0
	Mirror	Espejo	*Maja squinado*	19
TOTAL	**21** ***métiers identified.*** **16** ***métiers currently active***		**11** ***target species linked to the active métiers***[d]	**14,254** ***fishing days***

[a]Métiers identified through the fishers' survey. [b]Artisanal métiers allowed by existing fishing modalities but that were not practiced by the artisanal fleet in 2009. [c]Unlike the "Rod-hook" métier targeting *Merluccius merluccius*, in the "Rod-hook" métier targeting *Scomber scombrus* artificial bait is used. [d]*Merluccius merluccius; Conger conger; Dicentrarchus labrax; Loligo vulgaris; Scomber scombrus; Mullus surmuletus; Nephrops norvegicus; Lophius spp.; Octopus vulgaris; Anguilla Anguilla; Gelidium spp.*

On the other hand, at least four of the potential métiers[15] were not carried out by the artisanal fleet during 2009 (Merluccius rod-hook, Maja gillnet, driftnet, purse seine), while some others were only used by few vessels (mirror, Nephrops trammel net and Anguilla pot). In terms of gear types, longlines were the most used gears (29% of fishing operations), followed by trammel nets (27%) and gillnets and traps (both with 16% of fishing operations).

In addition to the problem of low information and poor knowledge in artisanal fleets, these results confirm the heterogeneity of a sector (the artisanal one) that frequently has been globally treated "as a whole" and classified within an only and undifferentiated national census category, which usually has been taken as a reference for management and planning public policies at a national and European scale. As the present case study has revealed, artisanal fleets actually comprehend a miscellaneous group of fishing modalities mainly practiced by multi-gear vessels with high differences in terms of gears, target species and level of practice within the artisanal fleet itself. As result, their internal complexity and heterogeneous structure should be taken into account when management measures are designed and implemented in those regions where artisanal vessels represent a relevant percentage of the total fleet.

Nevertheless, information gathered through the interviewing process was useful to understand the conditions and fishing strategies that influence fishers' decisions on what métiers should be used on a short and long term perspective. Thus, fishers' survey revealed that artisanal fishing patterns have changed significantly over the last years, and it is expected that their evolution will continue through the coming years. Certain métiers such as rod-hook, seaweed recollection or purse-seine are currently only practiced by vessels from other census categories that request to operate with "artisanal gears" temporally, while other métiers are quickly reducing their levels of activity in Asturias. As result, only 76% of the identified métiers can be considered active nowadays. Moreover, a trend of progressive reduction in the diversity of métiers practiced by this fleet in the last years has been confirmed: due to changes in fishing strategies and variations in fishers' rational behavior to maximize activity's profitability, in recent times net métiers (as for example Lophius tramelnet, Mullus trammel net and Merluccius gillnet) are getting higher presence and replacing other traditional ones.

Finally, in addition to the identified fishing métiers it must be also mentioned the goose barnacle (Pollicipes pollicipes) recollection, as an important complementary fishing activity for artisanal fishermen in the central and west coast of Asturias, especially from October to April, when other fishing possibilities are scarce. Even though this activity cannot be considered as a métier (it is a harvesting activity carried out through a personal seafood recollection license), a large number of vessels crewmembers alternate goose barnacle recollection with other fishing métiers considered compatible[16]. As a consequence, the vessel remains inactive for certain periods of time when all the crewmembers are dedicated to the recollection activity and the obtained incomes are distributed among the crew[17].

3.2. Métiers Alternation and Spatio-Temporal Distribution

Artisanal vessels have shown a high degree of alternation between different métiers along the year. Results in Figure 2 confirm the multi-specific and multi-gear character of the Asturian artisanal fleet, as in 2009 the 55% of artisanal vessels practiced 3 or more different métiers. This high flexibility is also observed in terms of fishing gears alternation, with 18.4% of boats using métiers corresponding to 3 or more different gear types. The annual mean number of métiers practiced was 2.89 (with a standard deviation of ±1.23). These findings reflect that annual fishing effort is distributed among an "optimal" combination of métiers for each vessel, and a certain degree of "substitution effect" among métiers exists.

Analyzing the spatio-temporal distribution of the métiers, it can be observed (Table 2) that the main métiers in terms of annual number of fishing days (Lophius spp. trammel net and Merluccius longline) are present along the majority of the Asturian fishing ports during most part of the year. In general terms, most ports use a highly diversified array of métiers, although some of them appear to be "specialized" (Figure 3) in certain types of fishing gears (e.g. in Bañugues only pot gears are utilized, while 60% of the fishing operations in Puerto de Vega are carried out using longlines). On the other hand, métiers carried out in shallow waters close to the shore and in estuaries targeting crustaceans (pots and Mirror), cephalopods (Octopus pot) or anguilliformes (Anguilla pot and glass eel fishing) show a highly seasonal pattern (Figure 4), being generally located in a few ports with a long tradition in the utilization of these type of métiers.

These latter métiers (and the mentioned goose barnacle recollection) are not only a clear expression of the traditional fishing gears performed by the regional fleet, but they also play a fundamental role as a complementary fishing activity during certain periods of the year when it is not possible to carry out other métiers. This is especially important between January and March, when longlines use to be inactive as the adverse weather conditions do not allow the gear to be operated efficiently.

3.3. Output Analysis: Weight, Income and Species Composition per Métier

In the 14,254 cases analyzed, a total of 115 commercial species were registered, even though only 11 have been identified as target for the 16 active métiers (Table 1). Although most of them were fishes, crustaceans (Maja squinado, Homarus gammarus, Necora puber, etc.) and cephalopods (Octopus vulgaris, Loligo vulgaris) were also recorded. The species with the highest market value was Anguilla anguilla (glass eel), followed by several crustaceans (Scyllarus arctus, Palinurus elephas and Palaemon serratus); although, landings of such species are relatively low in terms of weight, crustaceans total (real) landings are likely higher because a part is commercialized out of fish markets (direct sale). Other species with lower market prices contributed in a higher proportion to the total income of the artisanal fleet (Merluccius merluccius, Scomber scombrus, Mullus surmuletus).

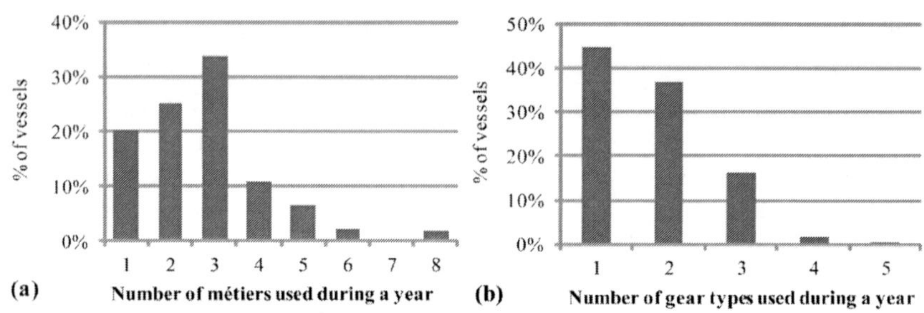

Figure 2. Number of métiers (a) and gears types (b) practiced per vessels over a one-year period (in percentage).

Average weight per sale bill was 99.19 kg (s.d. ± 382.78) in 2009, while the average income per fishing day was €396.65 (with a standard deviation of €454.82). The analyzed métiers significantly differed in the average values of both parameters (catches Kruskal-Wallis, $\chi^2 = 4246.78$, df = 15, $P < 0.001$; incomes Kruskal-Wallis, $\chi^2 = 862.69$, df = 15, $P < 0.001$). Given that under the null hypothesis of the Kruskal-Wallis test the mean ranks of the groups are the same, p-values obtained (under 0.05) lead to reject this assumption and confirm that significant differences exist between the analyzed métiers in terms of landings and incomes. As daily sale bills do not include discards or unsold captures, weight values presented through this section cannot be taken as a precise indicator of the total fishing effort exerted by the artisanal fisheries over the marine resources; notwithstanding, it is a highly valuable information regarding the economic activity of the artisanal vessels and the expected returns of the performed métiers in terms of species composition retained for sale, weight and incomes.

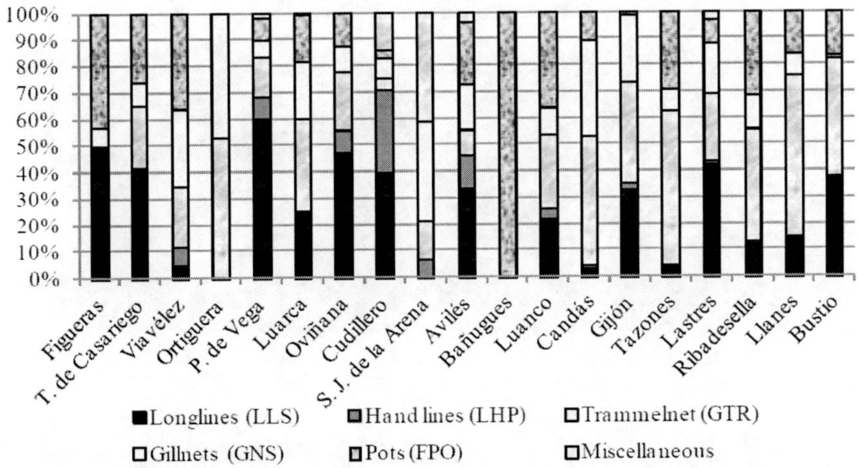

Figure 3. Degree of utilization of the different gear types by fishing port (in percentage).

The rod-hook targeting Scomber scombrus is the métier that generates the highest weight landed and income per fishing day, significantly greater than the rest of métiers (Table 2). Notwithstanding, its use is reduced to a few months (or even a few weeks) between February and May, depending on the migratory patterns of the objective species (Scomber scombrus) and, mainly, the existence of Total Allowable Catch (TAC) established at a EU level. On the other hand, the versatility (possibility of been carried out at least 7 months per year), high- catch and income profile of the Lophius trammel net and the Merluccius longline métiers explain their high degree of utilization by the Asturian artisanal vessels. The significant commercial value (first sale prices) of Lophius spp. and M. merluccius (about 6 €/kg both species) assures a relatively good income to vessels operating with Lophius trammel net and Merluccius longline (an average income of €422.93 and €433.97 respectively per fishing day). Other métiers used through most of the year are the Conger and Dicentrarchus longline and the Merluccius gillnet, that also show relatively high average incomes per fishing day, ranging from €300 to €425 (about 2.4 and 3.6 €/kg respectively).

Figure 4. Artisanal métiers practiced in the Asturian coastline. From the top to the bottom, Glass eel, Rod-hook targeting Merluccius merluccius and Mirror.

Table 2. Main characteristics of the artisanal active métiers in 2009.

Métier	Type of bottom[a]	Average catch (kg) per sale bill	Average income (€) per sale bill	% of fishing days	No of fishing vessels using it	No of fishing ports	Seasonality											
							J	F	M	A	M	J	J	A	S	O	N	D
Merluccius longline	R, M	70.08	433.97	18.75%	66	16												
Conger longline	R	126.26	304.29	3.94%	33	11												
Dicentrarchus longline	R	75.15	359.26	5.94%	20	8												
Squid jig-hook	R, S	34.23	300.14	5.13%	36	8												
Scomber rod-hook	A	1468.23	885.69	1.95%	13	7												
Mullus trammel net	R, RT	67.25	240.50	6.33%	22	9												
Nephrops trammel net	M	55.70	387.94	0.04%	2	1												
Lophius trammel net	S, M	72.28	422.93	21.13%	61	16												
Mullus gillnet	S	41.61	240.50	4.41%	16	8												
Merluccius gillnet	S, M	116.24	425.35	11.63%	40	14												
Octopus pot	A	56.42	275.34	6.17%	38	7												
Maja pot	R	31.56	295.65	8.44%	32	16												
Homarus pot	R, RT	29.80	428.19	1.14%	28	12												
Anguilla pot[b]	-	54.66	458.66	0.02%	1	1												
Glass eel[b]	-	4.75	323.85	4.83%	40	5												
Mirror	R	8.64	185.63	0.13%	1	1												

[a]Type of bottom (R, rocky bottoms; S, sandy bottoms; M, muddy bottoms; RT, mixture of rocky and sandy bottoms; A, All types of bottoms); Seasonality (no shading, no use or residual use; light green shading, moderate use; dark green shading, intensive use). [b]Fished on estuaries.

Although for most of métiers the largest sales in terms of weight correspond to the declared target species, for certain métiers (such as Maja and Homarus pot, Glass eel or Nephrops trammel net), the target species (with high first-sale prices) also constituted the highest contribution in terms of income. The species composition for the 5 most abundant commercial species of the 4 main métiers are presented in Figure 5. As it can be observed, while the three first métiers show a more even distribution of landings, 90.8% of the Octopus pot catches retained for sale correspond to the target species. Although significant differences do exist in terms of selectivity among the métiers performed by the Asturian artisanal fleet, in all cases the five most important species represented at least the 50% of the total annual weight landed or income; nevertheless, results achieved can be used for stakeholders to work in favour of gears selectivity improvements for the main species retained for commercial purposes.

3.4. Technical and Workforce Demands Associated to Each Métier

In order to better understand artisanal activity and exploitation patterns, technical and workforce demands per Achieved results allow more accurate inferences about costs structures and output models for this fleet at a métier level: both technical and workforce (number of crewmembers) characteristics significantly differ depending on the métiers (Table 3). Rod-hook targeting Scomber scombrus presents the higher demands (Figure 6) in terms of vessels total length (12.32 m), engine power (98.9 kW) and gross tonnage (11.55 GT), as a direct consequence of this fishery exploitation patterns: longer distance travels and large captures per fishing day (an average of 1468.23 kg, Table 2).

Vessels targeting mainly crustaceans and glass eel present the opposite typology: small vessels with low gross tonnage, engine power and crewmembers that develop their activity in shallow waters close to their respective fishing ports. Such small vessels may find difficulties to assure a stable income level throughout the year, as they focused on highly seasonal métiers and vessels' characteristics may limit the capacity of performing other métiers for which different technical characteristics are required. In general terms, it can be observed that vessels using trolling lines show the highest technical parameters, followed by vessels using trammel net métiers, gillnets, longlines and, finally, pots and other gears.

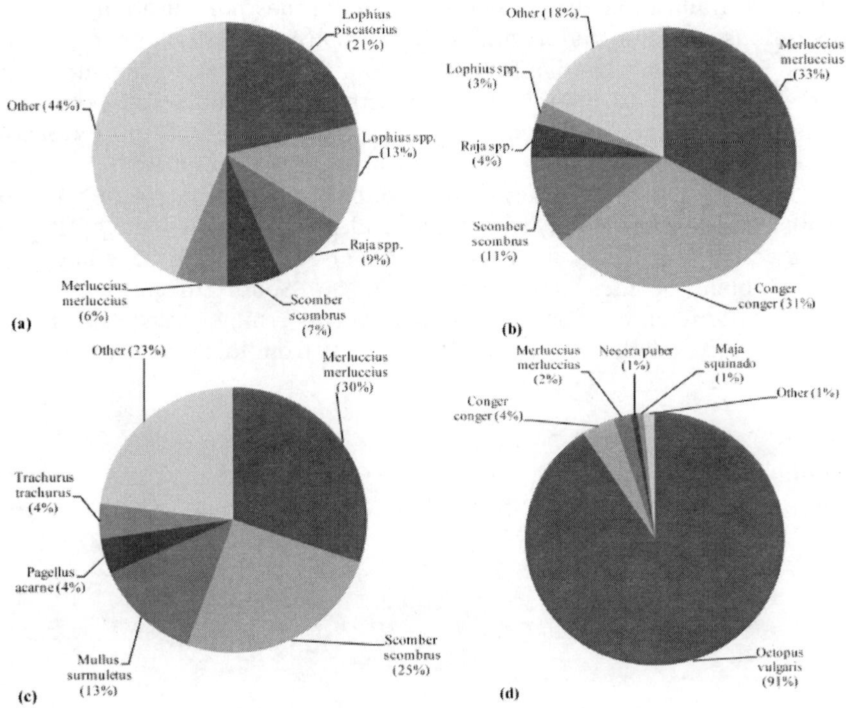

Figure 5. Species composition for the 5 most abundant commercial species of the 4 main identified métiers corresponding to 4 different gear types: (a) trammel net ("miño"), (b) Merluccius longline ("palangrillo"), (c) Merluccius gillnet ("beta") and (d) Octopus pot ("nasa de pulpo").

4. DISCUSSION

The concept of artisanal métiers actually embraces a highly diverse group of fishing modalities practiced by multi-gear vessels with a high variety of fishing operations, landings, incomes and spatio-temporal patterns. Many South-European artisanal fleets are mainly integrated by vessels under 12 and 10 metres of total length[18]; according to Regulation (CE) 1224/2009, these vessels are no subjects to comply with VMS and logbooks fill-in requirements respectively, which derives in a lack of monitoring systems that provide accurate, reliable and quantitative information on spatial and temporal fishing patterns for most of artisanal vessels[19]. This situation has traditionally complicated a sustainable management of these boats, as basic and detailed data on their activity is not available. Usually in this context, first-sales notes become the only systematic source of information, although it allows analyzing neither spatial issues nor duration of fishing operations, so fishing effort and its location are mainly unknown for these fleets. On the other hand, information from specific monitoring surveys on vessels less than 12 metres of total length and voluntary collaboration of fishermen in data collection processes also represent an interesting complementary data source for artisanal vessels, but with important limitations (lack of accuracy and reliability of quantitative information, spatial information mainly qualitative and poorly detailed, etc.). Although some research projects, as IBERMIX [28] or TECTAC [29] have addressed the problem of identifying métiers in the South-European area, the aforementioned limitations on fishing data concerning vessels less than 12 metres have excluded most of artisanal boats from these relevant analyses so far.

Table 3. Summary of the one-way ANOVA for technical characteristics among the artisanal métiers identifieda.

Vessel characteristics	Transformation		Sum of squares	d.f.	Mean squares	F
		Among groups	15,645.24	15	1043.02	315.37***
Length (m)	-	Within groups	47,089.49	14,238	3.31	
		Total	62,734.73	14,253		
		Among groups	1,796,675.25	15	119,778.35	105.91***
Engine power (kW)	-	Within groups	16,102,276.83	14,238	1130.94	
		Total	17,898,952.08	14,253		
		Among groups	56,611.49	15	3774.10	152.42***
Tonnage (GT)	ln(x + 1)	Within groups	352,543.27	14,238	24.76	
		Total	409,154.76	14,253		
		Among groups	2520.76	15	168.05	230.58***
Crewmembers	ln(x + 1)	Within groups	10,022.15	13,751	0.73	
		Total	12,542.91	13,766		

[a]n = 14.254 (13,766 for crew members' analysis); d.f., degrees of freedom; F, F-Ratio, P, level of significance (***$p < 0.001$). A dash (-) indicates no transformation.

This situation has traditionally hampered the design and implementation of specific management measures based on a good knowledge on the characteristics and problems of each métier. At the same time, South-European Atlantic regions as Asturias are registering a progressive disappearance of artisanal métiers with low economic profitability (although with high level of environmental and social sustainability), in parallel to a concentration process of artisanal fishing activities on a few main métiers.

Although previous studies also used the information from daily sales and direct interviews to fishermen in order to identify métiers [3] [10] [18] [24] [26] [32] [38] - [40] , this research proposes a methodological novelty based on the detailed monitoring of the regional administrative register of artisanal fishing licences used per day and vessel in the region of study.

(a)

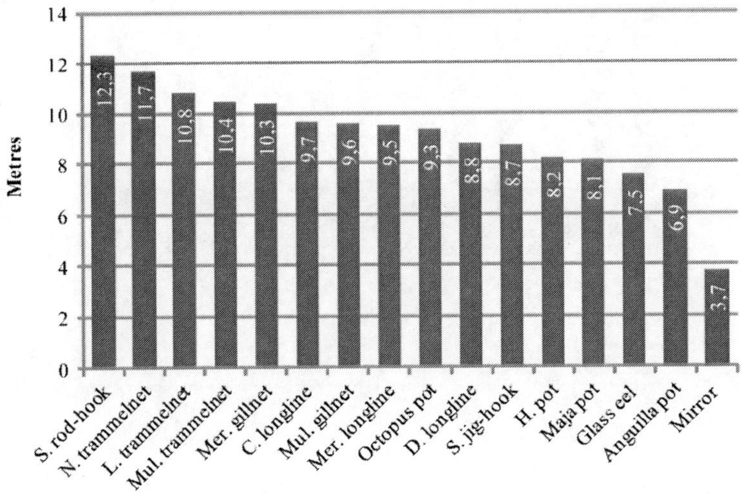

(b)

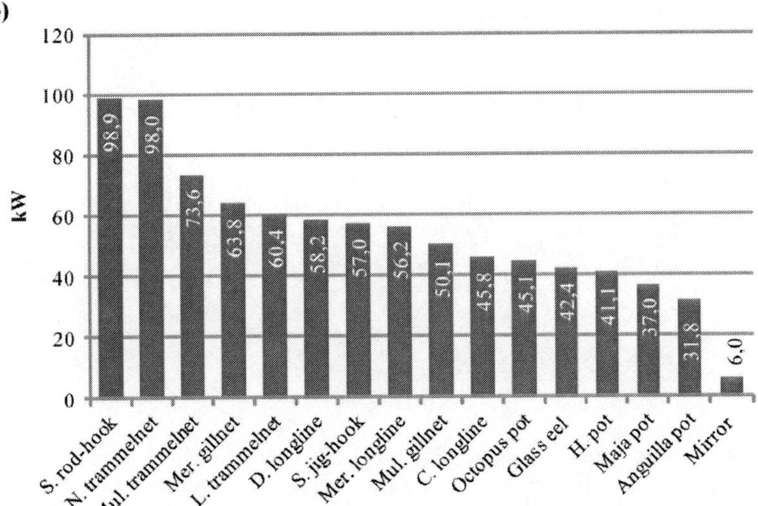

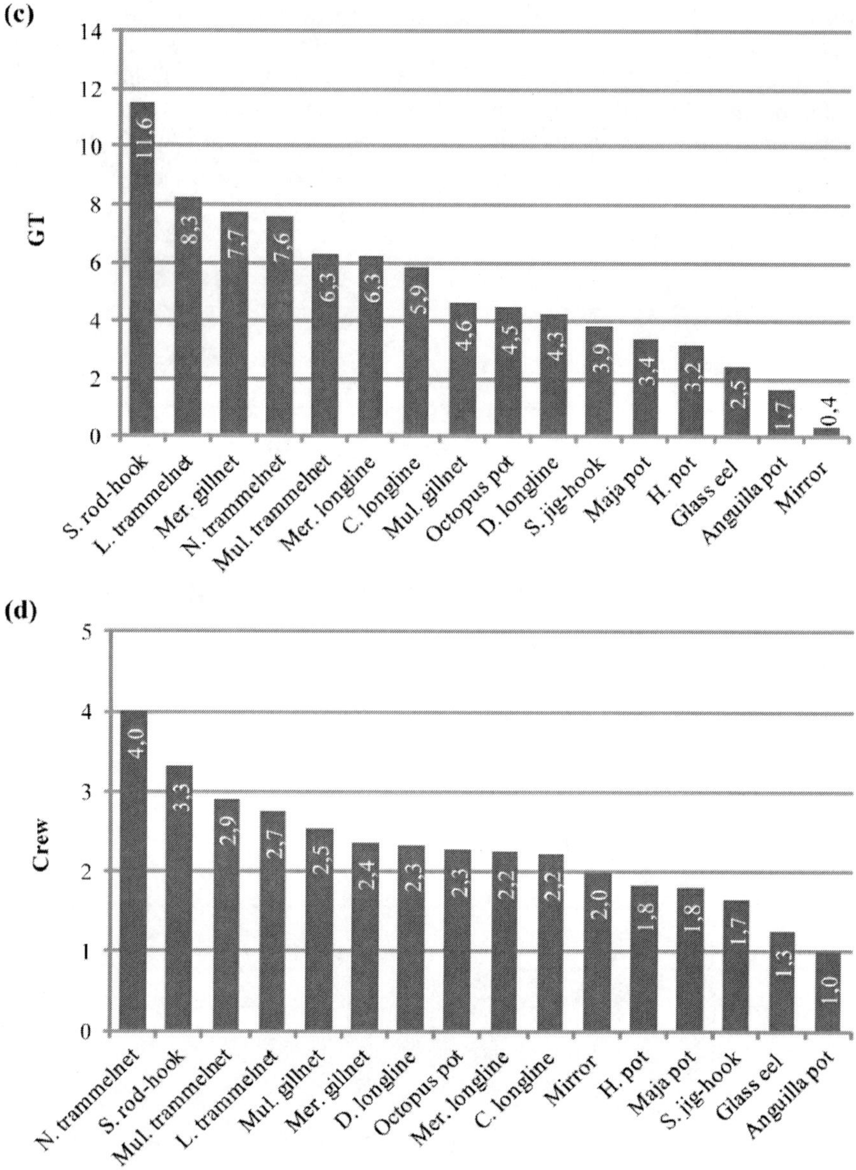

Figure 6. Average technical characteristics of artisanal vessels: (a) total length, (b) engine power, (c) tonnage and (d) crewmembers] according to the métiers performed.

A value added of this approach, in poor-data contexts, is that it allows reaching a highly-disaggregated and detailed artisanal metier identification and classification, that comprehends all the richness and diversity of métiers practiced by these type of boats, and provides higher resolution than other classification frameworks as the European Data Collection Framework (DCF-2008/949/EC).

Data mining of daily changes in fishing licences used by the artisanal regional fleet presents analytic advantages linked to the richness and high level of detailed data (per day and vessel) recorded, that avoids the lack of precision frequently derived from fishermen interviews and provides a large panel of accurate information that allows a quick preliminary identification of the most important métiers, as well as their main characteristics in terms of target species, fishing gears, monthly temporal pattern and level of rotation. After that, complementary information from daily sale bills, official fleet census and interviews was used to check, adjust and extent results in terms of catches retained for commercial purposes (species composition and weight), incomes, technical demands and fishing strategies per métier.

Although a precedent study carried out by Punzón et al. [41] provided a basic description of the artisanal fisheries in this area, it was focused on fishing gears, catches and their seasonality using information dated in 1996. From an empirical point of view, the present study provides an in-depth and updated characterization of the metiers practiced by the artisanal fleet in the region studied (Asturias, Northwest Spain), with high level of resolution and showing how artisanal activity is changing in recent times (trend of concentration in some métiers and cease of another).

The Asturian fleet is mainly artisanal, involving over 230 vessels (205 vessels with sales in 2009) that represent around 70% of the whole fleet. This sector has a great socioeconomic, cultural and historical importance given the strong bonds between the coastal communities, the artisanal fisheries and the marine environment [1] - [4] [12] . The large number of métiers identified in this study (21 métiers, 16 of which were active in 2009) is a direct consequence of the coastal, close-to-port character of the Asturian artisanal fisheries. As several studies have highlighted, the use of fixed fishing gears in coastal areas, where seasonal environmental variability and spatial heterogeneity are high, promotes the diversification of fishing tactics [10] [11] . Results from this case study confirm the real heterogeneity of vessels that usually are grouped together within the same category (named "Artes menores" in Spain, "Only passive gears" at a European scale, etc.), which is treated as a "uniform" census modality by management and planning public policies. As a consequence, in those regions where artisanal vessels belong to these census categories and represent a relevant part of the total fleet, management measures should be designed and implemented taking into account métiers heterogeneity and the high level of internal complexity of artisanal fleets.

As it occurs in another artisanal and small-scale fisheries in South Europe [10] [11] [31] [32] [38] [39] [42] , findings concerning the high annual rotation of fishing métiers and gears types confirm that polyvalence and versatility[20] are important features of artisanal fleets, representing the fishermen's ability to adapt, up to some extent, to dynamic environmental conditions and to different presence of resources in order to ensure stability in landings and sustain returns throughout the whole year.

Two interesting conclusions arise from these findings. Firstly, when métiers rotation exists, a certain degree of "substitution effect" among métiers exists, which must be interpreted as a strength and taken into account for artisanal fleets subsidies design; in these cases, a percentage of negative effects from changes in fish stocks, limitations of captures or increases in costs can be softened. In opposition to this, the practice of métiers with low possibilities of replacement among each other introduces rigidity in fishing strategies and reduces adaptation possibilities for the fleet, which should be considered by managers. This research has confirmed that some smaller vessels have constraints to redistribute their fishing effort, especially in those periods of worst weather conditions: some artisanal métiers identified, as the glass eel fishing, and other fishing activities as the goose barnacle (Pollicipes pollicipes) recollection, play a fundamental role as alternative fishing activities during certain periods when it is difficult and unprofitable to carry out another métiers (i.e. first months of the year when longlines use to be inactive because of bad weather conditions). In terms of sustainable management of artisanal fisheries, this kind of métiers should be prioritized by sectoral support measures, as they have strong socioeconomic effects at a local scale, play an important subsidiary role in such periods (when other métiers cannot be practiced) and contribute, as a result, to the stability of artisanal fishermen revenues along the year.

Nonetheless, despite their polyvalent character, certain métiers are used extensively by the artisanal fleet producing the major part of landings and revenues of these vessels: Lophius trammel net and Merluccius longline generates the 28.6% of landings and 43.1% of the total income generated by artisanal fisheries in 2009. It is worth highlighting that trammel net is one of the most representative artisanal gears across European small-scale fisheries [2] [10] [11] [32] [39] [40] [43] - [45] . However, while most Mediterranean trammel nets artisanal métiers target species such as scorpion fishes (Scorpaena spp.), red mullets (M. surmuletus), European hake (M. merluccius) or cuttlefish (S. officinalis), the trammel net mostly used by Asturian artisanal vessels target mainly Lophius spp. and Raja spp., although landings of M. surmuletus and M. merluccius are also recorded.

Identified Mullus trammel net and Merluccius gillnet are also widely used by the aforementioned Mediterranean artisanal fisheries.

On the other hand, although the use of longlines have been reported for several Mediterranean fisheries [10] [11] [38] [46] targeting mainly sea breams (Sparidae), their degree of use is clearly lower than shown for the Asturian artisanal fleet (used in 28.63% of the total fishing days). This specific gear targeting European hake (M. merluccius) and conger (C. conger) seems to be more common among Atlantic artisanal fisheries [47] .

In terms of spatial distribution, the identified métiers are located in several fishing ports covering broad geographical areas. In fact, 12 of the active métiers (75%) are present in 7 or more different fishing ports. These results contrast with those concerning artisanal fisheries of SW Spain [26] , where the identified métiers are located in a few fishing ports. In Asturias, only those traditional métiers carried out in shallow waters close to the shore and targeting mainly crustaceans and anguilliformes (Anguilla pot and glass eel fishing) are restricted to some specific locations: these métiers (mostly traps) also present the highest seasonal pattern (whereas most of the net and longline gears are used during larger periods of the year), so specific appraisals must be carried out in case of Marine Protected Areas declaration or catches limits setting (TAC's, resource exploitation plans, etc.) affecting them, as negative impacts and economic losses at a local scale may be important.

Analyses of vessels' typologies have confirmed the existence of significant differences in terms of technical characteristics among vessels depending on the performed métiers. Although other South-European regional studies have found out that métiers are independent from vessel's size [32] , some other have highlighted the existing relationships between vessels' size, other technical characteristics and the type of métiers used by small- scale boats [10] [11] [48] [49] . Findings in this research are in line with these studies and stress out that smaller vessels may be limited to exploit relatively seasonal fisheries surrounding the home ports, especially between January and March when adverse weather may put boats at risk. In addition to this, specific results on technical demands in this study provide researchers and managers useful information to carry out more accurate inferences about costs structures and output models for artisanal fleets, taking into account the technical demands of each métier. In terms of public policies design and budget provision, they help to assess up to what extent each métier is sensitive and need support from subsidies, training activities or security and qualification requirements.

Artisanal vessels have to decide on the métiers' choice in two time scales: daily (short-term) and annually (long-term), when they have to make investments in fishing gears. As it has been underlined by Tzanatos et al. [11] , in the short-term, the choice of fishing tactics or métiers depends on multitude of factors: recent yield and income, market demand and price, seasonal availability of fishing grounds, tradition, skippers' experience, rumors about other fishermen yields, etc. Hence, changes in the biological or economic conditions results in a rapid redistribution of the fishing effort of artisanal fisheries [50] , as it has been also shown in this case study throughout the high alternation between different métiers and gear types along one fishing season. Although the average landings weight was 99.19 kg and the average income was €396.65 per sale bill in 2009, Asturian artisanal métiers showed significant differences: the rod-hook targeting Scomber scombrus is the métier that generates the highest weight (1468 kg) and income (€886), being also important the Merluccius longline and gillnet métier (an average income of €434 and €425 respectively per fishing day) and the Lophius trammel net métier (€423), whereas 50% of Asturian active artisanal métiers registered an average income per fishing day under €350 in 2009. These results provide useful information that could guide artisanal fisheries managers and decision makers in the future allocation of ITQ's or the current distribution of TAC's, as well as they constitute valuable data to assess potential losses (and compensation measures) for vessels affected by certain management measures or environmental disasters.

These findings reflect that annual fishing effort is distributed among an "optimal" combination of métiers for each vessel, according to boats characteristics, expected annual profitability and fishing possibilities in the area, given the existing regulatory framework, so there is some margin of flexibility to adapt fishing decisions to changes in fish stocks, limitations of catches or increases in costs. This information constitutes a valuable baseline for further research focused on the assessment of multi-output models and productivity magnitudes, or the modeling of fishermen profit maximization.

The long-term decisions, what can be considered as fishing strategies, are affected by different considerations. As it was revealed in this case study, artisanal fishing patterns within Asturian fleet are changing significantly in the last years, and some of the identified métiers are nowadays in disuse or barely employed. Although the final choice of available gears may be limited by the lack of experience in some of them [51] , ship-owners face mainly a cost-benefit question regarding what would be the best investment in fishing gears. Answers obtained through the interviewing process have shown that the diversity of métiers performed by the Asturian artisanal fleet nowadays are being reduced and tend to concentrate in those ones with higher net profit rates: as García- de-la-Fuente et al. [12] have recently shown, the estimated profitability of nets clearly surpasses hooks, lines and traps artisanal métiers incomes. As an example, ship-owners have stated that despite that high yield and income traditionally generated by longlines, the bait costs have increased greatly over the last years, pushing down the final profitability of such métiers. As a consequence, many vessels have recently invested in trammel net gears (Lophius trammel net) as they present relatively lower exploitation costs. It constitutes a clear example of how fishing tactics and effort may vary over time, which suggests a new sustainable management challenge: it must be evaluated the social and environmental role of artisanal métiers with low economic profitability in order to support them through the current fishing policy. At the same time, environmental sustainability of artisanal net métiers should be increasingly monitored, as they are becoming more and more used.

Two final issues arise from these results. On the one hand, public authorities and fishing guilds should pay special attention to find and implement management and commercial/diversification strategies to support those artisanal métiers with low economic profitability but high level of environmental and social sustainability. On the other hand, although analyses in other Spanish artisanal fisheries have shown a low level of discards (averaging just 4.13% of the total weight according to Forcada et al. [10] , and some experiments specifically carried out with Asturian artisanal métiers as the Mullus gillnet [52] have shown that only 7.9% of the total catch weight is discarded, an increasing attention must be paid to artisanal net métiers selectivity, given that they become even more important among the Asturian artisanal fleet, so evaluation (and regulation) of "good practices" for an environmentally sustainable fishing activity could be recommendable.

It is hoped the present identification and characterization of Asturian artisanal métiers helps future management measures for artisanal fisheries. Nevertheless, exploring the pattern of métier rotation throughout longer time series would provide useful insights of the artisanal fishing activity and changes on fishing effort distribution [10] [48] . As well, improving knowledge on variable fishing tactics over the years, reasons and motivations explaining switching between métiers are considered key issues that deserve further research for the forecast and sustainable management of fishing effects on exploited stocks. In addition to this, it is necessary to highlight that future research will have to take into consideration total catches (including discards and unsold captures) in order to properly assess the fishing effort exerted by artisanal vessels at a regional scale. Moreover, it will be desirable to have available other kind of artisanal fisheries data in the near future, provided by logbooks or vessels monitoring systems (VMS) [16] that considerably increase accuracy and information about métiers effort in this region.

5. CONCLUSIONS

A métier-based approach is very useful to understand the spatio-temporal patterns of fishing effort allocation and catches, as well as to design and implement specific management measures based upon a good knowledge on the characteristics and problems of each métier. Although the problem of identifying métiers in the South- European area has been traditionally addressed in fisheries research, limitations on fishing data concerning vessels under 12 metres have excluded most of artisanal boats from many relevant analyses. Thus, when accurate, reliable and quantitative information on spatial and temporal fishing patterns for most of artisanal vessels is not available, many studies have applied alternative approaches as onboard samples or surveying methods to identify and characterize métiers, even though these sources mainly provide qualitative information and also have reliability problems. In this context, there is no precedent in exploring the potential of other sources of information to identify and characterize métiers in poor-data artisanal fisheries.

This study was aimed at achieving a detailed and updated characterization of the main technical, biological and economic issues concerning the métiers currently used by the artisanal fishing fleet of Asturias, an Atlantic region located in Northwest Spain. To reach this goal, a novel approach using administrative data on fishing licenses combined with daily sales registers and information collected through a surveying process has been applied. A value added of this approach, in poor-data contexts, is that it allows reaching highly disaggregated and detailed artisanal metier identification and classification, which comprehend all the richness and diversity of métiers practiced by these types of boats, and provides higher resolution than other classification frameworks. In the present case study, where the regional fleet is mainly artisanal and involved over 230 vessels in 2009 (70% of the whole fleet), applying this method required a detailed analysis of the regional administrative register of artisanal fishing licences used per day and vessel in the region during a whole year (more than 14000 fishing operations).

As a result, 21 métiers were identified, 16 of which were fully active in 2009. Findings have outlined the strong coastal, close-to-port character of the Asturian artisanal fleet, as well as a high annual rotation of fishing métiers and gears types that confirms its polyvalence and versatility. Nonetheless, certain métiers are used extensively by the artisanal fleet, producing the major part of landings and revenues of these vessels: Lophius trammel net and Merluccius longline generate the 28.6% of landings and 43.1% of the total income generated by artisanal fisheries in 2009. In terms of spatial distribution, the identified métiers are located in several fishing ports covering broad geographical areas: 12 of the active métiers (75%) are present in 7 or more different fishing ports. Only a few traditional métiers carried out in shallow waters close to the shore (as Anguilla pot and glass eel fishing) are restricted to some specific locations: these métiers (mostly traps) also present the highest seasonal pattern (whereas most of the net and longline gears are used during larger periods of the year). From an economic point of view, results by vessel typology have confirmed the existence of significant differences in terms of technical characteristics and workforce demands among vessels depending on the performed métiers. Additionally, although the average landings weight was 99.19 kg and the average income was €396.65 per sale bill in 2009, Asturian artisanal métiers showed significant differences: the rod-hook targeting Scomber scombrus is the métier that generates the highest weight (1468 kg) and income (€886), being also important the Merluccius longline and gillnet métier (an average income of €434 and €425 respectively per fishing day) and the Lophius trammel net métier (€423). Another interesting output of this research is that artisanal fishing patterns within Asturian fleet have changed significantly in the last years, and some of the identified métiers are nowadays in disuse or barely employed: as nets register lower exploitation costs than hooks, lines and traps, many vessels have recently invested in trammel net gears (Lophius trammel net).

The discussion of some findings in this article lead to other questions linked to the design of specific sustainable management measures in the context of South-European artisanal fisheries. Firstly, when highly heterogeneous artisanal vessels represent a relevant part of the whole fleets, management and planning measures should be designed and implemented taking into account métiers heterogeneity and the high level of internal complexity of artisanal fleets, avoiding grouping these vessels together within the same "uniform" category (i.e. named "Artes menores" in Spain, "Only passive gears" at a European scale, etc.). Secondly, it was confirmed that some vessels are limited to exploit relatively seasonal fisheries surrounding the home ports, especially when adverse weather may put boats at risk; given their constraints to redistribute their fishing effort, certain artisanal métiers (as glass eel fishing) and other fishing activities (as goose barnacle recollection, Pollicipes pollicipes), play in these periods a fundamental role as alternative/subsidiary fishing activities, when it is difficult and unprofitable to carry out another métiers. They become strategic at a local scale, as they provide income stability along the year to many artisanal fishermen. Finally, as part of the artisanal fleet is concentrating its activity on trammel net gears, a métier-diversity reduction is happening in parallel to a progressive cease of certain métiers; this changes suggest new sustainable management challenges: it should be evaluated the social and environmental role of artisanal métiers with low economic profitability in order to support them throughout the current fishing policy, whereas artisanal net métiers should be increasingly monitored in order to preserve the environmental sustainability of artisanal coastal fisheries.

ACKNOWLEDGEMENTS

This study was carried out within the PRESPO Project and co-funded by the European Regional Development Fund (ERDF) and the General Directorate of Maritime Fisheries of Asturias (Regional Ministry). Additionally, authors would like to thank Fishermen Guilds for their active collaboration and participation. Finally, we would like to thank reviewers for their detailed comments and insightful suggestions for the manuscript.

REFERENCES

1. FAO (2005) Increasing the Contribution of Small-Scale Fisheries to Poverty Alleviation and FOOD Security. FAO Technical Guidelines for Responsible Fisheries 10, Rome.
2. Stergiou, K.I., Moutopoulos, D.K., Soriguer, M.C., Puente, E., Lino, P.G., Zabala, C., Monteiro, P., Errazkin, L.A. and Erzini, K. (2006) Trammel Net Catch Species Composition, Catch Rates and Métiers in Southern European Waters: A Multivariate Approach. Fisheries Research, 79, 170-182. http://dx.doi.org/10.1016/j.fishres.2006.03.003
3. Tzanatos, E., Dimitriou, E., Papaharisis, L., Roussi, A., Somarakis, S. and Koutsikopoulos, C. (2006) Principal Socio-Economic Characteristics of the Greek Small-Scale Coastal Fishermen. Ocean and Coastal Management, 49, 511- 527. http://dx.doi.org/10.1016/j.ocecoaman.2006.04.002
4. European Parliament (2012) On Small-Scale Coastal Fishing, Artisanal Fishing and the Reform of the Common Fisheries Policy. Fisheries Committee Report A7-0291/2012, Rapporteur João Ferreira.
5. Florido del Corral, D. (2008) Focusing on Artisanal Fleets in a New Scenario: The Case of Andalusia (Spain). Marine Policy, 32, 1004-1012. http://dx.doi.org/10. 1016/j.marpol.2008.02.007
6. Gómez, S., Lloret, J., Riera, V. and Demestre, M. (2006) The Decline of the Artisanal Fisheries in Mediterranean Coastal Areas: The Case of Cap de Creus (Cape Creus). Coastal Management, 34, 217-232. http://dx.doi.org/10.1080/08920750500531389
7. European Commission (2001) Green Paper on the Future of the Common Fisheries Policy. COM (2001) 135, Brussels.
8. Farrugio, H., Oliver, P. and Biagi, F. (1993) An Overview of the History, Knowledge, Recent and Future Research Trends in Mediterranean Fisheries. Scientia Marina, 57, 105-119.
9. García-Flórez, L., Morales, J., Gaspar, M.B., Castilla, D., Mugerza, E., Berthou, P., García de la Fuente, L., Oliveira, M., Moreno, O., García del Hoyo, J.J., Arregi, L., Vignot, C., Chapela, R. and Murillas, A. (2014) A Novel and Simple Approach to Define Artisanal Fisheries in Europe. Marine Policy, 44, 152-159. http://dx.doi.org/10.1016/j.marpol.2013.08.021
10. Forcada, A., Valle, C., Sánchez-Lisazo, J.L., Bayle-Sempere, J.T. and Corsi, F. (2010) Structure and Spatio-Temporal Dynamics of Artisanal Fisheries around a Mediterranean Marine Protected Area. ICES Journal of Marine Science, 67, 191-203. http://dx.doi.org/10.1093/icesjms/fsp234
11. Tzanatos, E., Somarakis, S., Tserpes, G. and Koutsikopoulos, C. (2006) Identifying and Classifying Small-Scale Fisheries Métiers in the Mediterranean: A Case Study in the

Patraikos Gulf, Greece. Fisheries Research, 81, 158-168. http://dx.doi.org/10.1016/j.fishres.2006.07.007
12. García-de-la-Fuente, L., González-&AACUTE;lvarez, J., García-Flórez, L., Fernández-Rueda, P. and Alcázar-&AACUTE;lvarez, J. (2013) Relevance of Socioeconomic Information for the Sustainable Management of Artisanal Fisheries in South Europe. A Characterization Study of the Asturian Artisanal Fleet (Northern Spain). Ocean and Coastal Management, 86, 61-71. http://dx.doi.org/10.1016/j.ocecoaman.2013.05.007
13. Biseau, A. (1998) Definition of a Directed Fishing Effort in a Mixed-Species Trawl Fishery, and Its Impact on Stock Assessments. Aquatic Living Resources, 11, 119-136. http://dx.doi.org/10.1016/S0990-7440(98)80109-5
14. Laurec, A., Biseau, A. and Charuau, A. (1991) Modelling Technical Interactions. ICES Marine Science Symposia, 193, 225-236.
15. Mesnil, B. and Shepherd, J.G. (1990) A Hybrid Age- and Length- Structured Model for Assessing Regulatory Measures in Multiple-Species, Multiple-Fleet Fisheries. Journal du Conseil International Pour l'Exploration de la Mer, 47, 115-132. http://dx.doi.org/10.1093/icesjms/47.2.115
16. Russo, T., Parisi, A., Prorgi, M., Boccoli, F., Cignini, I., Tordoni, M. and Cataudella, S. (2011) When Behaviour Reveals Activity: Assigning Fishing Effort to Métiers Based on VMS Data Using Artificial Neural Networks. Fisheries Research, 111, 53-64. http://dx.doi.org/10.1016/j.fishres.2011.06.011
17. Ulrich, C. and Andersen, B.S. (2004) Dynamics of Fisheries, and the Flexibility of Vessel Activity in Denmark between 1989 and 2001. ICES Journal of Marine Science, 61, 308-322. http://dx.doi.org/10.1016/j.icesjms.2004.02.006
18. Cabrera, J.L. and Defeo, O. (2001) Daily Bioeconomic Analysis in a Multispecific Artisanal Fishery in Yucatan, Mexico. Aquatic Living Resources, 14, 19-28. http://dx.doi.org/10.1016/S0990-7440(00)01094-9
19. Oliveira, M.M., Camanho, A.S. and Gaspar, M.B. (2014) Enhancing the Performance of Quota Managed Fisheries Using Seasonality Information: The Case of the Portuguese Artisanal Dredge Fleet. Marine Policy, 45, 114-120. http://dx.doi.org/ 10. 1016/j.marpol.2013.11.014
20. Pascual, M., Borja, A., Galparsoro, I., Ruiz, J., Mugerza, E., Quincoces, I., Murillas, A. and Arregi, L. (2013) Total Fishing Pressure Produced by Artisanal Fisheries, from a Marine Spatial Planning Perspective: A Case Study from the Basque Country (Bay of Biscay). Fisheries Research, 147, 240-252. http://dx.doi.org/10.1016/j.fishres.2013.06.010
21. Castro, J., Punzón, A., Pierce, G.J., Marín, M. and Abad, E. (2010) Identification of Métiers of the Northern Spanish Coastal Bottom Pair Trawl Fleet by Using the Partitioning Method CLARA. Fisheries Research, 102, 184-190. http://dx.doi.org/10.1016/ j.fishres. 2009.11.011
22. Castro, J., Marín, M., Pierce, G.J. and Punzón, A. (2011) Identification of Métiers of the Spanish Set-Longline Fleet Operating in Non-Spanish European Waters. Fisheries Research, 107, 100-111. http://dx.doi.org/10.1016/j.fishres.2010.10.013
23. Holley, J. and Marchal, P. (2004) Fishing Strategy Development under Changing Conditions: Examples from the French Offshore Fleet Fishing in the North Atlantic. ICES Journal of Marine Science, 61, 1410-1431. http://dx.doi.org/10.1016/ j.icesjms.2004.08.010
24. Jiménez, M.P., Sobrino, I. and Ramos, F. (2004) Objective Methods for Defining Mixed-Species Trawl Fisheries in Spanish Waters of the Gulf of Cádiz. Fisheries Research, 67, 195-206. http://dx.doi.org/10.1016/j.fishres.2003.09.048
25. Murawski, S.A., Lange, A.M., Sissenwine, M.P. and Mayo, R.K. (1983) Definition and Analysis of Multispecies Otter-Trawl Fisheries off the Northeast Coast of the Unites States. Journal du Conseil International Pour l'Exploration de la Mer, 41, 13-27. http://dx.doi.org/10.1093/icesjms/41.1.13
26. Silva, L., Gil, J. and Sobrino, I. (2002) Definition of Fleet Components in the Spanish Artisanal Fishery of the Gulf of Cádiz (SW Spain ICES Division IXa). Fisheries Research, 59, 117-128. http://dx.doi.org/10.1016/S0165-7836(01)00420-9

27. Ulrich, C., Gascuel, D., Dunn, M.R., Le Gallic, B. and Dintheer, C. (2001) Estimation of Technical Interactions Due to the Competition for Resource in a Mixed-Species Fishery, and the Typology of Fleets and Métiers in the English Channel. Aquatic Living Resources, 14, 267-281. http://dx.doi.org/10.1016/S0990-7440(01)01132-9
28. Castro, J., Abad, E., Artetxe, I., Cardador, F., Duarte, R., García, D., Hernández, C., Marín, M., Murta, A., Punzón, A., Quincoces, I., Santurtún, M., Silva, C. and Silva, L. (2007) Identification and Segmentation of Mixed-Species Fisheries Operating in the Atlantic Iberian Peninsula Waters. IBERMIX Project. European Commission, Directorate-General for Fisheries and Maritime Affairs (Contract Ref.: FISH/2004/03-33).
29. Marchal, P., Ed. (2006) Technological Developments and Tactical Adaptations of Important EU Fleets (TECTAC No. QLK5-2001-01291). Final Project Report.
30. Guillou, A., Lespagnol, P. and Ruchon, F. (2002) La pêeche aux petits métiers en Languedoc-Roussillon en 2000-2001. Convention de Participation au Programme PESCA (PIC), DIRAM-IFREMER No. 00/3210040/YF.
31. Jabeur, C., Gobert, C. and Missaoui, H. (2000) Typologie de la flottille de pêche côtière dans le golfe de Gabès (Tunisie). Aquatic Living Resources, 13, 421-428. http://dx.doi.org/10.1016/S0990-7440(00)01069-X
32. Tzanatos, E., Dimitriou, E., Katselis, G., Georgiadis, M. and Koutsikopoulos, C. (2005) Composition, Temporal Dynamics and Regional Characteristics of Small-Scale Fisheries in Greece. Fisheries Research, 73, 147-158. http://dx.doi.org/10.1016/j.fishres.2004.12.006
33. O'Donnell, K.P., Molloy, P.P. and Vincent, A.C.J. (2012) Comparing Fisher Interviews, Logbooks, and Catch Landings Estimates of Extraction Rates in a Small-Scale Fishery. Coastal Management, 40, 594-611. http://dx.doi.org/10.1080/08920753.2012.727734
34. Neis, B., Schneider, D.C., Felt, L., Haedrich, R.L., Fisher, J. and Hutchins, J.A. (1999) Fisheries Assessment: What Can Be Learned from Interviewing Resource Users? Canadian Journal of Fisheries and Aquatic Sciences, 56, 1949- 1963. http://dx.doi.org/10.1139/f99-115
35. European Commission (2008) Commission Decision (2008/949/EC) of 6 November 2008. Adopting a Multiannual Community Programme Pursuant to Council Regulation (EC) No. 199/2008 Establishing a Community Framework for the Collection, Management and Use of Data in the Fisheries Sector and Support for Scientific Advice Regarding the Common Fisheries Policy. Official Journal of the European Union, L 346/37.
36. Martin, Q., Cabrero, M.T. and de Paz, Y. (2007) Tratamiento estadístico de datos con SPSS. Thomson Editores, Spain.
37. Gibbons, J.D. (1985) Nonparametric Methods for Quantitative Analysis. American Sciences Press, New York.
38. Battaglia, P., Romeo, T., Consoli, P., Scotti, G. and Andaloro, F. (2010) Characterization of the Artisanal Fishery and Its Socio-Economic Aspects in the Central Mediterranean Sea (Aeolian Islands, Italy). Fisheries Research, 102, 87-97. http://dx.doi.org/ 10.1016/j.fishres.2009.10.013
39. Colloca, F., Crespi, V., Cerasi, S. and Coppola, S.R. (2004) Structure and Evolution of the Artisanal Fishery in a Southern Italian Coastal Area. Fisheries Research, 69, 359-369. http://dx.doi.org/10.1016/j.fishres.2004.06.014
40. Merino, G., Morales-Nin, B., Maynou, F. and Grau, A.M. (2008) Assessment and Bioeconomic Analysis of the Majorca (NW Mediterranean) Trammel Net Fishery. Aquatic Living Resources, 107, 99-107. http://dx.doi.org/10.1051/alr:2008027
41. Punzón Merino, A. and Gancedo Crespo, R.M. (2000) Descripción de las pesquerías artesanales de Cantabria y Asturias (Norte de España). Informes Técnicos (Instituto Español de Oceanografía), 179, 3-43.
42. Freire, J. and García-Allut, A. (2000) Socioeconomic and Biological Causes of Management Failures in European Artisanal Fisheries: The Case of Galicia (NW Spain). Marine Policy, 24, 375-384. http://dx.doi.org/10.1016/S0308-597X(00)00013-0
43. Batista, M.I., Teixeira, C.M. and Cabral, H.N. (2009) Catches of Target Species and Bycatches of an Artisanal Fishery: The Case Study of a Trammel Net Fishery in the Portuguese Coast. Fisheries Research, 100, 167-177. http://dx.doi.org/10. 1016/j.fishres.2009.07.007

44. Erzini, K., Gonçalves, J.M.S., Bentes, L., Moutopoulos, D.K., Hernando, J.A., Soriguer, M.C., Puente, E., Errazkin, L.A. and Stergiou, K.I. (2006) Size Selectivity of Trammel Nets in Southern European Small-Scale Fisheries. Fisheries Research, 79, 183-201. http://dx.doi.org/10.1016/j.fishres.2006.03.004
45. Gonçalves, J.M.S., Stergiou, K.I., Hernando, J.A., Puente, E., Moutopoulos, D.K., Arregi, L., Soriguer, M.C., Vilas, C., Coelho, R. and Erzini, K. (2007) Discards from Experimental Trammel Nets in Southern European Small-Scale Fisheries. Fisheries Research, 88, 5-14. http://dx.doi.org/10.1016/j.fishres.2007.06.017
46. García-Rodríguez, M., Fernández, &AACUTE;.M. and Esteban, A. (2006) Characterization, Analysis and Catch Rates of the Small-Scale Fisheries of the Alicante Gulf (SE Spain) over a 10 Years Time Series. Fisheries Research, 77, 226-238. http://dx.doi.org/ 10.1016/j.fishres.2005.09.002
47. Puente, E. and Astorkiza, I. (2002) Estudio técnico-pesquero y socioeconómico de las pesquerías artesanales del País Vasco. Colección ITSASO 25, AZTI-Tecnalia.
48. Le Pape, O. and Vigneau, J. (2001) The Influence of Vessel Size and Fishing Strategy on the Fishing Effort for Multispecies Fisheries in Northwestern France. ICES Journal of Marine Science, 58, 1232-1242. http://dx.doi.org/10.1006/jmsc.2001.1121
49. Guyader, O., Berthou, P., Koutsikopoulos, C., Alban, F., Demanèche, S., Gaspar, M.B., Eschbaum, R., Fahy, E., Tully, O., Reynal, L., Curtil, O., Frangoudes, K. and Maynou, F. (2013) Small Scale Fisheries in Europe: A Comparative Analysis Based on a Selection of Case Studies. Fisheries Research, 140, 1-13. http://dx.doi.org/10. 1016/j. fishres. 2012.11.008
50. Holland, D.S. and Sutinen, J.G. (1999) An Empirical Model of Fleet Dynamics in New England Trawl Fisheries. Canadian Journal of Fisheries and Aquatic Sciences, 56, 253-264. http://dx.doi.org/10.1139/f98-169
51. Salas, S., Sumaila, U.R. and Pitcher, T. (2004) Short-Term Decisions of Small-Scale Fishers Selecting Alternative Target Species: A Choice Model. Canadian Journal of Fisheries and Aquatic Sciences, 61, 374-383.
52. Fernández-Rueda, M.P., Alcázar, J.L., Jiménez, F., Herrador, R., Muñoz, A. and García-Flórez, L. (2012) Catch Composition, Discards and Selectivity in the Red Mullet (Mullus surmuletus) Gillnet Fishery of the Asturias Coast. 13th International Symposium on Oceanography of the Bay of Biscay (ISOBAY 13), Santander, 11-13 April 2012, 150. http://www.ieo-santander.net/isobay13/documentos/posters/poster150_ mpfernandezrueda.pdf

ENDNOTES

[1]Achieving a standard definition of artisanal fisheries is a complex topic that exceeds the objective of this paper. A recent discussion in the South-European context can be found in García-Flórez et al. [9] .

[2]Decision 2008/949/EC has been repealed with effect from 1 January 2011 by the Commission Decision 2010/93 of 18 December 2009, in which a multiannual Community Programme for the collection, management and use of data in the fisheries sector for the period 2011-2013 is adopted.

[3]Mainly, vessels registered as “artes menores” in the National Census of the Operating Fishing Fleet. This category embraces multi-gear vessels and it may be mainly assimilated to the “Only passive gears” European census category, although it may also include some other minor gears with a strong artisanal character and not necessary passive.

[4]Vessels selected consisted of Danish seiners, gill netters, long-lines, otter trawlers and beam trawlers.

[5]Updated Asturian fishing grounds map, carried out in 2011 by the Regional Ministry with powers on fishing affairs. Available on line at http://tematico.asturias.es/dgpesca/din/divulgacion.php?tipo=caladeros.

[6]Data referred to June 2010.

[7]Commercial information per vessel concerning daily sales at local first-sale fish markets.

[8]Some of the supposedly active vessels in 2009-2010 (233) did not have sales in 2009.

[9]For more details, see the description in García-de-la-Fuente et al. [12] .

[10]See Appendix IV, North Atlantic Region: ICES areas V-XIV and NAFO areas.

[11]Although they were allowed to fish using the same administrative license.

[12]See Gibbons [37] for a detailed analogy and comparisons between non-parametric and parametric methods.

[13]Another three métiers linked to bait capture have been identified. However, these métiers have not been included in the analysis as they do not have any commercial interest.

[14]Although trammel net is considered a multi-species métier, Lophius spp. represents one of the main groups of target species according to the artisanal fishermen interviewed. Additionally, 2009 registered a high amount of Lophius spp. sales, with an important relative weight within the species composition of this métier. Taking these reasons into account, a specific reference to Lophius spp. trammel net is made in several parts of this article.

[15]It is impossible to assess the number of sale notes considered as fishing days for the seaweed recollection métier as these sales are not considered as a fishing resource and, consequently, are no registered on the sales database. However, it is known that just one artisanal vessel operated with such fishing modality in 2009.

[16]Those daily set and hauled, and always practiced in different days as goose barnacle recollection.

[17]Nevertheless, crewmembers keep the same retribution pattern and benefits share that when they fish using the vessel.

[18]As example, in this case study, 212 of the total 233 artisanal vessels belonging to the "Artes menores" National Census category were under 12 metres of total length (91%), and 157 were under 10 metres (67%). Similar situations can be found in other South-European regions with large

artisanal fishing fleets, as it has been recently outlined in García-Flórez et al. [9] .

[19]In the case of Spain, during the last years many regions and fishing authorities have been making an effort to install positioning systems on board vessels less than 12 metres of total length, but current coverage is already partial and highly variable depending on the area and métier. Additionally, these data are not directly available for research objectives and need to be previously processed.

[20]Real possibility in the short term of practicing a certain métier instead of another during the same season.

CHAPTER 4

Effect of dietary protein and lipid level on growth, feed utilization, and muscle composition in golden mandarin fish *Siniperca scherzeri*

Zohreh Sankian[1], Sanaz Khosravi[1], Yi-Oh Kim[2] and Sang-Min Lee[1*]

[1]*Department of Marine Biotechnology, Gangneung-Wonju National University, Gangneung 25457, South Korea.*

[2]*Department of Inland Fisheries Research Institute, Chungju 27329, South Korea.*

ABSTRACT

A feeding trial was designed to assess the effects of dietary protein and lipid content on growth, feed utilization efficiency, and muscle proximate composition of juvenile mandarin fish, Siniperca scherzeri. Six experimental diets were formulated with a combination of three protein (35, 45, and 55%) and two dietary lipid levels (7 and 14%). Each diet was fed to triplicate groups of fish (8.3 ± 0.1 g) to apparent satiation for 8 weeks. The results showed that growth performance in terms of weight gain (WG) and specific growth rate (SGR) increased with increasing dietary protein level from 35 to 55% at the same dietary lipid level. At the same dietary lipid content, WG and SGR obtained with diets containing 55% protein was significantly higher than those obtained with diets containing 45 and 35% protein. No significant effect on growth rate was found when the dietary level of lipid was increased from 7 to 14%. While the levels of protein and lipid in the diets had no significant effect on feed intake, other nutrient utilization efficiency parameters including daily protein intake (DPI), feed efficiency (FE), and protein efficiency ratio (PER) showed a similar trend to that of growth rates, with the highest values obtained with diets containing 55% protein. Muscle chemical composition was not significantly affected by the different dietary treatments for each dietary lipid or protein level tested. These findings may suggest that a practical diet containing 55% protein and 7% lipid provides sufficient nutrient and energy to support the acceptable growth rates and nutrient utilization of mandarin fish juveniles.

Keywords: Growth performance, Feed utilization, efficiency, Protein/lipid ratio, Mandarin, Siniperca scherzeri

1. BACKGROUND

Since protein is often the most costly component of the formulated fish feeds, numerous studies have been undertaken to define the optimum dietary protein requirement for developing more cost-effective, nutritionally balanced practical diets for various fish species. However, the optimum dietary protein requirements are known to be affected by several factors including the fish species and size, the quality of the protein source, and the amount of non-protein energy in the diet (NRC 2011). When insufficient non-protein energy is available in the diet, part of the dietary protein will be catabolized to supply energy, which is wasteful. Thus, dietary supplementation of energy-yielding nutrients, mainly lipids, has been suggested as a strategy to spare or improve the efficiency of protein utilization by fish, thereby enhancing economic returns and reducing water pollution. Conversely, supply of dietary lipid in excess of requirement can limit feed consumption, thereby reducing the intake of the necessary amount of protein and other essential nutrients for maximal growth rates of fish while enhancing body fat deposition. Therefore, it is very important to determine the optimal dietary balance between protein and lipid for achieving maximum growth and efficient feed utilization of fish.

Golden mandarin fish, Siniperca scherzeri, is one of the most commercially important freshwater species endemic to East Asia, mainly distributed in China, Korea, and Northern Vietnam (Zhou et al. 1988). Increased market demands combined with the dramatic decline in wild stocks as a result of over-catching and habitat destruction (Liang 1996; Wu et al. 1997) has created considerable interest in the development and improvement of culture practices for commercial production of this species. In fact, mandarin fish has become one of the most promising target species with high potential for aquaculture due to its excellent taste, high market value, rapid growth, and high resistance against disease. However, despite the commercial importance of mandarin fish, no commercial formulated feed is yet available for this species, and fish reared in commercial pens are usually fed live feed. Since live feed production is not cost-effective, feeding live prey to mandarin fish may hinder the prospective development of extensive commercial production of this species. Therefore, it is necessary to develop formulated feeds for mandarin fish culture to be far more practical and efficient in terms of commercial operation cost compared with the present practice of using live feed as the rearing diet. In fact, the present study is believed to be the first attempt to evaluate the effects of dietary protein and lipid levels in practical feeds on growth performance, feed utilization, and muscle composition of juvenile mandarin fish, S. scherzeri. The results of the current study could be helpful in formulating a cost-effective and nutritionally sound practical diet for this species.

2. METHODS

2.1 Experimental diets

Formulation and proximate composition of the experimental diets are provided in Table 1. Six experimental diets were formulated to contain three protein levels (35, 45, and 55% crude protein) each at two lipid levels (7 and 14% crude lipid). Anchovy fish meal served as the main protein source and equal proportions of squid liver oil and soybean oil as lipid sources in the experimental diets. All dried ingredients were well-mixed and, after addition of oil and double-distilled water, pelleted through a meat chopper machine. The pellets were dried overnight at room temperature, crushed into suitable size (3 mm in diameter), and stored at −30 °C until used.

2.2 Fish and feeding trial

Juvenile mandarin fish were kindly provided by Dr. Yi Oh Kim (Inland Fisheries Research Institute, Chungcheongbuk-do, South Korea). A commercial feed (50% crude protein and 13% lipid; Woosung, Daejeon, South Korea) was ground to a fine powder and sieved through a 2–3-mm mesh screen. Then it was pelletized into 1.5–1.8-mm long cylindrical pellets having a diameter of 3 mm.

Table 1 Formulation and nutrient contents (%) of the six experimental diets

Ingredients (%)	Diets					
	P35L7	P45L7	P55L7	P35L14	P45L14	P55L14
Anchovy fish meal[a]	42.0	59.0	76.0	42.0	59.0	76.0
Wheat flour[b]	40.1	25.1	10.1	30.0	15.0	0.0
Corn gluten meal[c]	0.0	0.0	0.0	2.8	2.8	2.8
Potato starch	3.5	7.0	10.5	3.5	7.0	10.5
Squid liver oil + soybean oil	2.2	1.5	0.8	9.5	8.8	8.1
Cellulose	9.6	4.8	00	9.6	4.8	0.0
Vitamin premix[d]	1.0	1.0	1.0	1.0	1.0	1.0
Mineral premix[e]	1.0	1.0	1.0	1.0	1.0	1.0
Vitamin C	0.3	0.3	0.3	0.3	0.3	0.3
Vitamin E	0.2	0.2	0.2	0.2	0.2	0.2
Choline salt	0.1	0.1	0.1	0.1	0.1	0.1
Proximate composition (% dry matter)						
Moisture	7.45	8.92	7.58	7.74	8.46	7.99
Crude protein	36.8	45.7	54.6	36.0	45.3	55.8
Crude lipid	6.5	6.7	6.8	13.7	13.9	14.5
Ash	9.7	12.2	14.2	10.0	11.7	11.8

[a]Pesquera Bahia Caldera, Caldera, Chile. Fishmeal composition (% dry matter): crude protein, 67.3; crude lipid, 8.6
[b]CJ CheilJedang Corporation, Seoul, South Korea. Wheat flour composition (% dry matter): crude protein, 16.4; crude lipid, 3.9
[c]WooSung Feed Corporation, Daejeon, South Korea. Corn gluten meal composition (% dry matter): crude protein, 66.1; crude lipid, 2.8
[d]Vitamin premix contained the following ingredients (g/kg premix), which were diluted in cellulose: thiamin hydrochloride, 2.7; riboflavin, 9.1; pyridoxine hydrochloride, 1.8; niacin, 36.4; Ca-D-pantothenate, 12.7; myo-inositol, 181.8; D-biotin, 0.27; folic acid, 0.68; p-aminobenzoic acid, 18.2; menadione, 1.8; retinyl acetate, 0.73; cholecalciferol, 0.003; cyanocobalamin, 0.003
[e]Mineral premix contained the following ingredients (g/kg premix): $MgSO_4{\cdot}7H_2O$, 80.0; $NaH_2PO_4{\cdot}2H_2O$, 370.0; KCl, 130.0; Ferric citrate, 40.0; $ZnSO_4{\cdot}7H_2O$, 20.0; Ca-lactate, 356.5; CuCl, 0.2; $AlCl_3{\cdot}6H_2O$, 0.15; KI, 0.15; $Na_2Se_2O_3$, 0.01; $MnSO_4{\cdot}H_2O$, 2.0; $CoCl_2{\cdot}6H_2O$, 1.0

Fish were fed the repelleted commercial diet for 2 weeks to be acclimated to the experimental conditions and facilities. Following the acclimation period, fish (initial mean body weight, 8.3 ± 0.1 g) were distributed in a closed recirculation system equipped with 18 square glass aquaria of 65 L capacity at a density of 20 fish per aquarium. Each aquarium was supplied with dechlorinated freshwater at a flow rate of 1.5 L min−1 and continuous aeration. The photoperiod was maintained on a 12:12-h (light/dark) schedule. The average water temperature during the feeding trial was 23 ± 0.7 °C. Triplicate groups of fish were fed with one of the test diets to visual satiation twice a day (09:00 and 17:00 h) for 8 weeks. The uneaten feed was collected, dried, and weighted to determine the feed intake level.

At the end of the feeding trial, all the fish in each tank were counted and bulk-weighed for calculation of survival, growth, and feed utilization parameters including weight gain (WG), specific growth rate (SGR), feed efficiency (EF), daily feed intake (DFI), daily protein intake (DEI), and protein efficiency ratio (PER) through the following formulas:

Weight gain=[(final body weight - initial body weight)÷initial body weight]×100

Specific growth rate=[(ln final body weight - ln initial body weight)÷ days]×100

Feed efficiency=(fish wet weight gain/feed intake)×100

Daily feed intake={feed intake/[(initial fish weight+final fish weight+dead fish weight)×days fed/2]}×100

Daily protein intake={protein intake/[(initial fish weight+final fish weight+dead fish weight)×days fed/2]}×100

Protein efficiency ratio=fish wet weight gain/protein intake

Five fish per tank were randomly sampled and stored at −45 °C for muscle proximate composition analyses. The proximate composition of the experimental diets and muscle samples of fish were analyzed according to standard methods (AOAC 1997). The crude protein content was determined using the Auto Kjeldahl System (Buchi, Flawil, Switzerland), the crude lipid content by the ether-extraction method, using a Soxhlet extractor (VELP Scientifica, Milano, Italy), the moisture content by oven drying (105 °C for 6 h), and the ash content using a muffler furnace (600 °C for 4 h).

2.2 Statistical analysis

The data were subjected to two-way ANOVA to test for differences in the mean effects of dietary protein and lipid levels, using SPSS version 20.0 (SPSS Inc., Chicago, IL, USA). Statistical significance was determined at

$P < 0.05$. Data were presented as mean ± SE. Percentage data were arcsine transformed before statistical analysis.

3. RESULTS

The increase of dietary protein level from 35 to 55% induced a significant increase in fish growth performance in terms of WG and SGR (Table 2). Fish fed diets with 55% protein had significantly higher growth rate than those fed the 35 and 45% protein, regardless of dietary lipid level. Although fish growth performance was not significantly affected by dietary lipid content, numerically higher values were observed in fish offered the diet with the highest lipid content (14%). While DFI was not affected by the dietary treatment, significantly higher DPI was found in fish fed the diets containing 55% protein compared to those fed the 45% protein or less. Fish fed the 55% protein diet showed significantly higher feed efficiency than those given lower protein at both dietary lipid levels. PER was significantly increased with increasing dietary protein from 35 to 55%, and the highest value was recorded in fish fed the P55L14 diet. However, dietary lipid contents had no significant effect on feed utilization efficiency of juvenile mandarin fish at all the dietary protein levels.

Two-way ANOVA revealed that neither dietary protein and lipid levels alone nor their interactions had significant effect ($P > 0.05$) on muscle composition of juvenile mandarin fish after 8 weeks of feeding (Table 2).

4. DISCUSSION

In the present study, growth performance of juvenile mandarin fish, in terms of weight gain (WG) and specific growth rate (SGR), was significantly increased with the increase in dietary protein level and the highest values were observed in those fed the highest dietary protein level of 55% (Table 3). This value fits within the range of those reported in previous studies for other strictly carnivorous fish species, such as yellow snapper, Lutjanus argentiventris (Peters 1869) (Maldonado-García et al. 2012), Senegalese sole, Solea senegalensis (Rema et al. 2008), Atlantic halibut (Hippoglossus hippoglossus, L.) (Hamre et al. 2003), common dentex, Dentex dentex (Espinos et al. 2003), Murray cod, Maccullochella peelii peelii (De Silva et al. 2002), and Mediterranean yellowtail, Seriola dumerilii (Jover et al. 1999) that generally have high dietary requirements for protein (Wilson 2002; NRC 2011).

Table 2 Growth performance and feed utilization of mandarin fish fed the six experimental diets for 8 weeks

Diets	WG[1]	SGR[2]	DFI[3]	DPI[4]	FE[5]	PER[6]	Survival
P35L7	51.4 ± 7.3^{a}	0.74 ± 0.08^{a}	2.30 ± 0.03	0.81 ± 0.01^{a}	31.5 ± 3.6^{a}	0.90 ± 0.10^{a}	99.0 ± 1.1
P45L7	79.3 ± 11.6^{b}	1.04 ± 0.11^{b}	2.49 ± 0.05	1.01 ± 0.02^{b}	39.9 ± 3.8^{ab}	0.98 ± 0.09^{ab}	96.7 ± 1.9
P55L7	117.7 ± 4.7^{c}	1.39 ± 0.04^{c}	2.41 ± 0.12	1.11 ± 0.04^{c}	55.0 ± 3.4^{c}	1.19 ± 0.05^{b}	97.0 ± 0.0
P35L14	62.4 ± 1.8^{ab}	0.87 ± 0.02^{ab}	2.38 ± 0.06	0.83 ± 0.02^{a}	35.3 ± 0.4^{ab}	1.01 ± 0.01^{ab}	97.7 ± 2.2
P45L14	83.2 ± 10.9^{b}	1.08 ± 0.10^{b}	2.33 ± 0.08	0.94 ± 0.03^{b}	44.7 ± 2.9^{b}	1.11 ± 0.07^{ab}	96.7 ± 1.9
P55L14	133.9 ± 17.7^{c}	1.51 ± 0.13^{c}	2.53 ± 0.03	1.16 ± 0.03^{c}	56.1 ± 3.3^{c}	1.23 ± 0.07^{b}	95.7 ± 1.3
Two-way ANOVA (*P* value)							
Protein	<0.001	<0.001	0.163	<0.001	<0.001	0.019	0.619
Lipid	0.216	0.199	0.890	0.912	0.196	0.160	0.983
Protein × lipid	0.860	0.819	0.094	0.103	0.788	0.775	0.262

Values are mean of triplicate groups and presented as mean ± SE. Values with different superscripts in the same column are significantly different ($P < 0.05$). The lack of superscript letter indicates no significant differences among treatments

[1]Weight gain (%) = [(final body weight − initial body weight)/initial body weight] × 100

[2]Specific growth rate (% day^{-1}) = [(ln final body weight − ln initial body weight)/days] × 100

[3]Daily feed intake (%) = {feed intake/[(initial body weight + final body weight + dead fish weight) × days fed/2]} × 100

[4]Daily protein intake (%) = {protein intake/[(initial body weight + final body weight + dead fish weight) × days fed/2]} × 100

[5]Feed efficiency (%) = (wet weight gain/feed intake) × 100

[6]Protein efficiency ratio = wet weight gain/protein intake

Table 3 Muscle proximate composition (%, wt basis) of mandarin fish fed the six experimental diets for 8 weeks

Diets	Moisture	Protein	Lipid	Ash
P35L7	76.6 ± 0.4	19.2 ± 0.3	0.8 ± 0.1	1.2 ± 0.1
P45L7	77.9 ± 0.6	19.3 ± 0.2	1.2 ± 0.4	1.1 ± 0.1
P55L7	76.5 ± 0.2	20.1 ± 0.5	1.1 ± 0.5	1.2 ± 0.2
P35L14	76.5 ± 0.3	20.0 ± 0.4	0.6 ± 0.2	1.1 ± 0.1
P45L14	77.2 ± 0.3	19.8 ± 0.1	1.3 ± 0.5	1.1 ± 0.0
P55L14	76.4 ± 0.7	19.7 ± 0.1	1.4 ± 0.3	1.0 ± 0.1
Two-way ANOVA (*P* value)				
Protein	0.197	0.249	0.324	0.956
Lipid	0.512	0.114	0.561	0.724
Protein × lipid	0.584	0.538	0.670	0.617

Values are mean of triplicate groups and presented as mean ± SE. Values with different superscripts in the same column are significantly different ($P < 0.05$). The lack of superscript letter indicates no significant differences among treatments

Since growth performance did not reach a plateau over the tested protein levels in the present study and fish fed the diets containing 55% protein exhibited higher growth rates than those fed the diet containing dietary protein of 35 and 45%, it could be suggested that mandarin fish juveniles require dietary protein of at least 55% to sustain their fast growth. Nevertheless, further research is needed to precisely define the optimal dietary protein requirement for desirable rate of growth using diets containing higher levels of protein than 55%.

In addition, although increased dietary protein levels had no significant effect on DFI in the present study, it resulted in significantly higher FE and PER values. This meant that fish fed the high-protein diets (45 to 55% dietary protein) used dietary protein more efficiently than fish fed the low-protein diet (35%). A similar trend has also been found by different authors for other freshwater carnivorous fish species including pikeperch, Sander lucioperca (Nyina-wamwiza et al. 2005), and snakehead (Aliyu-Paiko et al. 2010). Higher pressure on body protein in order to satisfy the dietary needs for tissue building, repair, and metabolism has been suggested as the reason for poor growth and dietary utilization observed in fish fed sub-optimal dietary protein levels (Mohanta et al. 2013).

Dietary energy has a major impact on the dietary protein requirements of fish, and by proper use of non-protein energy sources, particularly lipid, dietary protein in fish feed can be spared (Mohanta et al. 2013). Nevertheless, a protein-sparing effect, where supplementation of dietary lipid improves fish performance and feed utilization efficiency, was not evident in this study at all protein levels. Although numerically higher values were observed in those of fish fed 14% dietary lipid, there was no significant difference in weight gain and PER of fish fed diets containing 7 to 14% lipid. These results may suggest that 7% dietary lipid is probably sufficient to meet the minimal requirement of this fish while the amount of dietary lipid needed to achieve maximum growth seems to be at or close to 14%. Limited or no obvious protein-sparing effect was also observed in various other fish species including murray cod, Maccullochella peelii peelii (De Silva et al. 2002), grass carp, Ctenopharyngodon idella (Du et al. 2005), white seabream, Diplodus sargus (Ozorio et al. 2006), and tiger puffer, Takifugu rubripes (Kikuchi et al. 2009), where increasing levels of dietary lipid had no beneficial effects on growth and feed utilization efficiency.

In the present study, muscle chemical composition was not affected by dietary treatment. Similar results were recorded for Senegalese sole, Solea senegalensis Kaup, (Valente et al. 2011) Mediterranean yellowtail, Seriola dumerili (Vidal et al. 2008), Atlantic cod, Gadus morhua L. (Morais et al. 2001), and red drum, Sciaenops ocellatus L. (McGoogan and Gatlin 1999). This finding is in contrast to those reported in other studies for Totoaba, Totoaba macdonaldi (Rueda-López et al. 2011), topmouth culter, Culter alburnus Basilewsky (Zhang et al. 2015), and red porgy, Pagrus pagrus, (Schuchardt et al. 2008) where fish muscle composition was significantly affected by dietary protein/lipid ratios. This discrepancy may be attributed to the fish species variation or a difference in experimental condition particularly dietary protein/energy ratios.

5. CONCLUSION

Despite the growing importance of the mandarin fish as a promising target species with high potential for aquaculture, there is no information concerning the nutritional requirement of this freshwater finfish species. To the best of our knowledge, this is the first attempt to evaluate the protein and lipid requirements of this fish indicating that diets for juvenile mandarin fish should include at least 55% protein and 7% lipid to maintain a good performance. These findings may provide useful context for developing a more cost-effective and nutritionally balanced feed for the culture of mandarin fish.

ACKNOWLEDGEMENTS

This study was supported by a grant from the Fishery Commercialization Technology Development Program (D11524615H480000120) funded by Korea Institute of Marine Science & Technology Promotion.

REFERENCES

1. Aliyu-Paiko M, Hashim R, Shu-Chien AC. Influence of dietary lipid/protein ratio on survival, growth, body indices and digestive lipase activity in snakehead (Channa striatus, Bloch 1793) fry reared in re-circulating water system. Aquacult Nutr. 2010;16:466–74.
2. AOAC (Association of Official Analytical Chemists). Official Methods of Analysis. 16th ed. Arlington: Association of Official analytical chemists; 1997.
3. De Silva SS, Gunasekera RM, Collins RA, Ingram BA. Performance of juvenile Murray cod, Maccullochella peelii peelii (Mitchell), fed with diets of different protein to energy ratio. Aquacult Nutr. 2002;8:79–85.
4. Du ZY, Liu YJ, Tian LX, Wang JT, Wang Y, Liang GY. Effects of dietary lipid level on growth, feed utilization and body composition by juvenile grass carp (Ctenopharyngodon idella). Aquacult Nutr. 2005;11:139–46.
5. Espinos FJ, Tomas A, Perez LM, Balasch S, Jover M. Growth of dentex fingerlings (Dentex dentex) fed diets containing different levels of protein and lipid. Aquaculture. 2003; 218: 479–490.
6. Hamre K, Øfsti A, Næss T, Nortvedt R, Holm JC. Macronutrient composition of formulated diets for Atlantic halibut (Hippoglossus hippoglossus, L.) juveniles. Aquaculture. 2003;227:233–44.

7. Jover M, Garcıa-Gomez A, Tomas A, De la Gandara F, Perez L. Growth of mediterranean yellowtail (Seriola dumerilii) fed extruded diets containing different levels of protein and lipid. Aquaculture. 1999; 179: 25–33.
8. Kikuchi K, Furuta T, Iwata N, Onuki K, Noguchi T. Effect of dietary lipid levels on the growth, feed utilization, body composition and blood characteristics of tiger puffer Takifugu rubripes. Aquaculture. 2009;298:111–7.
9. Liang XF. Study on mandarin fish and its culture home and abroad. Fisheries Sci Tech Inform. 1996;23:13–7 (in Chinese).
10. Maldonado-García M, Rodríguez-Romero J, Reyes-Becerril M, Álvarez-González CA, Civera-Cerecedo R, Spanopoulos M. Effect of varying dietary protein levels on growth, feeding efficiency, and proximate composition of yellow snapper Lutjanus argentiventris (Peters, 1869). Lat Am J Aquat Res. 2012;40:1017–25.
11. McGoogan BB, Gatlin DM. Dietary manipulations affecting growth and nitrogenous waste production of red drum, Sciaenops ocellatus I Effects of dietary protein and energy levels. Aquaculture. 1999;178:333–48.
12. Mohanta KN, Subramanian S, Korikanthimath VS. Effect of dietary protein and lipid levels on growth, nutrient utilization and whole-body composition of blue gourami, Trichogaster trichopterus fingerlings. J Anim Physiol An N. 2013;97:126–36.
13. Morais S, Bell JG, Robertson DA, Roy WJ, Morris PC. Protein/lipid ratios in extruded diets for Atlantic cod Gadus morhua L.: effects on growth, feed utilisation, muscle composition and liver histology. Aquaculture. 2001;203:101–19.
14. NRC (National Research Council). Nutrient requirements of fish and shrimp. Washington, DC: National Academy Press; 2011. 392 pp.
15. Nyina-wamwiza L, Xu XL, Blanchard G, Kestemont P. Effect of dietary protein, lipid and carbohydrate ratio on growth, feed efficiency and body composition of pikeperch Sander lucioperca fingerlings. Aquac Res. 2005;36:486–62.
16. Ozorio ROA, Valente LMP, Pousao-Ferreira P, Oliva-Teles A. Growth performance and body composition of white seabream (Diplodus sargus) juveniles fed diets with different protein and lipid levels. Aquac Res. 2006;37:255–63.
17. Rema P, Conceicao LEC, Evers F, Castro-Cunha M, Dinis MT, Dias J. Optimal dietary protein levels in juvenile Senegalese sole (Solea senegalensis). Aquacult Nutr. 2008;14:263–9.
18. Rueda-López S, Lazo JP, Reyes GC, Viana MT. Effect of dietary protein and energy levels on growth, survival and body composition of juvenile Totoaba macdonaldi. Aquaculture. 2011;319:385–90.
19. Schuchardt D, Vergara JM, Fernandez-Palacios H, Kalinowski CT, Hernandez- Cruz CM, Izquierdo MS, Robaina L. Effects of different dietary protein and lipid levels on growth, feed utilization and body composition of red porgy (Pagrus pagrus) fingerlings. Aquacult Nutr. 2008;14:1–9.
20. Valente LMP, Linares F, Villanueva JLR, Silva JMG, Espe M, Escórcio C, Pires MA, Saavedra MJ, Borges P, Medale F, Alvárez-Blázquez B, Peleteiro JB. Dietary protein source or energy levels have no major impact on growth performance, nutrient utilization or flesh fatty acids composition of market-sized Senegalese sole. Aquaculture. 2011;318:128–37.
21. Vidal AT, Garcia FDG, Gomez AG, Cerda MJ. Effect of the protein/energy ratio on the growth of Mediterranean yellowtail (Seriola dumerili). Aquac Res. 2008;39:1141–8.
22. Wilson RP. Amino acid and proteins. In: Halver JE, Hardy RW, editors. Fish nutrition. New York: Academic Press; 2002. p. 143–79.
23. Wu LX, Jiang ZQ, Qin KJ. Feeding habit and fishery utilization of Siniperca scherzeri in Biliuhe reservoir. J Fish Sci Chin. 1997;44:25–9 (in Chinese with English abstract).
24. Zhang YL, Song L, Liu RP, Zhao ZB, He H, Fan QX, Shen ZG. Effects of dietary protein and lipid levels on growth, body composition and flesh quality of juvenile topmouth culter, Culter alburnus Basilewsky. Aquac Res. 2015;47:2633–41.
25. Zhou CW, Yang Q, Cai DL. On the classification and distribution of the Sinipercinae fishes (Family Serranidae). Zool Res. 1988;9:113–26 (in Chinese with English abstract).

CHAPTER 5

Vitellogenesis and choriogenesis in fishes

Akihiko Hara[1], Naoshi Hiramatsu1[1], Toshiaki Fujita[2]

[1] *Faculty of Fisheries Sciences, Hokkaido University, Hakodate, Hokkaido 041-8611, Japan*

[2] *Faculty of Engineering, Hachinohe Institute of Technology, Hachinohe, Aomori 031-8501, Japan*

ABSTRACT

In studies of sex discrimination in fish in the early 1900s, a specific antigen in the blood of gravid females was identified using immunological methods. At present, this specific antigen is known as vitellogenin, the major precursor of egg yolk protein that is synthesized in the female liver and is secreted into the blood to be incorporated into the egg. Recently, protein and gene analyses have revealed the presence of several vitellogenin variants. In addition, in the 1980s, choriogenin was identified as a novel precursor of egg envelope proteins that is secreted into the blood in response to stimulation by estrogen, similarly to vitellogenin. These two proteins not only play key roles in the process of oogenesis, but they are also used as effective biomarkers for assessing the impact of estrogen-like endocrine-disrupting chemicals (environmental hormones) in aquatic ecosystems.

Keywords: Vitellogenin, Choriogenin, Yolk protein, Estrogen, Biomarker, Endocrine disrupting chemicals, Fish

1. INTRODUCTION

The presence of a protein specifically found in female blood serum during oocyte growth in fishes was first reported in 1914 by Uhlenhuth and Kodama [1] in a paper entitled 'A study of sexual distinction reaction'. In this study, the authors used an immunological method (precipitation reaction) with antisera from rabbits immunized with carp *Cyprinus carpio* egg extract to detect a substance reactive with this antisera in the blood of carp carrying mature eggs. They named this substance ovumin. Later, in 1923, Sakuma [2] reported in his paper entitled 'On the precipitin reaction of the reproductive cells and the sexual differences' that he studied sexual differences in 14 marine fish species using a similar method and found that antisera against fish eggs indeed, as observed in the case of carp, reacted with blood serum from mature females.

In the 1940s, using the Tiselius method of electrophoresis, Deutsch et al. [3, 4] and Moore [5] analyzed the electrophoretic patterns of serum proteins from several fish species and reported species differences. In addition, in 1957, Saito [6, 7] used the Tiselius method along with paper electrophoresis, and reported on the existence of both qualitative and quantitative characteristic differences in serum proteins between teleosts and elasmobranchs, that also depended upon fish habitat. Later, fish serum proteins became widely used as biochemical tools among researchers in various fields such as taxonomy, genetics, physiology and biochemistry (review in Hara [8]: see references therein).

Since 1961, the presence of proteins specifically expressed in the blood of maturing females (serum vitelline [9], Sm antigen [10], HM factor [11], serum lipovitellin [12], FSPP [13], lipophosphoprotein [14]) has been observed in many fish species using various analytical methods, including newly developed variations of electrophoresis, gel-filtration, ion-exchange chromatography, ultracentrifugation and immunological methods. We also discovered a female-specific iron-binding serum protein (FS) in rainbow trout *Oncorhynchus mykiss* and chum salmon *O. keta* [15] while analyzing interspecies differences in electrophoretic patterns of serum proteins. In 1969, in the field of entomology, such female-specific proteins expressed in the blood (body fluid) during oogenesis were named 'vitellogenin (Vtg)' (vitelline + genin, meaning source of egg yolk) [16], and the name has been used ever since for proteins specifically expressed in the blood of maturing females in oviparous vertebrates. In general, Vtg in oviparous vertebrates is a lipoglycophosphoprotein and serves as a major precursor of egg yolk proteins stored as essential nutrients for future embryogenesis. The process or period in which this Vtg-derived egg yolk accumulates within oocytes is called vitellogenesis. During this process, estrogen secreted from the ovarian follicles triggers the synthesis of Vtg in the liver, which is then secreted into the blood, incorporated into oocytes and cleaved to generate multiple egg yolk proteins such as lipovitellin (Lv) and phosvitin (Pv) that will be stored in the ooplasm (reviews in Wallace [17], Mommsen and Walsh [18], Selman and Wallace [19], Specker and Sullivan [20], Hiramatsu et al. [21, 22], Patiño and Sullivan [23], Reading and Sullivan [24]).

In the 1970s, multiple female-specific serum proteins were reported by Plack et al. [25] and Aida et al. [13] in Atlantic cod *Gadus morhua* and ayu *Plecoglossus altivelis*, respectively. We also detected three female-specific serum proteins components in medaka *Oryzias latipes* [26], and four to five components in white-edged rockfish *Sebastes taczanowskii* [27]. In the era from the 1970s to the 1990s, the presence of female-specific serum proteins other than Vtg was shown. Some of these proteins were expressed prior to the main phase of yolk deposition, indicating the possibility that they differed from Vtg [13, 27, 28].

An estrogen-inducible female-specific protein different from Vtg with a relatively small molecular weight was discovered in 1984 in medaka, and was later revealed to be the precursor of egg envelope protein [29]. This substance was named 'choriogenin (Chg)' (chorion + genin, meaning 'source of egg envelope') [30]. Similar to Vtg, Chg is an estrogen-induced maternal protein that functions as an oogenic protein forming the inner layer of the egg envelope.

Since the mid-1990s, Vtg has attracted attention as a biomarker for evaluating the effects of endocrine system-disrupting chemicals present in various waters. Environmental induction of Vtg was first reported in male rainbow trout stocked near a wastewater treatment plant in the UK [31]. Later, many studies linking Vtg expression with estrogenic-like endocrine disrupting chemicals (EEDCs) were reported mainly from Europe and North America (reviews; Sumpter [32], Arcand-Hoy and Benson [33], Giesy and Snyder [34], Kime [35], Pait and Nelson [36], Arukwe and Goksøyr [37], Hiramatsu et al. [22]).

Over time, we acquired basic knowledge about Vtg in fish and established various immunoassays to detect the compound. Our laboratory has also conducted studies of Chg, mainly using Salmonidae species and focusing on estrogen-induction ('choriogenesis'), because this protein is a key biomarker for assessing EEDCs. Here, we review our studies on Vtg and Chg in fish and discuss some recent developments on this topic.

2. FISH VITELLOGENIN AND ITS RELATED EGG YOLK PROTEINS

Chicken oogenesis has been studied since the beginning of the 1900s, when oocyte growth was attributed to the uptake of low-molecular-weight substances such as amino acids, which were thought to be responsible for forming egg yolk proteins. A new idea that egg yolk proteins are first synthesized in the maternal liver and then transferred to growing oocytes via the blood was proposed, and it was first proven in 1974 that Vtg was the precursor of egg yolk proteins based upon a biochemical study using African clawed frogs *Xenopuslaevis* [38]. In this study, Vtg double-labeled with ^{3}H and ^{32}P was taken up by eggs after the administration of estrogen and cleaved into two egg yolk proteins with different molecular weights: Lv with ^{3}H radioactivity and Pv with ^{3}H and ^{32}P activity. This transformation from Vtg to egg yolk proteins was also demonstrated in chickens *Gallus domesticus*, using a similar method [39].

During the study of fish species, we isolated and purified Vtg and three egg yolk proteins, Lv, Pv and β′-component (β′-c), from fish in the family Salmonidae, including rainbow trout [40] and Sakhalin taimen *Hucho perryi* [41]. Using various antibodies against the purified proteins along with immunodiffusion and Western blotting methods, we visualized molecular cleavage of Vtg to produce Lv, Pv and β′-c. In the Sakhalin taimen, Vtg is a 540 kDa dimer consisting of two 240 kDa subunits containing 330 kDa Lv composed of two heavy-chains (92 kDa) and two light-chains (29 kDa), 34 kDa β′-c protein composed of a 17 kDa peptide, and 23 kDa Pv. Using biotin-labeled Vtg, we also observed its degradation into egg yolk proteins by bovine cathepsin D, suggesting specific degradation of Vtg in eggs by a cathepsin D-like enzyme [42, 43]. On the other hand, with regard to specific uptake of Vtg into eggs, a Vtg receptor was discovered, leading to the conclusion that the Lv domain within the Vtg molecule serves as a receptor-binding site [44].

For the purification of fish Vtg, we established a novel method in 1978 using rainbow trout by precipitating the serum of vitellogenic females under low ionic strength conditions (cold distilled water) and immediately loading samples onto a gel filtration column for isolation [40]. This simple and quick procedure is suitable for Vtg, which is prone to degradation during the purification process. We also developed another purification method for proteins that are difficult to precipitate or for Vtg isolation from other fish species by using hydroxylapatite chromatography [45]. Many other methods such as ion-exchange chromatography have been used to purify Vtg from various fish, including rainbow trout [46], Atlantic salmon *Salmo salar* [47], goldfish *Carassius auratus* [48, 49], Japanese eel

Anguilla japonica [50], catfish *Heteropneustes fossilis* [51], mummichog *Fundulus heteroclitus* [52] and coho salmon *Oncorhynchus kisutch* [53, 54] (review in Specker and Sullivan [20]).

Lv, a major degradation product of Vtg, is the main component of egg yolk, and its lipid content is about 20 % by mass. In structure, Lv is a dimer consisting of two different polypeptides, a heavy chain (LvH) and a light chain (LvL). Lv is abundant in various types of amino acids and lipids that are essential nutritional sources for embryogenesis. In contrast, Pv is a phosphoprotein with approximately 10 % phosphorus and with serine (Ser) occupying 50 % of the amino acid content, making Pv difficult to detect with normal staining techniques, and giving it extremely weak antigenicity because of its high phosphorus content. Minerals bound to the Ser residues in this phosphoprotein are thought to be essential components for osteogenesis and metabolic functions. The third component of the egg yolk protein, β′-c, which is devoid of any lipid and phosphorus, was discovered in the rainbow trout by Jared and Wallace [55], and identified as a kind of serum protein. Later, Markert and Vanstone [56] isolated a similar protein from the eggs of coho salmon and named the protein β′-c. It has also been isolated from fish other than salmonids, including barfin flounder *Verasper moseri* [57], white perch *Morone americana* [58] and grey mullet *Mugil cephalus* [59]. β′-c is thus considered to be a ubiquitous egg yolk protein in teleosts. In the elasmobranch, we reported for the first time that β′-c is also present in the eggs of cloudy catshark *Scyliorhinus torazame* [60]. Most of its function remains unclear, other than being a source of amino acids necessary in embryogenesis, but β′-c contains large amounts of cysteine (Cys) residues within the molecule. Recently, Shimizu et al. [61] discovered that β′-c is one of the allergens that cause fish egg allergy symptoms. Similarly, 40 kDa glycoprotein (YG40) containing abundant Cys residues was found in chicken eggs [62], corresponding to fish β′-c. Recently, from a study of the barfin flounder, the fourth egg yolk protein component considered to have originated from the furthest C-terminus of the Vtg polypeptide (C-terminal coding domain) was found [63]. Based on these results, the molecular sequence of the egg yolk proteins in Vtg is considered to be expressed, from the N-terminus, as NH_2–(LvH)–(Pv)–(LvL)–(β′-c)–(C-terminal coding domain)–COOH (see Fig. 2) [21, 22, 64, 65, 66].

3. VITELLOGENESIS IN FISH

The general characteristics of Vtg in oviparous vertebrates can be summarized as follows: (1) Vtg is a protein specifically expressed in female blood serum during vitellogenesis; (2) The administration of estrogen to male or juvenile fish induces the production of plasma Vtg; (3) Vtg is a high-molecular-mass complex protein consisting of sugar, lipid and phosphorus and binding other elements such as calcium, iron and zinc; (4) Vtg is a precursor of egg yolk proteins that react with antibodies raised against egg extracts.

Figure 1 depicts an oogenesis model of Salmonidae involving Vtg and the substance Chg, which is described later. The process of vitellogenesis in fish is understood as follows: (1) Follicle-stimulating hormone (FSH) secreted into the bloodstream by the pituitary gland acts on follicle cells surrounding developing oocytes to induce synthesis of the sex steroid hormone, estrogen (estradiol-17β), which induces vitellogenesis. The estrogen links to sex steroid hormone-binding globulin in the blood and is transferred to hepatocytes to act on the Vtg gene in the nucleus via estrogen receptors. The complex of estrogen and the estrogen receptor in hepatocytes binds to the promoter region of the Vtg gene, which activates the gene to initiate and accelerate transcription. (2) Then, translation of the Vtg transcript products begins, after which Vtg proteins undergo modification processes such as lipidation, phosphorylation and glycosylation before being secreted into the blood. (3) Blood Vtg binds to the Vtg receptor on the oocyte plasma membrane and is taken into the cell. (4) Vtg that enters an oocyte is degraded specifically by a cathepsin D-like enzyme and undergoes molecular cleavage to produce Lv, Pv, and β′-c, which are stored in the cell. This is the ‘single Vtg model’, which was formerly proposed when Vtg was considered to be composed of a single molecular species of protein. However, as a result of the recent discovery of multiple species of Vtg, the ‘multiple Vtg model’ with more complex processes is now accepted as a more accurate description of vitellogenesis [22].

One of the most interesting features of Vtg is that, through exposure to estrogen, Vtg can be induced in systems where its synthesis would not normally occur, such as in male liver or in vitro liver tissue, and cultured hepatocytes. In general, Vtg can be detected in any culture system, regardless of fish species, when induced with $\geq 10^{-8}$M estradiol-17β [67]. In vivo Vtg induction studies using estrogen treatment have demonstrated that a single treatment with 10 μg estradiol-17β/kg body weight resulted in blood Vtg detection within 24 h, regardless of fish species. The amount of induced Vtg in the blood reached the order of μg/ml in 12 h, depending upon conditions such as the dose of estradiol-17β and the water temperature. It has also been suggested that, judging from observations that Vtg was hardly detected in the blood when estradiol-17β was administered at concentrations less than 10 μg/kg body weight, even with multiple injections, there exists a certain threshold in the estrogen concentration required to induce Vtg synthesis.

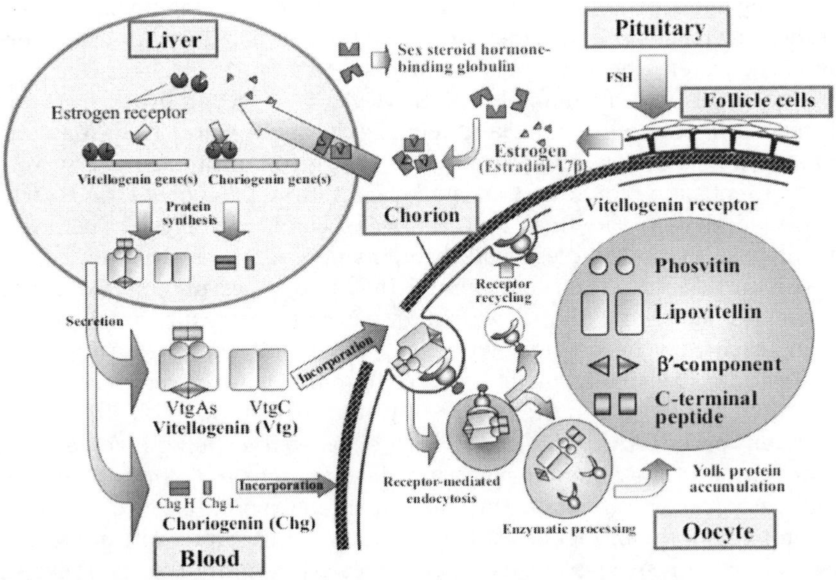

Figure 1 Model of vitellogenesis and choriogenesis in salmonid fish. See text for details and abbreviations. Redrawn from Hiramatsu et al. [22]

The Vtg level also changes as fish progress through oogenesis. We observed an annual change in the Vtg level in Sakhalin taimen (Salmonidae), as measured by an enzyme linked immunosorbent assay and the Mancini method (mentioned later), demonstrating fluctuations in the blood Vtg concentration in accordance with fish maturation [68]. In pond cultures, female Sakhalin taimen generally become sexually mature and sufficiently gravid for egg collection within 6–7 years. In 3-year-old fish, the level of Vtg in blood serum ranges from 1 to 10 μg/ml. As fish approach sexual maturity, Vtg levels gradually increase to 1 mg/ml in the year prior to egg collection, followed by a rapid increase reaching several tens of mg/ml a few months prior to egg collection. As indicated by this study, blood serum levels of Vtg in females fluctuate over a relatively wide range (from 1 μg/ml to several tens of mg/ml), compared with those of other serum proteins.

As described above, Vtg is subjected to limited degradation during vitellogenesis to produce at least three types of egg yolk proteins (Lv, Pv and β′-c), which are stored in eggs. The initial degradation during vitellogenesis is termed the first proteolysis, and later degradations during final maturation and embryogenesis are called the second and third proteolysis, respectively [21]. In addition to the first proteolysis described earlier, second proteolysis was first reported in mummichogs [69, 70], where egg yolk proteins were further degraded during the final maturation of the oocytes. This phenomenon is observed in fish living in seawater or brackish water whose oocytes absorb large amounts of water during the final stages of maturation [71, 72]. However, this second proteolysis is not observed in salmonids that lay their eggs in fresh water [43]. Matsubara and Sawano [73] described three types of egg yolk proteins (Lv, Pv and β′-c) in the barfin flounder that underwent first proteolysis and further degraded during the final maturation (second proteolysis), where the majority of Pv and β′-c egg yolk proteins degraded into free amino acids (FAA). In addition, they determined that two types of Vtgs exist in the barfin flounder, and that the dual Lv variants correspondingly derived from each of these Vtgs were degraded differently during oocyte maturation. The two forms of Lv were found to be involved in controlling the buoyancy of eggs in the course of water absorption, with a differential contribution of FAA to the osmolyte pool, as well as being involved in the distribution of nutrients during early development [63]. In contrast, little has been reported about the third proteolysis of egg yolk proteins accompanying embryogenesis. In our study using salmonids, in which no second proteolysis was observed, we revealed that three types of egg yolk proteins (Lv, Pv and β′-c) were subjected to separate degradations after fertilization, and that the degradation of LvH to smaller products and dephosphorylation of Pv occurred after the eyed embryo stage, while β′-c remained intact throughout embryo development [43].

4. MULTIPLICITY OF FISH VITELLOGENINS

Recently, several Vtg gene transcripts were discovered in the African clawed frog (VtgA1, A2, B1, B2) [74, 75] and chicken (Vtg I, II, III) [76, 77], along with fish species such as mummichog [78, 79], barfin flounder [64], haddock *Melanogrammus aeglefinus*, a species of cod [80], tilapia *Oreochromis aureus* [81] and sheepshead minnow *Cyprinodon variegatus*, an estuarine-species of pupfish [82]. Among fishes, a gene encoding a special form of Vtg lacking the Pv region (phosvitinless Vtg, Pv-less Vtg) was first discovered in zebrafish *Danio rerio* [83]. In addition, genomic analysis of rainbow trout [84] revealed the presence of 20 active genes for Vtg as well as ten pseudogenes. The Vtg genes in rainbow trout had ≤3 % sequence heterogeneity, and they likely encode nearly identical proteins. This gene redundancy probably occurs so that large amounts of Vtg can be synthesized by this species, which spawns comparatively large eggs.

In 2002, we successfully purified three different types of Vtgs from the blood plasma of white perch for the first time [58] and proposed a temporary classification for the three Vtg forms. Fish Vtgs can be divided into two main types, complete and incomplete [66] (Fig. 2). The complete Vtg has a deduced primary amino acid sequence composed of the five egg yolk protein regions described earlier (LvH, Pv, LvL, β′-c, C-terminal coding domain). As a result of the homology analysis, the complete Vtg was further divided into two subgroups, type A (VtgA) and type B (VtgB). The second major group, the incomplete Vtg, mainly consists of LvH and LvL, the main components of lipoprotein in fish eggs, and it has very low homology with the complete forms of Vtg. Pv-less Vtg in zebrafish (mentioned above) is considered to be an incomplete Vtg. Using the same classification system as the complete Vtg, we named this incomplete Vtg as 'Vtg type C' (VtgC). Because VtgA and VtgB have highly homologous primary structures and share similar characteristics such as molecular weight, the purification of individual proteins is difficult. In contrast, because VtgC has a lower molecular weight than the other two types, it is relatively easy to purify. To date, the complete Vtg (VtgA and/or VtgB) and incomplete Vtg (VtgC) have been successfully isolated and purified as two different proteins with distinct molecular weights and antigenicity in various fish species, including tilapia (*O. aureus* [85] and *O. mossambicus* [86]), medaka [87], Japanese goby *Acanthogobius flavimanus* [88] and Sakhalin taimen [89]. In addition, three distinct types of Vtg have been purified or detected in mosquitofish *Gambusia affinis* [90], red seabream *Pagrus major* [91] and grey mullet [92]. Two or three types of Vtgs have been detected as proteins with distinct antigenicity in various existing fish species [66]. Table 1 shows a summary of multiple Vtg genes and proteins reported in the literature [22].

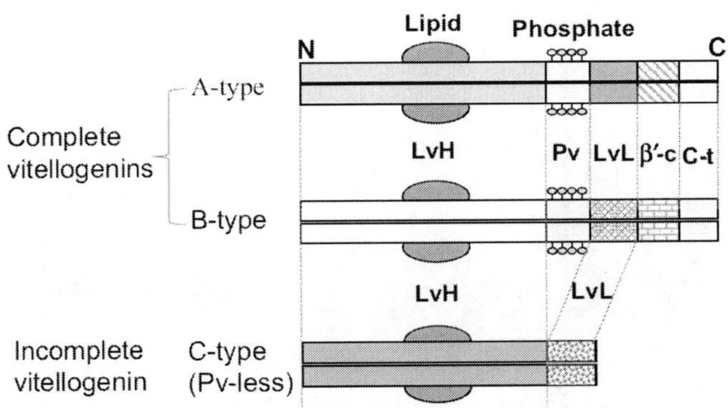

Figure 2 Domain structures of fish vitellogenins with regard to their derived yolk protein products, lipovitellin (Lv), phosvitin (Pv), β′-component (β′-c) and C-terminal coding domain (C-t). Monomeric Lv is composed of two subunits: a heavy chain (LvH) and a light chain (LvL). This figure is reproduced from the one published in Hiramatsu et al. [66]

Table 1 Multiple vitellogenin (Vtg) cDNA and protein in various teleosts

Species	cDNA			Protein		
	VtgA (VtgAa)*	VtgB (VtgAb)*	VtgC (Pv-less)	VtgA	VtgB	VtgC (Pv-less)
Barfin flounder	○	○		500–520 kDa	530–550 kDa	
Red seabream	○	○	○	610 kDa	610 kDa	340 kDa
Japanese common goby	○		○	530 kDa		320 kDa
Mosquitofish	○	○	○	600 kDa	600 kDa	400 kDa
Mummichog	○	○	○			
Walleye pollock	○	○				
Medaka	○	○	○	570 kDa		460 kDa
White perch	○	○	○	532 kDa	532 kDa	426 kDa
Grey mullet	○	○	○	570 kDa	580 kDa	335 KDa
	Vtg*					
Zebrafish	○ (VtgAo)		○	VtgAs*		
Pacific herring	○ (VtgAc)		○	540 kDa		380 kDa
White spotted charr	○ (VtgAs)		○			

The structure and function of fish Vtgs has diversified during the evolution. At present, fish Vtg genes are considered to have diversified through whole genome duplication (WGD) events, and also via lineage specific tandem gene duplication (TGD), followed by neofunctionalization. Recently, Finn and Kristoffersen [93] conducted detailed analyses of deduced amino acid sequences of Vtg gene transcripts from birds, amphibians and fish, and proposed a classification method based on the molecular evolution of Vtgs. According to their nomenclature, the Vtgs formerly named VtgA and VtgB become VtgAa and VtgAb paralogs, respectively. This nomenclature suggests that a chordate Vtg (VtgABCD), considered to be the ancestor of vertebrate Vtg, appeared during the first round of whole genome duplication (1R-WGD). For example, silver lamprey *Ichthyomyzon unicuspis* Vtg is classified as VtgABCD. In the subsequent 2R-WGD, Vtgs were differentiated into VtgAB (chondrostean vitellogenin; found in chondrostean fish, amphibians and birds) and VtgCD. Later, during the 3R-WGD, VtgAB allegedly differentiated to yield VtgA (found in all teleosts) and VtgB, and VtgCD was further divided into VtgC (fish VtgC or Pv-less Vtg) and VtgD. The presence of genes encoding VtgB and VtgD still remains unconfirmed among extant teleosts, suggesting that these genes may have been deleted after the 3R-WGD. VtgA supposedly experienced further lineage-specific gene duplication within teleosts, resulting in various paralogous and orthologous Vtg subtypes (e.g. VtgAa and VtgAb). More specifically, fish belonging to Paracanthopterygii and Acanthopterygii, considered to be more evolved groups of fish, are supposed to possess all three Vtg orthologues (VtgAa, VtgAb and VtgC). In contrast, paralogous Vtg variants found in some other fish cannot be categorised as VtgAa, VtgAb or VtgC. For example, the Salmonidae family in Protacanthoperygii differentiated relatively early in phylogeny and contains two types of Vtgs, one being a complete form of VtgA (salmonid-type A-type Vtg: VtgAs) that is not differentiated as VtgAa or VtgAb, and the other, VtgC. With regard to the complete forms of Vtgs, VtgAo1 and VtgAo2 are present in Ostariophysi as Ostariophysian-type A-type Vtg, and VtgAe1, VtgAe2 and VtgAe3 are present in Elopomorpha as Elopomoruph-type A-type Vtgs. In both cases, VtgC orthologues exist in addition to these paralogous VtgAs [24]. Figure 3 shows a molecular phylogenetic tree of amino acid sequences deduced from Vtg gene transcripts constructed by Finn and Kristoffersen [93], with some additional insights obtained through our research [60, 94, 95, 96]. It should be noted that more recent analyses of Vtg gene synteny have raised new questions concerning the relative importance of WGD versus TGD in the evolution of fish Vtgs and the timing of these duplications [97, 98], and indicate that the A-type and C-type Vtgs have a much longer evolutionary history (BP ~ 425 million) than previously thought. Nonetheless, the Vtg nomenclature proposed by Finn and Kristoffersen [93] remains applicable and useful for keeping track of Vtg gene orthology.

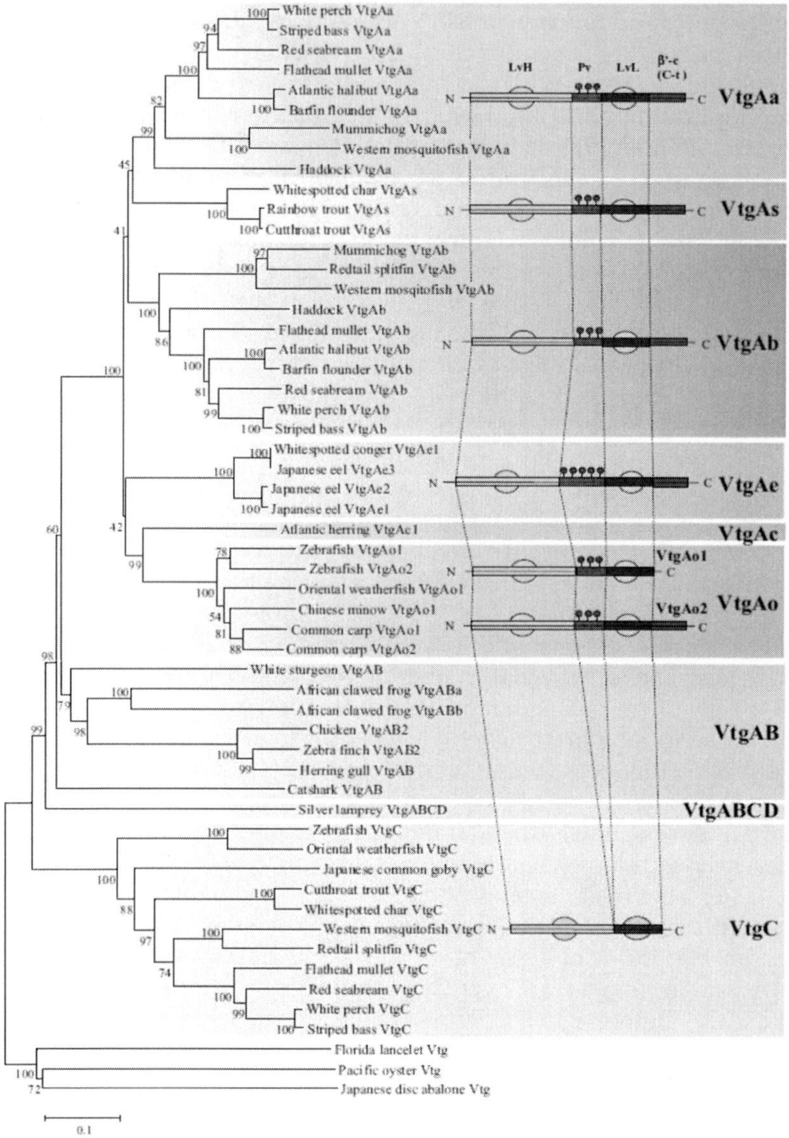

Figure 3 Left Phylogenetic distribution of vitellogenin (Vtg) polypeptide sequences from selected egg-laying animals. Nomenclature of Vtg sequences follows a classification scheme proposed by Finn and Kristoffersen [93]. The analysis was performed using the neighbour-joining algorithm, and sequences were rooted using invertebrate Vtg sequences as an outgroup. Numbers besides nodes indicate the percent of bootstrap values for each branch of the tree in the 1000 bootstrap trials. Right Models representing the linearized domain structures of selected types of Vtgs

The naming scheme for Vtgs should become more precise in the future as genetic and protein analyses become more advanced.

5. DETECTION AND MEASUREMENT OF FISH VITELLOGENIN

Because fish Vtg is a protein containing approximately 1 % phosphorous, Vtg in blood serum can be detected by staining with phosphorus dye after electrophoresis. In addition, Vtg can be indirectly quantified by measuring alkaline-labile phosphorus in blood serum. Furthermore, because of its high molecular weight, it can simply be detected by observing high-molecular-weight peaks that appear during gel-filtration chromatography [54]. However, qualitative and quantitative analyses based on immunological approaches using antibodies are generally superior. Because Vtgs are specifically found in females and induced by estrogen, and function as precursors of egg yolk proteins, three materials have been used as immunogens in the preparation of antibodies: (1) vitellogenic female blood serum, (2) blood serum of estrogen-treated fish, and (3) extracts of egg obtained during vitellogenesis. Female-specific antisera are prepared by subjecting antisera from animals (e.g. rabbits) immunized with these antigens to absorption procedures with male fish blood serum. Antisera prepared in this manner react strongly with Vtgs in female blood serum without reacting with male serum. Purified Vtg or egg yolk proteins (Lv, β′-c) can also be used as immunogens. Pv reportedly has no antigenicity under normal circumstances; however, antibodies can be obtained by immunizing with dephosphorylated Pv [41]. The preparation of universal antibodies that react regardless of species is now under development using a well-conserved region of the Vtg molecule as a peptide antigen [22]. Here, we present a brief summary of Vtg immunoassay.

6. OUCHTERLONY DOUBLE IMMUNODIFFUSION (AGAR GEL IMMUNODIFFUSION, PASSIVE DOUBLE IMMUNODIFFUSION)

This is the simplest detection method. Samples such as blood serum and antibodies are placed in different holes (wells) on an agar gel plate and left from several hours to overnight to develop. Afterwards, a precipitation line generated by antibody–antigen complexes (precipitation reaction) appears. This method is semiquantitative with a detection sensitivity of approximately 10 μg/ml. It requires relatively large amounts of antibodies compared with other methods introduced below and is low in Vtg detection sensitivity. However, it has advantages in its simplicity both in manipulation and required equipment [8].

7. MANCINI METHOD (SINGLE RADIAL IMMUNODIFFUSION)

This quantitative method has a similar sensitivity to the Ouchterlony double immunodiffusion. Agar gel plates containing antibodies are prepared in advance, and Vtg concentrations are measured by calculating the area of the precipitation ring that forms around the sample well. Because Vtg levels in the blood at the peak of vitellogenesis are quite high (approximately 40 mg/ml) in Salmonidae, this method of detection and quantification is sufficient for these species under normal conditions [8, 68].

8. SENSITIVE HIGH-THROUGHPUT IMMUNOASSAYS

These are quantitative assays with high sensitivity: radio immunoassay (RIA), enzyme immunoassay (EIA), fluorescent immunoassay (FIA) and chemiluminescent immunoassay (CLIA). These methods share a common detection principle, but use different labeling methods. Using tracers (radioactive material, enzymes, fluorescent material and chemiluminescent material), these methods exhibit >10,000-fold higher sensitivity as compared to agar plate precipitation methods and can detect concentrations on the order of several tens of pg/ml. We introduce a CLIA method for the measurement of Vtg in Salmonidae, a sandwich method using no radiolabeled materials [22, 99]. Antibodies (anti-β′-c antibodies) against Vtg were absorbed in a microtiter plate with 96 wells. Following this, known concentrations of Vtgs were added to the plate as standards, and samples were also added. Then, antibodies (anti-Lv antibodies) labeled with chemiluminescent acridinium were added to sandwich the Vtgs. Later, by measuring the luminescence of the labeled antibodies bound to the Vtgs, Vtg concentrations were calculated on the basis of a calibration curve (60 pg/ml–500 ng/ml) prepared in advance.

9. SIMPLE DETECTION BY IMMUNOCHROMATOGRAPHY

In collaboration with a private company, we established a quick salmonid Vtg detection method by developing an immunochromatography-based method generally used in pregnancy tests and influenza tests. Immunochromatography results can be obtained within 2–15 min, with a detection sensitivity of 30 ng/ml. Because this method offers a quick and more sensitive option than other conventional methods, it is useful for primary screening of proteins distributing in as wide a range of concentrations as Vtgs [22].

10. CHORIOGENESIS AND CHORIOGENIN

The egg membrane of teleosts (vitelline envelope, zona radiata or zona pellucida) consists of an extremely thin, high-density outer layer and a thick, low-density inner layer making up most of the structure. Some fish species have outer layers that are further divided into two parts. The timing of the formation of egg envelope components (choriogenesis) differs by layer. In general, outer layer formation starts around the perinucleolar oocyte stage, followed by the formation of the inner layer during the previtellogenic oocyte stage. Up until the mid-1980s, it was believed that proteins constituting the egg envelope were synthesized exclusively within the oocytes. However, a study of medaka [29] reported that a vitelline coat protein precursor synthesized in the liver after induction by estradiol-17β was secreted into the blood and deposited on the inner layer of the egg envelope. This liver-derived precursor of egg envelope protein was named Chg in mummichog and medaka [30]. To date, Chg has been isolated and purified in several fish species including medaka [100, 101], Sakhalin taimen [102], masu salmon *Oncorhynchus masou* [103], Atlantic cod [104] and red lip mullet *Chelon haematocheilus* [105], and the presence of two different Chgs with distinct molecular weights and antigenicities, namely Chg high molecular weight (Chg H) and Chg low molecular weight (Chg L), has been confirmed in these species (see Fig. 1).

Three Chg cDNAs that encode Chg H, Chg L and Chg H minor (Chg Hm) were cloned from the medaka liver, and their high homology to zona pellucida glycoprotein (ZP: ZPB and ZPC), which is synthesized in the mammalian ovary, was reported as well [106, 107]. In addition, three genes encoding egg envelope proteins, *zp1a,zp1b* and *zp3,* were cloned from gilthead seabream *Sparus aurata*, and were reported to be highly expressed in the liver [108]. In addition to such liver-derived egg envelope-related genes, the presence of egg envelope-related genes exclusively expressed in the ovary was also found. In carp [109, 110] and zebrafish [111], the expression of ZP2 and ZP3 mRNA was only observed in the ovary, and based on deduced amino acid sequences, ZP2 and ZP3 reportedly have high homology to mammalian ZPB and ZPC, respectively. However, none of the liver-derived egg envelope-related genes mentioned above have been found in Cyprinidae. Based on these results, it has been concluded that the site(s) of egg envelope protein synthesis depends on the fish species, and that the mechanism of teleost choriogenesis can be divided into two major groups: 'liver type' (Chg type) in Oryziatidae and Salmonidae, in which egg envelope production is solely based on Chg, and 'ovary type' (ZP type) in Cyprinidae, in which the egg envelope is made from ZP proteins only. Recently, however, this classification was disproved by some new reports. According to these studies, *zp* was obtained by cloning from the ovary of medaka [112] as well as from the liver. In addition, in masu salmon, three *chg* cDNAs encoding Chg Hα, β and Chg L were cloned from the liver of mature females [113], and two *zp* cDNAs were cloned from the ovary (ZPB [114] and ZPC [Fujita and Hara, unpubl. data]), revealing the presence of five egg envelope protein-related genes in total. Therefore, it appears that both Chg and ZP are generally involved in choriogenesis in fish that was formerly categorised as 'liver type', and both proteins can be expected to contribute to the complex structure of the egg envelope of these species. However, among fish categorized as having 'liver type' egg envelope precursors, the presence of ZP has been reported only in the two aforementioned species along with gilthead seabream, and the role of ZP in choriogenesis remains unclear. This suggests the possibility that, in general, both Chg and ZP are involved in choriogenesis in fish and that the variations in the mechanism of egg envelope formation arise from structural differences between Chg and ZP as well as the amounts of each protein that are synthesized.

Because the formation of the egg envelope starts before the accumulation of egg yolk, Chg synthesis may occur at an earlier stage of sexual maturation than Vtg synthesis. Such differences in the timing of synthesis may involve different sensitivities of the transcriptional response of each gene to estrogen stimulation. To date, we have purified two types of Chg (Chg H and Chg L) from salmonid species using methods similar to those used to purify medaka Chgs, and we have established a Mancini assay and EIA using antibodies raised against these proteins [115]. In masu salmon treated with estrogen, blood concentrations of two types of Chg exceeded that of Vtg when a low concentration of estrogen was used; the Vtg level surpassed the Chg levels when the estrogen concentration was increased. A similar phenomenon was also observed in female fish during sexual maturation at the transition between previtellogenic and vitellogenic stages. As observed in such cases, when induced by low concentrations of estrogen, blood levels of Chg exceeded those of Vtg, which suggests that Chg is a suitable marker of the effect of estrogen-like substances present at low concentrations [116]. Although the absence of Chg in the fish with 'ovary-type' egg envelope precursors needs to be taken into consideration, Chg is expected to be a useful biomarker for assessing the effect of estrogen-like substances in aquatic environments when examining fish with 'liver type' egg envelope precursors, perhaps in combination with Vtg measurements.

11. APPLICATIONS OF VITELLOGENIN AS BIOMARKERS FOR DETECTION OF ESTROGENIC ACTIVITY

Various chemicals associated with human activities are present in aquatic environments. Such chemicals include compounds that demonstrate estrogen-like activity, called estrogenic endocrine-disrupting chemicals (EEDCs). The effects of EEDCs on aquatic organisms have been of growing concern, particularly in Europe and North America. More recently, evidence of the widespread presence of EEDCs has been reported in Japan as well, and similar situations are presumed to be occurring in neighbouring Asian countries [22].

At present, the most reliable methods available for examining EEDC activity in water include the detection of Vtg induced in the blood of male or juvenile fish. As described previously, Vtg is a precursor of egg yolk proteins; its synthesis is controlled by estradiol-17β in the liver of female fish during maturation, and it is taken into oocytes after being transferred to the ovary via the blood (see Fig. 1). Vtg can also be induced in the blood of male or juvenile fish exposed to exogenous estrogen, and the presence of Vtg in the blood of males or juveniles has been widely accepted as a biomarker of their exposure to EEDCs [21, 22, 31, 32].

Various immunoassays for Vtg have been established in many fish species for detecting their exposure to EEDCs. Most of these methods have been developed under the premise of the 'single Vtg model'. More recently, however, the measurement of Vtg subtypes has been attempted in fish species in which Vtg polymorphisms were identified. In general, a complete form of Vtg is the major subtype and the incomplete Vtg (VtgC) is often a minor subtype [22]. This is particularly true for Salmonidae. In addition, in fish species where the complete form of Vtg has been further classified into VtgAa and VtgAb, the ratio of subtypes present in the blood depends on each species [117, 118]. Because of the large molecular weight differences, the ratio of complete to incomplete forms of Vtg can be readily determined by electrophoresis and similar methods. In contrast, multiple subtypes of Vtg often have identical or extremely similar molecular weights, and, as noted, they are difficult to separate or individually identify at the protein level. In such cases, major Vtg subtypes can be identified using quantitative real-time PCR, to measure their subtype-specific mRNA. The obtained Vtg gene expression profiles represent a valid index for the detection and assessment of EEDC activity in addition to Vtg protein expression. In general, after exposure to estrogen, Vtg protein in the blood can be detected for a longer period of time than Vtg mRNA in the liver, suggesting that each measurement reflects a distinct exposure history, with the elevation of Vtg mRNA levels in the liver being a temporary phenomenon reflecting more recent exposure than protein detection in the blood. Ideally, the detection of both products will enable chronological assessments in terms of EEDC exposure history [66].

12. EVALUATION ON EFFECTS OF ESTROGENIC ENDOCRINE DISRUPTING CHEMICALS

The detection of EEDC activity in the field using fish Vtg as an index was first reported from the UK in 1995 [31]. In that study, male rainbow trout were placed in a cage downstream from a drainage channel. Increased Vtg levels in their blood were documented and the values were compared with those of trout from a control site. Based on the results, the authors suggested estrogenic activity in wastewater drainage and identified nonylphenol as the causative agent. Since then, similar studies have been conducted in Salmonidae and Cyprinidae. In vitro exposures have also been performed in Salmonidae and small freshwater fish, and various insights have accumulated regarding the association between Vtg induction and abnormal reproductive functions. Because small freshwater fish such as medaka and zebrafish are easy to breed and have a relatively short lifespan and generation interval, many laboratory exposure studies have used such fish species [22].

Rainbow trout and masu salmon are small Salmonidae that can be cultured in ponds where they can complete their entire life cycle. We established CLIA for two types of Vtgs in pond-grown masu salmon; namely, a complete form of Vtg (VtgAs) and an incomplete form of Vtg (VtgC), and observed the changes in blood levels during sexual maturation. Concentrations of complete and incomplete forms of Vtgs in the female blood during vitellogenesis were around 100 and 50 ng/ml, respectively, demonstrating approximately twice the amount of the complete form compared to the incomplete form. In contrast, at the peak of vitellogenesis, blood levels of complete and incomplete forms of Vtg were approximately 20 and ≤0.5 mg/ml, respectively, yielding a relatively large ratio of 40:1. Because both types of Vtg are controlled by estradiol-17β, the fluctuation of this ratio is thought to be due to hormone sensitivity differences of each gene subtype [Nishikawa and Hara, unpubl. data]. At present, the mechanism that controls the Vtg synthesis ratios is unknown; however, a recent study suggested that the synthesis ratio of two types of Chgs, induced in a similar manner as Vtgs, was also dependent on the amount of estrogen administered [115]. These studies suggest that to conduct an accurate assessment of contamination, we need to consider the possibility that an incomplete form of Vtg (a minor component of female blood during vitellogenesis) is synthesized as a relatively major component in the presence of low concentrations of estrogen. In fact, although almost no complete forms of Vtgs were detected in the blood of pond-grown male masu salmon, trace amounts of incomplete Vtg (VtgC) [Fujita and Hara, unpubl. data] and Chg have been detected throughout the year [119].

Lately, the number of reports on assessment of the effects of EEDCs in marine fish has been growing. Estrogen activity has been mainly detected in Heterosomata and Gobiidae particularly in the coastal areas of Japan, Europe and North America [66]. To date, our research group has been conducting a study of EEDC activity along the coasts of Japan, Korea and China, targeting Mugilidae. In this study, we targeted two types of Mugilidae to expand the area of investigation: grey mullet that is widely distributed in the south, and red lip mullet in the north. For the grey mullet, various analyses have been conducted, including a qualitative analysis of multiple Vtgs and their product egg yolk proteins, gene cloning and characterization of multiple Vtg subtypes, and confirmation of estrogen response. Additionally, a Vtg measurement system specific for the various Vtg subtypes has been developed, establishing an assessment system that fully considers the 'multiple Vtg model' mentioned above [118]. Furthermore, in red lip mullet, we established identification and measurement systems for major Vtg subtypes and confirmed estrogen responsiveness. Moreover, we developed systems for qualitative analysis, identification, and measurement of Chg and attempted a preliminary investigation using a model river [105]. Until now, the investigation targeting both species of Mugilidae detected extremely high estrogen activity in urban coastal waters, and histopathological observations revealed some fish to have abnormal gonads, including ovotestes. However, the connection between blood Vtg levels and these reproductive abnormalities has not yet been confirmed [120].

Adding to in vivo experiments where small fish were exposed to estrogen and EEDCs, studies of free-ranging fishes in freshwater and marine waters contaminated with estrogenic compounds have generated a relatively large amount of useful information. However, our insight into the activity of EEDCs in the open ocean is still limited. Pink salmon *Oncorhynchus gorbuscha* and chum salmon, members of Salmonidae that undertake a feeding migration over long distances in the open ocean, are appropriate target species for investigation in northern oceanic areas. We collected pink salmon and chum salmon off the coast in the North Pacific and measured Vtg and Chg levels in the male blood. Although some Vtg and Chg were detected in male fish collected from the open sea, levels were within the normal range (baseline) for Salmonidae, suggesting that the open seawater is relatively clean and free from significant contamination with estrogenic compounds. In addition, when we measured blood Vtg levels of male salmon returning to major rivers in Hokkaido, the levels were even lower than those observed in males collected in the open ocean. Although the reason for this phenomenon is unclear, no EEDC activity has been detected to date in major rivers in Hokkaido, indicating that there is no effect on homing migration and early development of Salmonidae in the area [Takahashi and Hara, unpubl. data].

13. INTERPRETATION OF VITELLOGENIN INDUCTION

When using Vtg detected in male or immature fish blood as a biomarker of EEDC activity, care must be taken to ensure appropriate interpretation. For instance, in in vivo exposure experiments or caging experiments conducted in the field, the level of Vtg in each individual prior to the start of the experiment can be used as a baseline control, enabling unambiguous detection of the presence or absence of EEDC activity within the targeted area. However, no clear baseline Vtg level can be set for experiments with wild fish. Therefore, the presence of Vtg in the blood of male or immature fish does not necessarily indicate exposure to EEDCs, unless the presence of Vtg is correlated with other observations indicative of such exposure, such as gonadal or other physical reproductive abnormalities. Since the establishment of Vtg measurement systems with high sensitivity, low levels of Vtg have actually been detected in reproductively normal male individuals of several fish species [22]. Possible causes of the appearance of male fish with low levels of Vtg are: (1) low-level endogenous estrogen, (2) feed-derived estrogen, and (3) estrogen derived from excretion or secretions from female fish cohabiting in proximal areas during the spawning season. Furthermore, blood Vtg levels in normal male fish may vary depending on the species, season and maturity of the fish. Therefore, in every EEDC activity assessment in the aquatic environment, the normal range of the Vtg level (baseline) of the target male fish needs to be known beforehand. We have been investigating these baselines in several index fish species [22, 66]. For Salmonidae, using pond-grown masu salmon or Sakhalin taimen bred for many generations in captivity and confirmed to have normal reproductive functions, by measuring the Vtg levels through a whole year, we determined the species baseline Vtg levels.

Vtg levels in the blood of males that exceed the baseline suggest that the targeted individual is or has been exposed to EEDCs in the relatively recent past. However, even when temporary induction of Vtg exceeding the baseline level is observed in a male individual, its effects on reproduction (e.g. abnormalities in gonad morphology, reproductive behaviour, fertilization, development and sexual differentiation) largely depend on individual differences, including age. In general, the EEDC exposure level that is required for the induction of Vtg is lower than that associated with reproductive abnormalities. In contrast, Vtg induction occurs at or above the EEDC exposure level that triggers reproductive abnormalities in immature fish [22]. Therefore, when an abnormal Vtg level is detected in a wild mature male fish, it is highly likely that reproductive abnormality has already been progressing among other immature fish from the same area, even if no reproductive abnormality has been detected in the tested mature male individuals. As shown here, the results of field research not only demonstrate the current or relatively recent history of exposure of targeted individuals to EEDCs, but also predict the development of reproductive disorders in immature fish, including juveniles or fry, and can be considered as an early warning.

The discovery of multiple (polymorphous) Vtgs in fish may lead to new interpretations of the assessment results of EEDC contamination based on Vtg measurements. Our recent research on the physiological functions of each type of Vtg during oocyte maturation and embryogenesis following fertilization suggested that egg quality is directly dependent on the relative abundance (accumulation ratio) of each Vtg within the ovulated eggs [117, 118]. Assuming that this hypothesis is correct, when EEDCs disrupt the ratios of different forms of Vtgs found in the blood or accumulated in the egg, this suggests a potential direct impact on fertilization or survival rates. Until recently, only male or immature fish have been targeted in Vtg-based EEDC contamination studies. In the future, the ratios of multiple Vtgs in the blood of mature female fish and those accumulated in eggs are expected to function as novel biomarkers of endocrine disruption caused by EEDCs.

14. FUTURE CONCERNS

Having summarized information on Vtg and Chg as major proteins involved in oogenesis, we will now discuss a few ongoing studies. Until recently, fish Vtg protein has mainly been studied as an index for gender identification in farmed fish or for assessing estrogenic activity in fish in aquatic environments. Future studies should focus on analysis of Vtg polymorphisms from a physiological standpoint, following the entire process of vitellogenesis within a species, beginning with Vtg synthesis in the liver and ending with the migration of secreted proteins from the blood into the egg yolk. To generalise the concept of a 'multiple Vtg model', it is also essential to collect data regarding polymorphous Vtg from a wide variety of fish species. We have been conducting a detailed analysis of two types of Vtg in Salmonidae for the first time, analyzing the primary structure of yolk protein domains and their characteristic motifs and revealing the expression profile of each form of Vtg during ovarian development [95].

Even less information is known about Vtg receptors. To date, only a limited number of studies have examined the relationships between receptor proteins and their Vtg ligands. Recently, based on an experiment using a yeast 2-hybrid system (Y2H), the affinity between VtgAb and the Vtg receptor (an orthologue of mammalian very-low-density lipoprotein (VLDL) receptor) was confirmed in tilapia, and the binding regions were revealed to be located in the LvH region of VtgAb and on a ligand-binding repeat of the Vtg receptor [121]. This finding is in agreement with our experimental results, which biochemically demonstrated the presence of a binding region on Lv that binds to the Lr8-type Vtg receptor [65]. Moreover, Mizuta et al. [122] prepared a specific antibody to a recombinant homologue of the Lr8-type Vtg receptor in cutthroat trout *Oncorhynchus clarki* and immunobiochemically proved that an approximately 100 kDa receptor protein was the translation product of the gene encoding the Lr8-type Vtg receptor. Furthermore, immune-histochemical analysis using the same antibody demonstrated that the Lr8-type Vtg receptor protein is mainly located at the periphery of oocyte membranes within the ovary, supporting the hypothesis that the Lr8-type receptor binds to Vtg at this same site to be taken up into oocytes. Additional insights on the fish Lr8-type Vtg receptor have been accumulating; however, when considering the 'multiple Vtg model', almost no information is available regarding the binding of different Vtg subtypes with the Lr8-type Vtg receptor. In an experiment with white perch, Reading et al. [123, 124] confirmed the possibility that VtgAa and VtgAb may bind to different receptors and indicated that in addition to the Lr8-type Vtg receptor, a novel lipoprotein receptor (low-density lipoprotein receptor-related protein 13, Lrp13) may be involved in Vtg binding. Many unidentified lipoprotein receptors exist in fish, and the affinity of these receptors for Vtg will likely be revealed in the future.

Many fish eggs contain not only Vtg-derived lipids (mainly phospholipids), but also neutral lipids (mainly triglyceride or wax esters) called 'oil globules'. Because some fish species accumulate oil globules in a similar quantity as yolk granules, they may play an important role in the development of embryos and juvenile fish. However, their mechanism of formation, including their origins, remains unknown for the most part. We hypothesised that oil globules originate from a group of serum proteins rich in neutral lipids (e.g. VLDL), and we have attempted to construct a novel model for oil globule accumulation. At present, there are two proposed models: (1) a pathway where an oil globule is formed with free fatty acids released from VLDL by lipase in the blood and subsequently taken into the ovum and (2) a pathway where VLDL enters directly via a receptor and the release of fatty acids and synthesis of oil globules occur within the ovum. Experiments are currently in progress. To date, our focus has been on lipase and fatty acid transport molecules involved in the first pathway described above, and we have been investigating the location of expression of each molecule within follicles and the expression dynamics during oogenesis. For instance, when we analysed the dynamics and location of the expression of lipase in the ovary of cutthroat trout, we found that the protein was strongly expressed in the granulosa cell layer, and its expression increased in the oil droplet stage in which oil globule formation starts, followed by a decrease with the advance of vitellogenesis [125].

Fish oogenesis, particularly the formation of the yolk globules, egg envelope and oil globules, involves very precise regulatory mechanisms and is of direct relevance to final egg quality and seed integrity. This process, by which a primary oocyte grows by several orders of magnitude while accumulating or synthesizing everything necessary to eventually be fertilized and support complete development of a new life, is amongst the most intriguing of biological phenomena. Advances in our comprehensive understanding of this event will undoubtedly contribute to the development of applied technologies to improve the egg quality and seed integrity in farmed fishes and to protect reproductive processes in wild fish.

ACKNOWLEDGMENTS

This study started when one of the authors, Akihiko Hara, was under the supervision of late professor emeritus Hidematsu Hirai, along with several other professors in the Department of Biochemistry, Hokkaido University School of Medicine. We received generous cooperation from Professor Craig V. Sullivan at North Carolina State University, who has been a long-term collaborator. We would like to express our gratitude to Drs. Takahiro Matsubara, Kiyoshi Soyano, Takashi Todo, Munetaka Shimizu, Haruhisa

Fukada, Makiko Kitamura, Nobuyuki Okubo, Haruna Amano, Hong Lei, Yuka Morita, Yong-Woon Ryu, Hiroko Mizuta, Wu Meiqin, Osamu Nishimiya, Yuji Mushirobira and other graduate students, who made significant contributions to studies described in this article.

REFERENCES

1. Uhlenhuth P, Kodama T (1914) Studien über die Geschlechtsdifferenzierung. Kokka Igakkai Zashi 324-335:385–405 **(in Japanese)**
2. Sakuma K (1923) On the precipitin reaction of the reproductive cells and the sexual differences. Nippon Fujinka-Gakkai Zasshi 18:1101–1126 **(in Japanese)**
3. Deutsch HF, Goodloe MB (1945) An electrophoretic survey of various animal plasmas. J Biol Chem 161:1–20
4. Deutsch HF, McShan WH (1949) Biophysical studies of blood plasma proteins XII. Electrophoretic studies of the blood serum proteins of some lower animals. J Biol Chem 180:219–234
5. Moore DH (1945) Species differences in serum protein patterns. J Biol Chem 161:21–32
6. Saito K (1957) Biochemical studies on the fish blood-VIII. On the electrophoretic specificity of serum proteins. Nippon Suisan Gakkaishi 22:752–759 **(in Japanese with English abstract)**
7. Saito K (1957) Biochemical studies on the fish blood-IX. On the paper electrophoretic analysis of serum proteins. Nippon Suisan Gakkaishi 22:760–767 **(in Japanese with English abstract)**
8. Hara A (1987) Studies on female-specific serum proteins (vitellogenin) and egg yolk proteins in teleosts: immunochemical, physicochemical and structural studies. Mem Fac Fish Hokkaido Univ 34:1–59
9. Vanstone WE, Ho CW (1961) Plasma proteins of coho salmon, *Oncorhynchus kisutch*, as separated by zone electrophoresis. J Fish Res Bd Can 18:393–399
10. Krauel KK, Ridgway GJ (1963) Immunoelectrophoretic studies of red salmon (*Oncorhynchus nerka*) serum. Int Archs Allergy appl Immun 23:246–253
11. Utter FM, Ridgway GJ (1967) A serologically detected serum factor associated with maturity in English sole, *Parophrys vetulus*, and Pacific halibut, *Hippoglossus stenolepis*. Fish Bull Fish Wildl Serv US 66:47–58
12. Amirante GA (1972) Immunochemical studies on rainbow trout (*Salmo gairdnerii* Rich.) lipovitellin. Acta Embryol Exp Suppl:373–383
13. Aida K, Ngan PV, Hibiya T (1973) Physiological studies on gonadal maturation of fishes. I. Sexual difference in composition of plasma protein of ayu in relation to gonadal maturation. Nippon Suisan Gakkaishi 39:1091–1106
14. Emmersen BK, Petersen IM (1976) Natural occurrence, and experimental induction by estradiol-17-β, of a lipophosphoprotein (vitellogenin) in flounder (*Platichthys flesus*, L.). Comp Biochem Physiol 54B:443–446
15. Hara A (1976) Iron-binding activity of female-specific serum proteins of rainbow trout (*Salmo gairdneri*) and chum salmon (*Oncorhynchus keta*). Biochim Biophys Acta 427:549–557
16. Pan MJ, Bell WJ, Telfer WH (1969) Vitellogenic blood protein synthesis by insect fat body. Science 165:393–394
17. Wallace RA (1985) Vitellogenesis and oocyte growth in non-mammalian vertebrates. In: Browder LW (ed) Development Biology, vol 1. Plenum Press, New York, pp 127–177
18. Mommsen TP, Walsh PJ (1988) Vitellogenesis and oocyte assembly. In: Hoar WS, Randall DJ, Farrell AP (eds) Fish physiology, vol 11A. Academic Press, San Diego, New York, pp 347–406

19. Selman K, Wallace RA (1989) Cellular aspects of oocyte growth in teleost. Zool Sci 6:211–231
20. Specker JL, Sullivan CV (1994) Vitellogenesis in fishes: status and perspectives. In: Davey KG, Petter RE, Tobe SS (eds) Perspectives in comparative endocrinology. National Research Council, Ottawa, pp 304–315
21. Hiramatsu N, Matsubara T, Weber GM, Sullivan CV, Hara A (2002) Vitellogenesis in aquatic animals. Fish Sci 68(Suppl I):694–699
22. Hiramatsu N, Cheek AO, Sullivan CV, Matsubara T, Hara A (2005) Vitellogenesis and endrocrine disruption. In: Mommsen TP, Moon TW (eds) Biochemistry and molecular biology of fishes, vol 6. Elsevier Science, Amsterdam, pp 431–471
23. Patiño R, Sullivan CV (2002) Ovarian follicle growth, maturation, and ovulation in teleost fish. Fish Physiol Biochem 26:57–70
24. Reading BJ, Sullivan CV (2011) Vitellogenesis in fishes. In: Ferrell AP (ed) Encyclopedia of fish physiology: from genome to environment. The reproductive organs and processes. Elsevier, Maryland Heights, Missouri, pp 635–646
25. Plack PA, Pritchard DJ, Fraser NW (1971) Egg proteins in cod serum. Biochem J 121:847–856
26. Hara A, Takano K, Hirai H (1983) Immunochemical identification of female-specific serum protein, vitellogenin, in the medaka, *Oryzias latipes* (teleosts). Comp Biochem Physiol 74A:135–141
27. Hara A, Takemura A, Matsubara T, Takano K (1986) Immunochemical identification of female-specific serum proteins in a viviparous fish, the white-edged rockfish (*Sebastes taczanowskii*), during vitellogenesis and pregnancy, and after estrogen treatment. Bull Fac Fisheries Hokkaido Univ 37:101–110 **(in Japanese with English abstract)**
28. Matsubara T, Wada T, Hara A (1994) Purification and establishment of ELISA for vitellogenin of Japanese sardine (*Sardinops melanostictus*). Comp Biochem Physiol 109B:545–555
29. Hamazaki T, Iuchi I, Yamagami Y (1984) Chorion glycoprotein-like immunoreactivity in some tissues of adult female medaka. Zool Sci 1:148–150
30. Yamagami K (1996) Studies on the hatching enzyme (choriolysin) and its substrate, egg envelope, constructed of the precursors (choriogenins) in *Oryzias latipes*: a sequel to the information in 1991/1992. Zool Sci 13:331–340
31. Sumpter JP, Jobling S (1995) Vitellogenesis as a biomarker for estrogenic contamination of the aquatic environment. Environ Health Perspect 103:173–178
32. Sumpter JP (1997) Environmental control of fish reproduction: a different perspective. Fish Physiol Biochem 17:25–31
33. Arcand-Hoy LD, Benson WH (1998) Fish reproduction: an ecologically relevant indicator of endocrine disruption. Environ Toxicol Chem 17:49–57
34. Giesy JP, Snyder EM (1998) Xenobiotic modulation of endocrine function in fish. In: Kendall RJ, Dickerson RL, Giesy JP, Suk WA (eds) Principles and processes for evaluating endocrine disruption in wildlife. SETAC Press, Florida, pp 155–237
35. Kime DE (1998) Endocrine disruption in fish. Kluwer Academic Publishers, Boston **416 pp**
36. Pait AS, Nelson JO (2002) Endocrine disruption in fish: an assessment of recent research and results, NOAA Technical Memo. NOS NCCOS CCMA 149, Silver Spring, MD, NOAA, NOS, Center for Coastal Monitoring and Assessment
37. Arukwe A, Goksøyr A (2003) Eggshell and egg yolk proteins in fish: hepatic proteins for the next generation: oogenetic, population, and evolutionary implications of endocrine disruption. Comp Hepatol 2:1–21
38. Bergink EW, Wallace RW (1974) Precursor-product relationship between amphibian vitellogenin and yolk proteins, lipovitellin and phosvitin. J Biol Chem 249:2897–2903
39. Deeley RG, Mullinix KP, Wetekam W, Kronenberg HM, Meyers M, Eldridge JD, Goldberger RF (1975) Vitellogenin synthesis in the avian liver: vitellogenin is the precursor of the egg yolk phosphoproteins. J Biol Chem 250:9060–9066

40. Hara A, Hirai H (1978) Comparative studies on immunochemical properties of female-specific serum protein and egg yolk proteins in rainbow trout (*Salmo gairdneri*). Comp Biochem Physiol 59B:339–343
41. Hiramatsu N, Hara A (1996) Relationship between vitellogenin and its related egg yolk proteins in Sakhalin taimen (*Hucho perryi*). Comp Biochem Physiol 115A:243–251
42. Hiramatsu N, Hara A (1997) Specific proteolysis of vitellogenin to egg yolk proteins in white spotted-charr *Salvelinus leucomaenis*. Nippon Suisan Gakkaishi 63:701–708 **(in Jananese with English abstract)**
43. Hiramatsu N, Ichikawa N, Fukada H, Fujita T, Sullivan CV, Hara A (2002) Identification and characterization of proteases involved in specific proteolysis of vitellogenin and yolk proteins in salmonids. J Exp Zool 292:11–25
44. Hiramatsu N, Fukada H, Sullivan CV, Hara A (2001) Simple and sensitive detection of vitellogenin receptor(s) in Sakhalin taimen (*Hucho perryi*). Bull Fish Hokkaido Univ 52:5–9
45. Hara A (1978) Sexual differences in serum proteins of chum salmon and the purification of female-specific serum protein. Nippon Suisan Gakkaishi 44:689–693 **(in Japanese with English abstract)**
46. Campbell CM, Idler DR (1980) Characterization of an estradiol-induced protein from rainbow trout as vitellogenin by the cross reactivity to ovarian yolk fractions. Biol Reprod 22:605–617
47. Idler DR, Hwang SJ, Crim LW (1979) Quantification of vitellogenin in Atlantic salmon (*Salmo salar*) plasma by radioimmunoassay. J Fish Res Bd Can 36:574–578
48. Hori SH, Kodama T, Tanahashi K (1980) Induction of vitellogenin synthesis in goldfish by massive doses of androgens. Gen Comp Endocrinol 37:306–320
49. de Vlaming VL, Wiley HS, Delahuntry G, Wallace RA (1980) Goldfish (*Carassius auratus*) vitellogenin: Induction, isolation, properties and relationship to yolk proteins. Comp Biochem Physiol 67B:613–623
50. Hara A, Yamauchi K, Hirai H (1980) Studies on female-specific serum protein (vitellogenin) and egg yolk protein in Japanese eel (*Anguilla japonica*). Comp Biochem Physiol 65B:315–320
51. Nath P, Sundararaj BI (1981) Isolation and identification of female-specific serum lipophosphoprotein (vitellogenin) in the catfish, *Heteropneustes fossilis*. Gen Comp Endocrinol 43:184–190
52. Wallace RA, Selman K (1982) A new procedure for the isolation of intact vitellogenin from teleosts. In: Richter CJJ, Goos HJTh (eds) Proceedings of the international symposium on reproductive physiology of fish, Wageningen, The Netherlands, p 161
53. Gordon MR, Owen TG, Ternan TA, Hildebrand LD (1984) Measurement of a sex-specific protein in skin mucus of premature coho salmon (*Oncorhynchus kisutch*). Aquaculture 43:333–339
54. Hara A, Sullivan CV, Dickhoff WW (1993) Isolation and some characterization of vitellogenin and its related egg yolk proteins from coho salmon (*Oncorhynchus kisutch*). Zool Sci 10:245–256
55. Jared DW, Wallace RA (1968) Comparative chromatography of the yolk proteins of teleosts. Comp Biochem Physiol 24:437–443
56. Markert JP, Vanstone WE (1971) Egg proteins of coho salmon (*Oncorhynchus kisutch*). J Fish Res Bd Can 28:1853–1856
57. Matsubara T, Koya Y (1997) Course of proteolytic cleavage in three classes of yolk proteins during oocyte maturation in barfin flounder *Verasper moseri*, a marine teleost spawning pelagic eggs. J Exp Zool 272:34–45
58. Hiramatsu N, Matsubara T, Hara A, Donato DM, Hiramatsu K, Denslow ND, Sullivan CV (2002) Identification, purification and classification of multiple forms of vitellogenin from white perch (*Morone americana*). Fish Physiol Biochem 26:355–370
59. Amano H, Fujita T, Hiramatsu N, Shimizu M, Sawaguchi S, Matsubara T, Kagawa H, Nagae M, Sullivan CV, Hara A (2007) Egg yolk proteins in gray mullet (*Mugil cephalus*): purification and classification of multiple lipovitellins and other vitellogenin-derived

yolk proteins and molecular cloning of the parent vitellogenin genes. J Exp Zool 307A:324–341

60. Yamane K, Yagai T, Nishimiya O, Sugawara R, Amano H, Fujita T, Hiramatsu N, Todo T, Matsubara T, Hara A (2013) Characterization of vitellogenin and its derived yolk proteins in cloudy catshark (*Scyliorhinus torazame*). Fish Physiol Biochem 39:373–390
61. Shimizu Y, Nakamura A, Kishimura H, Hara A, Watanabe K, Saeki H (2009) Major allergen and its IgE cross-reactivity among salmonid fish roe allergy. J Agric Food Chem 57:2314–2319
62. Yamamura JI, Adachi T, Aoki N, Nakajima H, Nakamura R, Matsuda T (1995) Precursor-product relationship between chicken vitellogenin and the yolk proteins: the 40 kDa yolk plasma glycoprotein is derived from the C-terminal cysteine-rich domain of vitellogenin II. Biochim Biophys Acta 1244:384–394
63. Matsubara T, Nagae M, Ohkubo N, Andoh T, Sawaguchi S, Hiramatsu N, Sullivan CV, Hara A (2003) Multiple vitellogenins and their unique roles in marine teleosts. Fish Physiol Biochem 28:295–299
64. Matsubara T, Ohkubo N, Andoh T, Sullivan CV, Hara A (1999) Two forms of vitellogenin, yielding two distinct lipovitellins, play different roles during oocyte maturation and early development of barfin flounder, *Verasper moseri*, a marine teleost spawning pelagic eggs. Dev Biol 213:18–32
65. Hiramatsu N, Hara A, Hiramatsu K, Fukada H, Weber GM, Denslow ND, Sullivan CV (2002) Vitellogenin-derived yolk proteins of white perch, *Morone americana*: purification, characterization and vitellogenin-receptor binding. Biol Reprod 67:655–667
66. Hiramatsu N, Matsubara T, Fujita T, Sullivan CV, Hara A (2006) Multiple piscine vitellogenins: biomarkers of fish exposure to estrogenic endocrine disruptors in aquatic environments. Mar Biol 149:35–47
67. Kanetoshi A, Katsura E, Fujimoto T, Kojima H, Hori Y, Fukada H, Takahara S, Hara A (2004) Study on the screening test of endocrine disrupting chemicals using carp (*Cyprinus carpio*) hepatocyte culture: vitellogenin induction test by estrogens. Rep Hokkaido Inst Pub Health 54:1–6 **(in Japanese with English abstract)**
68. Hiramatsu N, Shimizu M, Fukada H, Kitamura M, Ura K, Fuda H, Hara A (1997) Transition of serum vitellogenin cycle in Sakhalin taimen (*Hucho perryi*). Comp Biochem Physiol 118C:149–157
69. Wallace RA, Begovac PC (1985) Phosvitins in *Fundulus* oocytes and eggs: preliminary chromatographic and electrophoretic analyses together with biological considerations. J Biol Chem 260:11268–11274
70. Wallace RA, Selman K (1985) Major protein changes during vitellogenesis and maturation of *Fundulus* oocytes. Dev Biol 110:492–498
71. Greeley MS Jr, Calder DR, Wallace RA (1986) Changes in teleost yolk proteins during oocyte maturation: correlation of yolk proteolysis with oocyte hydration. Comp Biochem Physiol 84B:1–9
72. Matsubara T, Adachi S, Ijiri S, Yamauchi K (1995) Changes of lipovitellin during *in vitro* oocyte maturation in Japanese flounder, *Paralichthys olivaceus*. Fish Sci 61:478–481
73. Matsubara T, Sawano K (1995) Proteolytic cleavage of vitellogenin and yolk proteins during vitellogenin uptake and oocyte maturation in barfin flounder (*Verasper moseri*). J Exp Zool 272:34–45
74. Wahli W, Dawid IB, Wyler T, Jaggi RB, Weber R, Ryffel GU (1979) Vitellogenin in *Xenopus laevis* is encoded in a small family of genes. Cell 16:535–549
75. Felber BK, Maurhofer S, Jaggi RB, Wyler T, Wahli W, Ryffel GU, Weber R (1980) Isolation and translation *in vitro* of four related vitellogenin mRNAs of estrogen-stimulated *Xenopus laevis*. Eur J Biochem 105:17–24
76. van-het Schip FD, Samallo J, Broos J, Ophuis J, Mojet M, Gruber M, Ab G (1987) Nucleotide sequences of a chicken vitellogenin gene and derived amino acid sequence of the enclosed yolk precursor protein. J Mol Biol 196:245–260

77. Silva R, Fischer AH, Burch JB (1989) The major and minor chicken vitellogenin genes are each adjacent to partially deleted pseudogene copies of the other. Mol Cell Biol 9:3557–3562
78. LaFleur GJ Jr, Byrne BM, Kanungo J, Nelson LD, Greenberg RM, Wallace RA (1995) *Fundulus heteroclitus* vitellogenin: the deduced primary structure of a piscine precursor to noncrystalline, liquid-phase yolk protein. J Mol Evol 41:505–521
79. LaFleur GJ Jr, Byrne BM, Haux C, Greenberg RM, Wallace RA (1995) Liver-derived cDNAs: vitellogenins and vitelline envelope protein precursors (choriogenins). In: Goetz FW, Thomas P (eds) Proceedings of the fifth international symposium on the reproductive physiology of fish, University of Texas, Austin, pp 336–338
80. Reith M, Munholland J, Kelly J, Finn RN, Fyhn HJ (2001) Lipovitellins derived from two forms of vitellogenin are defferentially processed during oocyte maturation in haddock (*Melanogrammus aeglefinus*). J Exp Zool 291:58–67
81. Lee BH, Lim EH, Lam TJ, Ding JL (1994) Two major groups of vitellogenin cDNA clones from *Oreochromis aureus* (Steindachner). Biochem Mol Biol Int 34:75–83
82. Bowman CJ, Kroll KJ, Hemmer MJ, Folmer LC, Denslow ND (2000) Estrogen-induced vitellogenin mRNA and protein in sheepshead minnow (*Cyprinodon variegatus*). Gen Comp Endocrinol 120:300–313
83. Wang H, Yan T, Tan JTT, Gong ZA (2000) Zebrafish vitellogenin gene (vg3) encodes a novel vitellogenin without a phosvitin domain and may represent a primitive vertebrate vitellogenin gene. Gene 256:303–310
84. Trichet V, Buisine N, Mouchel N, Moran P, Pendas AM, Le Pennec JP, Wolff J (2000) Genomic analysis of the vitellogenin locus in rainbow trout (*Oncorhynchus mykiss*) reveals a complex history of gene amplification and retroposon activity. Mol Gen Genet 263:828–837
85. Ding JL, Hee PL, Lam TJ (1989) Two forms of vitellogenin in the plasma and gonads of male *Oreochromis aureus*. Comp Biochem Physiol 93B:363–370
86. Kishida M, Specker JL (1993) Vitellogenin in tilapia (*Oreochromis mossambicus*): induction of two forms by estradiol, quantification in plasma and characterization in oocyte extract. Fish Physiol Biochem 12:171–182
87. Shimizu M, Fujiwara Y, Fukada H, Hara A (2002) Purification and identification of a second form of vitellogenin from ascites of medaka (*Oryzias latipes*) treated with estrogen. J Exp Zool 293:726–735
88. Ohkubo N, Mochida K, Adachi S, Hara A, Hotta K, Nakamura Y, Matsubara T (2003) Development of enzyme-linked immunosorbent assays for two forms of vitellogenin in Japanese common goby (*Acanthogobius flavimanus*). Gen Comp Endocrinol 131:353–364
89. Amano H, Mochizuki M, Fujita T, Hiramatsu N, Todo T, Hara A (2010) Purification and characterization of a novel incomplete-type vitellogenin protein (VgC) in Sakhalin taimen (*Hucho perryi*). Comp Biochem Physiol 157A:41–48
90. Sawaguchi S, Koya Y, Yoshizaki N, Ohkubo N, Andoh T, Hiramatsu N, Sullivan CV, Hara A, Matsubara T (2005) Multiple vitellogenins (Vgs) in mosquitofish (*Gambusia affinis*): identification and characterization of three functional Vg genes and their circulating and yolk proteins products. Biol Reprod 72:1045–1060
91. Sawaguchi S, Kagawa H, Ohkubo N, Hiramatsu N, Sullivan CV, Matsubara T (2006) Molecular characterization of three forms of vitellogenin and their yolk protein products during oocyte growth and maturation in red seabream (*Pagrus major*), a marine teleost spawning pelagic eggs. Mol Reprod Develop 73:719–736
92. Amano H, Fujita T, Hiramatsu N, Sawaguchi S, Matsubara T, Sullivan CV, Hara A (2007) Purification of multiple vitellogenins in grey mullet (*Mugil cephalus*). Mar Biol 152:1215–1225
93. Finn RN, Kristoffersen BA (2007) Vertebrate vitellogenin gene duplication in relation to the "3R hypothesis": correlation to the pelagic egg and the oceanic radiation of teleosts. PLoS One 2:e169
94. Williams VN, Reading BJ, Hiramatsu N, Amano H, Glassbrook N, Hara A, Sullivan CV (2014) Multiple vitellogenins and product yolk proteins in striped bass, *Morone*

saxatilis: molecular characterization and processing during oocyte growth and maturation. Fish Physiol Bichem 40:395–415

95. Mushirobira Y, Mizuta H, Luo W, Morita Y, Sawaguchi S, Matsubara T, Hiramatsu N, Todo T, Hara A (2013) Changes in levels of dual vitellogenin transcripts and proteins in cutthroat trout *Oncorhynchus clarki* during ovarian development. Nippon Suisan Gakkaishi 79:175–189 **(in Japanese with English abstract)**
96. Wu M, Nishimiya O, Nakamori M, Soyano K, Todo T, Hara A, Hiramatsu N (2014) Molecular cloning and characterization of the expression profiles of vitellogenin transcripts in the dojo loach (*Misgurnus anguillicaudatus*) in response to 17α-ethinylestradiol and 17β-estradiol administration. Zool Sci 31:202–212
97. Babin PJ (2008) Conservation of a vitellogenin gene cluster in oviparous vertebrates and identification of its traces in the platypus genome. Gene 413:76–82
98. Finn RN, Kolarevic J, Kongshaug H, Nilsen F (2009) Evolution and differential expression of a vertebrate vitellogenin gene cluster. BMC Evol Biol 9:2. doi:10.1186/1471-2148-9-2
99. Fukada H, Haga A, Fujita T, Hiramatsu N, Sullivan CV, Hara A (2001) Development and validation of chemiluminescent immunoassay for vitellogenin in five salmonid species. Comp Biochem Physiol 130A:163–170
100. Hamazaki TS, Iuchi I, Yamagami K (1987) Isolation and partial characterization of a "spawning female-specific substance" in the teleost, *Oryzias latipes*. J Exp Zool 242:343–349
101. Murata K, Iuchi I, Yamagami K (1993) Isolation of H-SF substance, the high-molecular-weight precursors of egg envelope proteins, from the ascites accumulated in the oestrogen-treated fish, *Oryzias latipes*. Zygote 1:315–324
102. Shimizu M, Fujita T, Hara A (1998) Purification of the precursors to vitelline envelope proteins from serum of Sakhalin taimen, *Hucho perryi*. J Exp Zool 282:385–395
103. Fujita T, Shimizu M, Hiramatsu N, Fukada H, Hara A (2002) Purification of serum precursor proteins to vitelline envelope (choriogenins) in masu salmon, *Oncorhynchus masou*. Comp Biochem Physiol 132B:599–610
104. Fujita T, Scott AP, Katsiadaki I, Amano H, Hong L, Hiramatsu N, Todo T, Hara A (2009) Purification of multiple precursors for egg chorion proteins in Atlantic cod (*Gadus morhua*). Zool Sci 26:870–877
105. Hong L, Fujita T, Wada T, Amano H, Hiramatsu N, Zhang X, Todo T, Hara A (2009) Choriogenin and vitellogenin in red lip mullet (*Chelon haematocheilus*): purification, characterization, and evaluation as potential biomarkers for detecting estrogenic activity. Comp Biochem Physiol 149C:9–17
106. Murata K, Sasaki T, Yasumasu S, Iuchi I, Enami J, Yasumasu I, Yamagami K (1995) Cloning of cDNAs for the precursor proteins of low-molecular-weight subunit of the inner layer of the egg envelope (chorion) of the fish *Oryzias latipes*. Dev Biol 167:9–17
107. Murata K, Sugiyama H, Yasumasu S, Iuchi I, Yasumasu I, Yamagami K (1997) Cloning of cDNA and estrogen-induced hepatic gene expression for chriogenin H, a precursor protein of the fish egg envelope (chorion). Proc Natl Acad Sci USA 94:2050–2055
108. Modig C, Modesto T, Canario A, Cerda J, von Hofsten J, Olsson PE (2006) Molecular characterization and expression pattern of zona pellucida proteins in gilthead seabream (*Sparus aurata*). Biol Reprod 75:717–725
109. Chang YS, Wang SC, Tsao CC, Huang FL (1996) Molecular cloning, structural analysis, and expression of carp ZP3 gene. Mol Reprod Dev 44:295–304
110. Chang YS, Hsu CC, Wang SC, Tsao CC, Huang FL (1997) Molecular cloning, structural analysis, and expression of carp ZP2 gene. Mol Reprod Dev 46:258–267
111. Wang H, Gong Z (1999) Characterization of two zebrafish cDNA clones encoding egg envelope proteins ZP2 and ZP3. Biochim Biophys Acta 1446:156–160
112. Kanamori A, Naruse K, Mitani H, Shima A, Hori H (2003) Genomic organization of ZP domain containing egg envelope genes in medaka (*Oryzias latipes*). Gene 305:35–45
113. Fujita T, Fukada H, Shimizu M, Hiramatsu N, Hara A (2008) Molecular cloning and characterization of three distinct choriogenins in masu salmon, *Oncorhynchus masou*. Mol Reprod Dev 75:1217–1228

114. Fujita T, Fukada H, Hara A (2003) Masu salmon (*Oncorhynchus masou*) zona pellucida protein B: cDNA cloning and tissue expression. Fish Physiol Biochem 28:333–334
115. Fujita T, Fukada H, Shimizu M, Hiramatsu N, Hara A (2004) Quantification of serum levels of precursors to vitelline envelope proteins (choriogenins) and vitellogenin in estrogen treated masu salmon, *Oncorhychus masou*. Gen Comp Endocrinol 136:49–57
116. Fujita T, Fukada H, Shimizu M, Hiramatsu N, Hara A (2005) Annual changes in serum levels of two choriogenins and vitellogenin in masu salmon, *Oncorhynchus masou*. Comp Biochem Physiol 141B:211–217
117. Williams VN, Reading BJ, Amano H, Hiramatsu N, Schilling J, Salger SA, Williams TI, Gross K, Sullivan CV (2014) Proportional accumulation of yolk proteins derived from multiple vitellogenins is precisely regulated during vitellogenesis in striped bass (*Morone saxatilis*). J Exp Zool 321A:301–315
118. Amano H, Fujita T, Hiramatsu N, Kagawa H, Matsubara T, Sullivan CV, Hara A (2008) Multiple vitellogenin-derived yolk proteins in gray mullet (*Mugil cephalus*): disparate proteolytic patterns associated with ovarian follicle maturation. Mol Reprod Dev 75:1307–1317
119. Fujita T, Fukada H, Shimizu M, Hiramatsu N, Hara A (2008) Monthly changes in the serum levels of two choriogenins and vitellogenin in male masu salmon *Oncorhynchus masou*. Fish Sci 74:1198–1200
120. Aoki J, Nagae M, Takao Y, Hara A, Lee YD, Yeo IK, Lim BS, Park CB, Soyano K (2010) Survey of contamination of estrogenic chemicals in Japanese and Korean coastal waters using the wild grey mullet (*Mugil cephalus*). Sci Total Environ 408:660–665
121. Li A, Sadasviam M, Ding JL (2003) Receptor-ligand interaction between vitellogenin receptor (VtgR) and vitellogenin (Vtg), implications on low density lipoprotein receptor and apolipoprotein B/E. J Biol Chem 278:2799–2806
122. Mizuta H, Luo W, Ito Y, Mushirobira Y, Todo T, Hara A, Reading BJ, Sullivan CV, Hiramatsu N (2013) Ovarian expression and localization of a vitellogenin receptor with eight ligand binding repeats in the cutthroat trout (*Oncorhynchus clarki*). Comp Biochem Physiol 166B:81–90
123. Reading BJ, Hiramatsu N, Sullivan CV (2011) Disparate binding of three types of vitellogenin to multiple forms of vitellogenin receptor in white perch. Biol Reprod 84:392–399
124. Reading BJ, Hiramatsu N, Schilling J, Molloy KT, Glassbrook N, Mizuta H, Luo W, Baltzegar DA, Williams VN, Todo T, Hara A, Sullivan CV (2014) Lrp13 is a novel vertebrate lipoprotein receptor that binds vitellogenins in teleost fishes. J Lipid Res 55:2287–2295
125. Ryu YW, Tanaka R, Kasahara A, Ito Y, Hiramatsu N, Todo T, Sullivan CV, Hara A (2013) Molecular cloning and transcript expression of genes encoding two types of lipoprotein lipase in the ovary of cutthroat trout, (*Oncorhynchus clarki*). Zool Sci 30:224–237

CHAPTER 6

Role of inland fishery and aquaculture for food and nutrition security in Nepal

Tek Bahadur Gurung*

National Animal Science Research Institute, Nepal Agricultural Research Council, Kathmandu, Nepal

ABSTRACT

Fish as a food is generally acceptable to all regardless of region, religion, race, gender and age across Nepal. The aquaculture and open-water capture fishery contributes about 2 % of agricultural gross domestic production; this share of fisheries sector is tiny but promising having a fastest 8–9 % annual growth rate. Nevertheless, the contribution of fisheries sector is big from nutritional security perspective to poor and marginalized communities and thus cannot be ignored. Generally, the role and importance of the inland water fisheries had been rarely elaborated with perspectives to food and nutritional security in Himalayan landlocked country. Therefore, the aim of this paper was to elucidate the role and potential of inland fisheries for food and nutrition security in Nepal. Information was collected from secondary sources to estimate per capita fish consumption. The per capita consumption was calculated dividing the total fish production by population of specific year, which showed that per capita fish consumption has increased from 125 g in 1975 to 2060 g in year 2013, at least nine times below compared with average global consumption. The value of per capita fish consumption would increase, if native shellfish (gastropod, crab, shrimp, and turtle), frogs and aquatic plants such as foxnut (Euryale ferox) and water chest nut (Trapa sp) which are consumed by ethnic communities are added. In the fish production among the five development regions of the country, Central Development Region was at the top, while Far-Western Development Region was at the bottom. However, on altitudinal basis southern plains contribute highest production, while least is in mountain areas although potentiality of cold-water aquaculture seems high. These imply that support services and innovation should be extended in the Western Development Region and mountainous regions in addition to the plains for improving fisheries-based food and nutrition security.

Keywords: Inland Fishery, Aquaculture, Nepal GDP, Per capita availability

1. BACKGROUND

Nepal's total fish production from captured and aquaculture together is estimated to be about 64,900 metric tons in year 2013/2014 with the share of 33.17 % to capture fisheries [40]. Aquaculture is the farming and husbandry of fish and aquatic organisms [6] where, with some intervention, increase in yield is expected with some form of ownership [8]; contrarily, the fish stocks in capture fisheries remain as a common property until harvested [5]. The term "fish" indicate fish, crustaceans, mollusks, frogs and turtles, excluding crocodiles, alligators, aquatic mammals and aquatic plants [13]. Captured fishery is one of the ancient approaches of food and nutrition supply contrarily to aquaculture which has been started lately. Almost all fish produced are consumed as food in Nepal.

Recent health benefit reports on ω-3 or N-3 fatty acids found in fish food have probably been one of the major triggers of increasing fish demand. Fish is a rich source of animal protein, amino acids, folic acids, ω-3, vitamins and micronutrients useful for health benefit to all age group [27, 37, 46, 50], especially for brain development, memory, mental and cardiac health [45, 48], working against Alzheimer disease and dementia [33, 36]. The ω-3 polyunsaturated fatty acid reduces free fatty acids, plasma triglyceride, cholesterol concentrations and platelet aggregation leading to decreasing blood pressure, blood clotting and atherosclerosis [37, 48].

The Government of Nepal has recommended at least 30 g per day fish or animal protein diet to each man, women, adolescent girls and boys [13]. Compared with the recommendation the fish intake is still far below than the recommended amount, suggesting that fish production needs to be promoted for productivity and production enhancement. The importance of fish over meat, poultry and eggs is due to low level of saturated fatty acids [7]. Thus, the role of fish products is immense in macro- and micronutritional security [44], maintaining the health and nutrition of the global community from brain cells development to reducing chances of cardiac failure and enormous other health benefits [13, 14, 29, 33].

Nepal underlies as one of the food and nutrition-deficient countries. It has been estimated that 41 % of Nepalese children suffer from stunting or chronic malnutrition [17, 31], having potential negative impact on the human population and economic development. Poor nutrition is cited as the major factor in more than half of all child deaths in Nepal—a significantly higher proportion than those claimed by other infectious diseases [17]. In early 1990s, the Nepalese standard food intake had very small amount of fish in their diet, but recently the amount of fish protein in common people diets is increasing [30]. This suggests that the fish production, availability, affordability, purchasing capacity and awareness might have led the increased consumption, implying that the role of fisheries in food and nutritional security is becoming indispensable and increasing. Coinciding with these facts, the Agriculture Development Strategy (2015–2030) has included fish as one of the prioritized commodities [2].

Recent technological innovation in cold-water aquaculture using rainbow trout (Oncorhynchus mykiss) has demonstrated that vast pristine Himalayan cold-water resources could be utilized for fish production [20]. Despite high potentiality for inland fisheries and aquaculture development, fish as a quality source of macro- and micronutritional food has yet to establish among a large section of population in Nepal. The role and importance of fisheries in Himalayan landlocked country, where fish symbolizes for fertility and prosperity, were rarely elaborated for food and nutritional security. Therefore, the aim of the present work is to transact and elucidate the per capita fish consumption for emphasizing the role and potential of fishery and aquaculture for food and nutritional security.

2. METHODS

Data on fish production were obtained from the Directorate of Fisheries Division, Government of Nepal, Fisheries Research Division, and Bureau of Agricultural Statistics Division [40]. The data on population of Nepal and different developmental regions were collected from Ministry of Population. Information on various fish products such as fish oil, fish meal, canned food and fresh fishes was available in various places and supermarkets in Nepal.

The per capita consumption of fish was calculated using the standard method derived from the total national production divided by the number of inhabitants:

Fish consumption per capita=total national production/number of inhabitants,

where the total national production is fin fish yield from aquaculture practices and capture fisheries yield.

Similarly, the annual growth of inland fish production was estimated using the formula:

Growthrate(%)=(Total fish production current year−total fish production previous year)×100/Total fish production current year.

Information on various aquatic products consumable in different agroecological regions of Nepal such as pila (Pila globosa), bivalve (Lammelidens marginelis), crabs, shrimp, frogs (Paa liebigii, Paa blanfordii), turtle, Makhan (Euryale ferox) and water chest nuts (Trapa spp.) was collected from various studies. The average value of major nutrients and micronutrients in fish products was derived from various publications.

3. RESULTS AND DISCUSSION

3.1 Fish production and consumption

Nepal is a country with 27 million people surrounded by India and China with the unique altitude ranging from 68 msl elevation to highest mountains peak of the world within an area of about 885 km long and 193 km width. Nepal has approximately 17 % of flat land located in southern end commonly known as "tarai," while toward the north 83 % is occupied by hills and mountains (Fig. 1). The vertical zonation for fish and fisheries activities and potential is shown in Fig. 2. The tarai is the site for carp, tilapia and catfish (Pangas, Clarias and others) production, while in lower mid-hills, carp along with cool-water fish species such as Mahseer (Tor spp) and Katle (Acrossocheilus hexagonolepis) could be cultivated. Asla or cold water cyprinids (Shizotharax spp) and rainbow trout (Oncorhynchus mykiss) in mid-hills can be produced in areas where glaciers waters with suitable low temperature are available, while upper regions represent the domains of the rainbow trout. In extremely cold waters above 2300–5000 meters range, probably arctic charr (Salvelinus alpinus) could perform well; however, studies require to be carried out on before reaching any conclusion (Fig. 2).

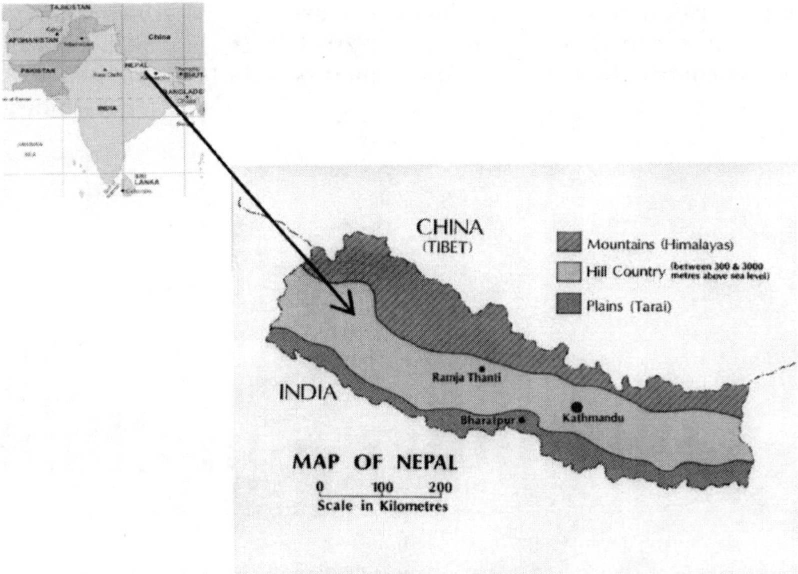

Figure. 1 Location of Nepal in surrounded by India and China

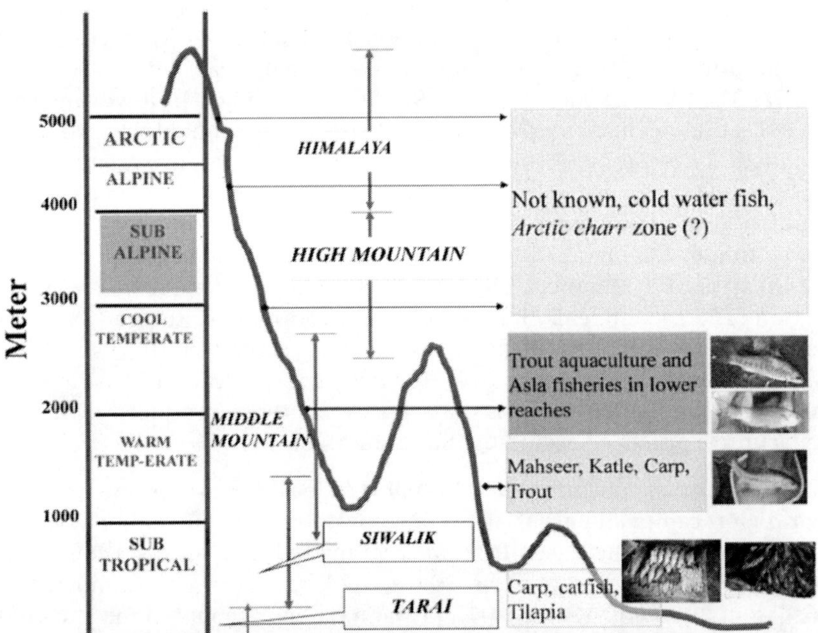

Figure. 2 Fisheries zoning in different altitudes of Nepal

Fish production in Nepal is gradually increasing with a growth rate of 8–9 % per year reaching 64,900 mt in year 2013/2014 (Fig. 3a, b). Out of the total, the contribution of the capture fishery was 33.17 % and rest 66.83 % from the culture practices. The value of fish production includes the quantity of fin fish produced in the country.

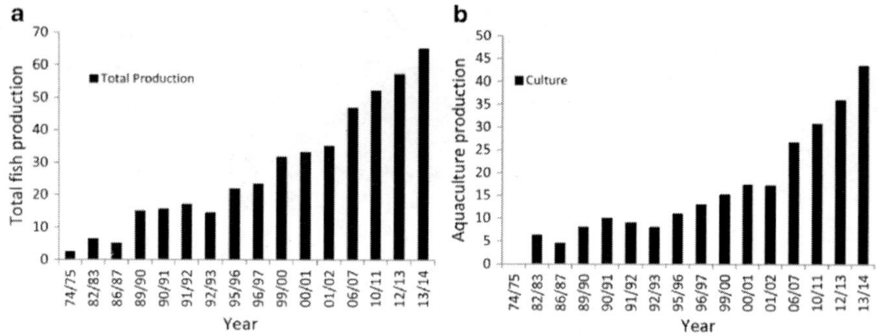

Figure. 3 Total fish production (aquaculture + captured fish) (a) and only aquaculture production (b) in Nepal (unit in x'000 metric ton)

The temporal changes in fish consumption showed steadily increasing despite the higher pace of the population growth (Fig. 4). The per capita fish consumption in 1975 was 194 g per day per year, after 39 years in 2013/2014 that has increased to 1983 g $d^{-1}yr^{-1}$ (Fig. 4). However, the fish consumption declined slightly in year 2013–2014, probably due to low fish production caused by unexpected devastating flood in tarai resulting heavy loss (Fig. 4). The per capita consumption of the fish in various parts of the country (Fig. 5) showed that highest fish consumption was in Central Development Region (1730 g), then in Eastern Development Region (1129 g), Western Development Region (1037 g), Mid-Western Development Region (762 g), and least in Far-Western Development Region (354 g). The per capita consumption of fish in Nepal represents only fin fish captured and cultured within the country. However, the per capita consumption in different development regions only represents those produced by means of aquaculture practices excluding the capture fishery.

The present fish consumption in Nepal is at least nine times lower than the world per capita apparent fish consumption. The per capita world fish consumption has increased from an average of 9.9 kg in the 1960s to 11.5 kg in the 1970s, 12.5 kg in the 1980s, 14.4 kg in the 1990s and reaching 16.4 kg in 2005 and 19.2 in 2012 [15]. China has accounted for most of the world growth with per capita fish supply of about 26.1 kg [15].

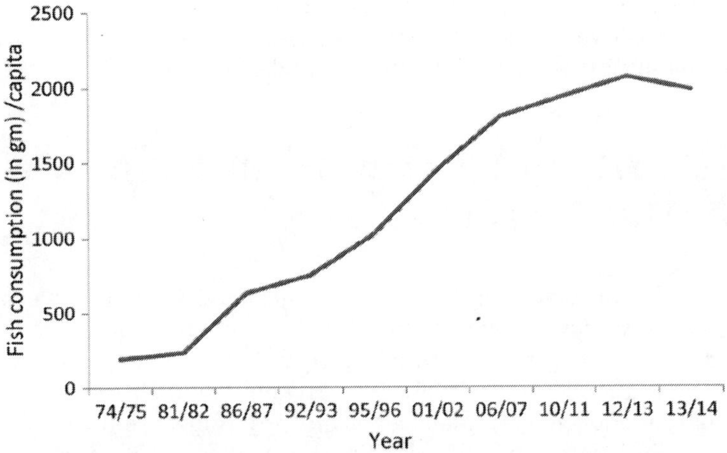

Figure. 4 Temporal changes in per capita fish consumption in Nepal

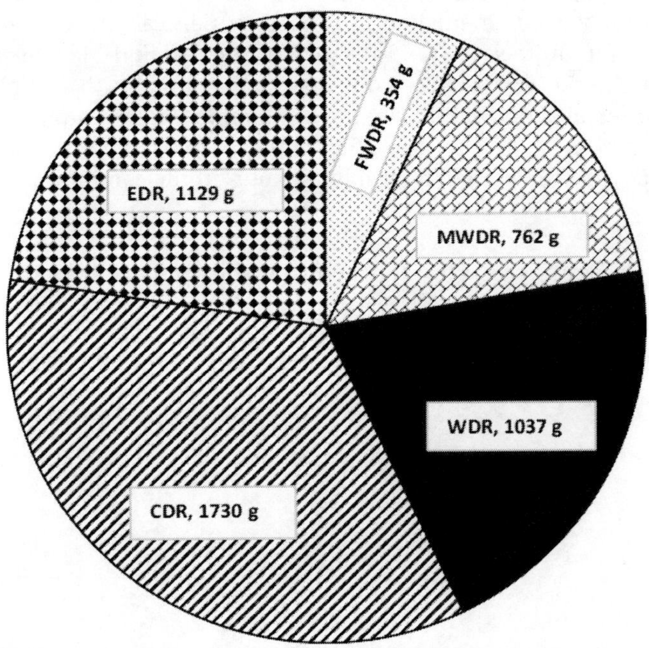

Figure. 5 Per capita fish consumption (in gram, g) in year 2014 in Far-Western Development Region (FWDR), Mid-Western Development Region (MWDR), Western Development Region (WDR), Central Development Region (CDR) and Eastern Development Region (EDR) of Nepal

Interestingly, the global per capita increasing trend of the fish consumption showed a positive trend with economic growth, and the countries with weak economic growth have low protein intake [15, 19, 42].

4. DIVERSITY OF FISH FOODS FOR FOOD AND NUTRITIONAL SECURITY

In this study, various aquatic products consumed in different ecological regions of Nepal have been compiled (Table 1). Fin fish such as indigenous major carp (Labeo rohita, Cirrhinus mrigala, Catla catla), Chinese major carp (Aristichthys nobilis, Hypophthalmichthys molitrix, Ctenopharyngodon idella), Mahseer (Tor spp), Snow trout or cold water cyprinids (Shizothorax spp), Gardi Labeo dero, Tilapia (Tilapia spp), Pangas (Pangasius sp), Common carp (Cyprinus carpio), Rainbow trout (Oncorhynchus mykiss) and other minor fishes are consumed all over the country. Frogs belonging Paa liebigii and Paa blanfordii are also known to be eaten for various medicinal purposes by Central Himalayan hill ethnic Gurung communities [22]. Crabs are eaten in mid-hills as well as in southern tarai. Turtle are consumed occasionally by certain ethnic communities in tarai [42]. Shrimp are consumed in mountains, hill in tarai. The gastropod such as Pila (Pila globosa) and bivalve (Lammelidens marginelis) are consumed in tarai by various ethnic communities [43].

Table 1 General fishery products consumable in agroecological regions of Nepal

Fishery products	High mountains	Mid-hills	Tarai
Fin fish	○	○	○
Frog	○	○	
Crab		○	○
Turtle		○	○
Shrimp	○	○	○
Gastropod			○
Bivalve			○
Nelumbo	○	○	
Trapa (water chest nut)		○	○
Makhan			○

Indigenous water chest nut (Trapa sp), Makhan or foxnut (Euryale ferox), mollusk (P. globosa and L. marginelis) crustaceans, imported frozen fish, canned products and dried fish are available in market [23, 24]. Water chest nut (Trapa sp) grown in wetlands of tarai and mid-hills are fondly consumed (Table 1). The makhan or foxnut (E. ferox) is highly valuable products used in holy occasions to worship the god. Varieties of processed fish products are sold, such as sea fish (sardine), fish pickle, fish oil capsule, fish fillet, dry fish, dry shrimp, fish meal, salted fish, fried fish and smoked fish (Table 2). But, these have not been included in the present estimate of per capita consumption (Tables 1, 2).

Table 2 Some processed fish food products available in Nepalese market

Products	Availability
Sea fish (sardines)	Canned
Fish oil	Capsule
Fish fillet (Vasa)	Plastic wrapped fillet
Dry fish	Plastic packed/open
Dry shrimp and fish meal	Plastic packed/open
Salted fish	Plastic packed/open
Fried fish	Plastic packed/open
Smoked fish	Open
Fish pickle	Plastic packed

However, the aquatic products those have not been envisaged into the per capita fish consumption, especially the indigenous products (Tables 1, 2), might need to be included in future estimates. Because, according to several definitions [6] and recognition by FAO [12] these commodities categorized as aquaculture products are harvested and consumed ensuring food and nutritional security and addresses the poverty alleviation. These commodities contain substantial nutritional values supporting women and children of marginalized and poor communities (Table 3). Therefore, in future research and scaling-up activities these commodities should be prioritized for production and productivity enhancement.

Table 3 Average value of major nutrients and micronutrients in fish products.

Product name	Macronutrients (%)				Elements (mg/100 g)					
	Protein	Moisture	Lipids	Carbohydrate	K	Na	Mg	Fe	Zn	Ca
Fresh water fish	14.8–22.8	68–78.5	0.6–19.4	2.3–6.9	49–501	158–780	22–187	0.31–15.95	0.4–40.2	–
Dry fish (sun dried)	64.9	13.96	4.21	16.9	–	–	–	–	–	–
Dried fish	58.5	19.7	11.24		–	–	–	–	–	–
Salted dry fish	62.2	13.4	6.03	18.4	–	–	–	–	–	–
Smoked dry fish	70.8	10.42	7.23	11.6	–	–	–	–	–	–
Crab small	11.2–13.48	65.3	9.8	9.1	5.20 ± 3.45	7.51 ± 4.10	0.81 ± 1.32	0.014 ± 1.13	0.15 ± 0.19	–
Turtle	16.4	–	13.4	–	150	69	10	0.9	1.6	18
Prawn	19.1	77.4	1	0.8	–	–	–	–	–	–
Snail (*Pila globosa*)	15.83 ± 1.03	71.61 ± 1.9	0.06 ± 0.02	5.54 ± 0.11	3.5 ± 0.23	13.38 ± 0.89	–	25 ± .98	–	1369.42 ± 91
Bivalve (*Lamellidens marginalis*)	13.39 ± 1.09	72.57 ± 2	4.71 ± 0.4	7.68 ± 0.56	0.33 ± .03	2.49 ± 0.19	–	98.75 ± 7.44	–	2130 ± 160
Water chest nut (Green variety)	0.275	62.5	0.84	8.7	5.22	0.64	90	200	600	0.25
Red variety	0.251	62.7	0.83	8.2	5.32	0.59	110	200	650	0.26
Makhan	15.6	12.1	1.35	61	–	–	–	–	–	–

Substantially large quantities of fresh fishes are imported especially from India and other countries to fulfill the local demand in the capital and many others urban destinations [23]. Besides, huge amount of fish meal is imported in animal and poultry feed manufacture. The dried fish for human consumption are also imported from abroad. These all imply that the fish production has immense market opportunities in Nepal.

5. THE ROLE OF FISHERIES AND AQUACULTURE IN MACRO- AND MICRONUTRITIONAL SECURITY

The importance of fish as a rich source of animal protein is well established to justify fish as a valuable food, whereas very little attention has been given to the role and understanding of fish in supplying vitamin A and minerals in the diet [35, 47, 48]. The fish contributes substantially on macro- and micronutrients supply to human food, especially vitamin A and minerals from the different types of species [13]. Macronutrients are the energy-giving caloric components such as starch, oil and structural proteins, while micronutrients comprised of vitamins, minerals, trace elements, phytochemicals and antioxidants essential for health. The present compilation of protein contents in freshwater fish revealed a range from 14.8 to 22.8 g/100 g (Table 3). High amount of protein from 58.5 to 70.8 % might occur in dried fish (sun-dried, slated dried and smoked dry fish). The protein in crab, turtle, prawn, snail, and bivalve is also comparative to the protein level in fin fish species (Table 3). In water chest nut, the Zn concentration is remarkable (Table 3).

The variation in vitamin A content in fish species is extreme [35]. The higher concentration of the vitamin A is found in the eyes and viscera of fish [13, 49]. These facts indicate that the cleaning process is highly important and might depend on the size and species of the fish for the retention of vitamin A [45]. Vitamin A in fish is found as retinol and dehydroretinol isomers [49].

Fish consumption contributes positively to adults and children's brain function, memory and improving eyesight [25, 33, 49]; particularly ω-3 consumption can improve mental health and behavior disorders [1, 36, 37]. ω-3 fatty acids also contribute to heart protection, Alzheimer disease [29, 33] and dimentia with a low ω-3 fatty acid status [36]. It is stated that in the blood, ω-3 polyunsaturated fatty acids produce vasodilation by reducing free fatty acids, plasma triglyceride, cholesterol concentrations and platelet aggregation [14, 49]. All these ultimately lead to decreasing blood pressure, clotting and atherosclerosis [32, 49]. Thus, the cardio-protective role of dietary fish intake includes prevention of coronary heart disease, reduced incidence of myocardial attack, protection against thrombosis, reduced risk of sudden death and overall mortality rates [36]. Fish oil is a rich source of the ω-3 fatty acids. Since our body does not efficiently produce some ω-3 fatty acids, it is necessary to obtain adequate amounts through fish and fish oil products [45].

Fish is also a good source of micronutrients required for human metabolism. Much is yet to be known about the level of micronutrients and essential elements of the Himalayan fishes. In present work, only little information has been reviewed from various sources (Table 3), and detail accounts will be beyond the scope of this paper. The level of calcium (Ca) is extremely high in snail (Pila globosa) and bivalve (Lamelliidens marginalis) compared with other organisms (Table 3) measuring up to 1369.4–2130 mg per 100 g wet weight [26, 43]. Relatively high elemental level of Mg, Fe and Zn was observed in water chest nut [18]. Such high or lowest elemental compositions of the commodities provide the specialty of the product and thus importance in human nutrition. A recent study found substantial support from the fish micronutrients protecting the community from mineral deficiency diseases [28].

In human nutrition, essential elements such as Ca, magnesium (Mg), sodium (Na), potassium (K), iron (Fe), zinc (Zn), copper (Cu) and manganese (Mn) are required for the normal maintenance of the human body for several biochemical reactions [10, 41]. The Ca, Mg and phosphorus (P) play crucial role in the formation of bones and teeth, while Na and K play role in electrolyte balance and transmission of nerve signals; Cu, Zn and Mn play the role in many enzyme reactions, and Fe constitutes the main parts of the hemoglobin for carrying oxygen from one place to another [9, 41]. These elements have the direct relationship with the malnutrition problems mostly associated with the children, lactating mothers, pregnant women and puberty-attaining girls in underdeveloping poor countries in the form of stunting, retardation, underweight, anemia, osteoporosis, etc.

6. GENERAL FISH CONSUMPTION PATTERN ACROSS NEPAL

In Nepal, the increase in fish consumption has not been uniform and adequate across regions. Nepal is a predominantly a mountainous country endowed with the tallest peak on the earth, the Mt. Everest. Earlier the people living in remote mountain villages had the poor accessibility to fish diets. People living in mountain and plain areas were known to suffer by high prevalence rate of goiter diseases [39], which may or may not be related to fish diet, but taken as the sign of lack of animal origin food and iodine deficiency in human diet. It was only lower reaches from where supply in the form of dried fish was possible to high mountain residents. Even in modern days the river valleys where rivers mouths are met, often traditional fisher's communities involving in capture fisheries supply smoked or dried hill stream fishes in small hill bazar's vegetable shop, grocery, hotel and restaurant. Contrarily, the people living in river valleys and southern plains where river water spreads over large areas provide ample opportunity for easy access to fish diets compared with mountain people. The depiction of carp paintings in walls and wood carving in door in houses in southern tarai might represent the affinity and importance of fish in their nutrition.

The analysis of fish consumption in rural areas versus urban areas would be a matter of interest; however, detail studies have yet to carry out. Generally, the major part of fresh fish sold in Kathmandu valley and other urban centers comprised of imported one from India [21]. The trends might suggest that fish consumption in urban areas is higher than in the rural.

In present study, the Central Development Region has been found to be consuming higher amount of fish (Fig. 5), while contrarily, it was the Far-Western Development Region where fish consumption was the least. This trend probably associated with the low fish production in the region and poor transportation in remote mountain districts. It is known that fish is more favorably received on the market when it is fresh, rather than processed owing to its perishable nature and limited shelf life [15].

The driving force behind the enormous surge in the consumption of animal products is a combination of population growth, rising incomes, increasing urbanization [15] and health awareness. Economic development and rising incomes usually lead to advances in the availability and quality of food, better overall nutritional status and the elimination of food shortages [51]. Growing urbanization usually modifies dietary patterns, both quantitatively and qualitatively, and changes the lifestyles of individuals [48]. Therefore, it is likely the role of fisheries and aquaculture production in the country would be increasing with societal dynamism in future.

7. CONCLUSIONS

Nepal predominantly a mountainous country is endowed with three river basins and one river system having about 6000 rivers of various lengths [21], offering opportunities for warm, cool as well as cold-water fish production. From water abundance perspectives, there could be substantial scaling up of the warm- and cold-water aquaculture. In general, the warm- and cold-water aquaculture requires rapid scaling up using plain tarai and mountain hill terraces, respectively, for fish production. In warmer tarai, aquaculture might have several agricultural enterprises in competition, but in mountain terraces, it is likely there would be rare competitor against the fish farming. If this argument is true, then it is likely that Nepal could be one of the global destinations for cold-water fish products, if feed and transportation constraints are resolved.

The standard food chart recommended 30 g of fish or meat protein for daily food intake [11], which is unlikely to be achieved soon with present pace of fisheries development in Nepal. Thus, more emphasis should be given to enhance the domestic fish production. Currently, Agriculture Development Perspective Plan (2015–2030) and Food and Nutrition Security Plan of Action (FNSP), the national programs for food and nutrition security [16], have emphasized the fisheries sector. Hopefully, with implementation of these national plans and policies, fisheries and aquaculture sector would further ensure the food and nutritional security in Nepal.

ACKNOWLEDGEMENTS

Thanks to all staffs of National Animal Science Research Institute (NASRI) for support and coordination during the preparation of this paper. The partial fund for this work has been provided by the Nepal Agricultural Research Council. The author also greatly depreciates the contribution of FAO, USAID, Asian Development Bank (ADB) and Japan International Cooperation Agency (JICA) for their supports to enhance fisheries research and development in Nepal.

REFERENCES

1. Addis PB. Omega-3 fatty acid content of Lake Superior Fish. Minneapolis: Minnesota extension service University of Minnesota, Home Economics HE-FO-561 8-b; 1990. p. 4.
2. ADS Agriculture Development Strategy, Ministry of Agriculture Development, Government of Nepal 2014, p. 285.
3. Anonymous. Food nutrition facts and count calories in food. 2016. http://slism.com/calorie/111243/.
4. Athiyaman R, Rajendran K. Nutritional value of freshwater Prawns Macrobrachium scabriculum (Heller, 1862) and Macrobrachium idellaidella (Hilgendorf, 1898). Int J Res Biol Sci. 2013;3(1):5–7. http://www.urpjournals.com.
5. Baby RL, Hasan I, Kabir KA, Naser MN. Nutrient analysis of some commercially important molluscs of Bangladesh. J Sci Res. 2010;2(2):390–6.
6. Bardach JE, Ryther JH, McLarney WO. Aquaculture, the farming and husbandry of freshwater and marine organisms. New York: Wiley-Interscience; 1972.
7. Bates B, Lennox A, Prentice A, Bates C, Swan G. National diet and nutrition survey; Headline results from years 1, 2 and 3 (combined) of the rolling programme (2008/09–2010/11). 2012. https://www.gov.uk/government/publications/national-diet-and-

nutrition-survey-headline-results-from-years-1-2-and-3-combined-of-the-rolling-programme-200809-201011.
8. Belton B, Thilsted SH. Fisheries in transition: food and nutrition security implications for the global South. Glob Food Secur. 2014;3:59–66.
9. Beveridge MCM, Little DC. Aquaculture in traditional societies. In: Costa-Pierce BA, editor. Ecological aquaculture. Oxford: Blackwell; 2002.
10. Bost M, Houdart S, Oberli M, Kalonji E, Huneau J-F, Margaritis I. Dietary copper and human health: current evidence and unresolved. J Trace Elem Med Biol. 2016;35:107–15.
11. Chai T-T, Ooh K-F, Quah Y, Wong F-C. Edible freshwater macrophytes: a source of anticancer and anti-oxidative natural products—a mini-review. Phytochem Rev. 2015;14:443–57. doi:10.1007/s11101-015-9399-z.
12. Department of Health, and Food Standards Agency, London; 2012.
13. DFTQC. Food composition table for Nepal. Kathmandu: Department of Food Technology and Quality Control Ministry of Agriculture Development, Government of Nepal; 2012. p. 80.
14. FAO. Human vitamin and mineral requirements, Report of a joint FAO/WHO expert consultation Bangkok, Thailand. Rome: Food and Agriculture Organization of the United Nations World Health Organization, Food and Nutrition Division, FAO; 2001. p. 289.
15. FAO. World review of fisheries and aquaculture. Rome: The State of World Fisheries and Aquaculture 2008, Food and Agriculture Organization Part 1, FAO.org; 2007. p. 84.
16. FAO. The state of world fisheries & aquaculture; 2014. p 96. http://www.fao.org/3/a-i3720e/i3720e01.pdf.
17. Faruk MO, Amin MZ, Sana NK, Shaha RK, Biswas KK. Biochemical analysis of two varieties of water chestnuts (Trapa sp.). Pak J Biol Sci. 2012;15(21):1019–26.
18. FNSP. Food and Nutrition Security Plan of Action (FNSP), A National Programme for Food and Nutrition Security, Ministry of Agriculture Development, Government of Nepal; 2013. p. 52. http://www.nnfsp.gov.np/PublicationFiles/d405d609-bb00-4708-831c-149dea4b4f49.pdf.
19. GoN. Multi-sector nutrition plan for accelerating the reduction of maternal and child under-nutrition in Nepal 2013–2017. Approved by the Cabinet Meeting of the GoN on May 15, 2012. GoN, NPC, p. 185. http://www.npc.gov.np/images/ download/ MSNP_english.pdf.
20. Gordon A, Finegold C, Crissman CC, Pulis A. Fish production, consumption, and trade in Sub-Saharan Africa: a review analysis. Prepared by WorldFish, with funding from the World Bank Fish to 2030: Sub-Saharan Africa Fish Trade in a Changing Climate; 2013. p. 48. http://www.worldfishcenter.org/content/fish-production-consumption-and-trade-sub-saharan-africa-review-analysis.
21. Gurung TB. Harnessing fisheries innovation for transformational impact in Nepal. Hydro Nepal. 2014;15:53–9.
22. Gurung TB, Gurung A, Doody TM. Connecting flows, fish diversity and ecology in the Koshi Basin, chap. 10. In: Doody TM, Cuddy SM, Bhatta LD, editors. Connecting flows and ecology in Nepal: current state of knowledge for the Koshi Basin. Sustainable Development Investment Portfolio (SDIP) project. Australia: CSIRO; 2016. p. 105–122. http://www.birdlifenepal.org/download-newsletter/41.
23. Gurung TB. Integrated aquaculture within agriculture irrigation for food security and adaptation to climate change. Hydro Nepal. 2012;11(1):73–7 (Special issue: conference proceedings). http://www.nepjol.info/index.php/HN/article/view/7214/5843.
24. Gurung TB, Pradhan N, Singh DM, Thapa TB. Prospects of trout (Oncorhynchus mykiss) farming commercialization in relation to global context in Nepal: In: Gurung TB, editors. Proceedings of the workshop on scaling-up of Rainbow trout (Oncorhynchus mykiss) farming strategies in Nepal, Kathmandu, Nepal; 2008. p 158.
25. HRC. Frog consumption trend in Manaslu conservation area; 2016. http:// hrcnepal.org.np/.
26. Kromhout D, Yasuda S, Geleijnse JM, Shimokawa H. Fish oil and omega-3 fatty acids in cardiovascular disease: do they really work? Eur Heart J. 2011. doi:10. 1093/eurheartj/ehr362.

27. Limbu KP, Subba BR, Rai BK. Nutritional status of three species of Molluscs in Nepal, Chapter: 10. In: Gupta VK, Verma AK, Singh GD, editors. Perspectives in animal ecology and reproduction, vol. 9. Delhi: Daya Publishing House; 2012. p. 232–9.
28. McKenzie JM, Wyness L. Review of Nutritional & Health Benefits for the British Trout Association, Queen Margaret University, Edinburgh, British Trout Association Client Report, Edinburg; 2013. p 30.
29. Mogobe O, Mosepele K, Masamba WRL. Essential mineral content of common fish species in Chanoga, Okavango Delta, Botswana, African. J Food Sci. 2015;9(9):480–6.
30. Nestel P, Clifton P, Colquhoun D, Noakes M, Mori TA, Sullivan D, Thomas B. Indications for omega-3 long chain polyunsaturated fatty acid in the prevention and treatment of cardiovascular disease. Heart Lung Circ. 2015;24:769–79.
31. NPC. Nepal thematic report on food security and nutrition 2013. Based on the findings of the Nepal living standards survey 2010/11. By, National Planning Commission, Central Bureau of Statistics in association with World Food Program, World Bank, AusAID, UNICEF; 2013 p 99.
32. Obande RA, Omeji S, Isiguzo I. Proximate composition and mineral content of the Fresh water snail (Pila ampullacea) from River Benue, Nigeria. 2013.
33. Philibert A, Fillion M, DeGuire J, Weiler H, Passos C, Lemire M, Mergler D. Plasma phospholipid omega-3 fatty acids and freshwater fish consumption in the Brazilian Amazon. Food Nutr Sci. 2013;4(9A):137–149. doi:10.4236/fns.2013.49A1021. http://www.scirp.org/journal/PaperInformation.aspx?PaperID=36488.
34. Praveen N, Singh AK. While effect of kinetic on protein concentration of Makhana (Euralye ferox Salisb.) during fruit development. Int J Basic Appl Sci Res. 2014;1(1):6–14.
35. Ratnayake W, Galli C. Fat and fatty acid terminology, methods of analysis and fat digestion and metabolism: a background, review paper. Ann Nutr Metab. 2009;55(1):8–43.
36. Roos N, Islamy MM, Thilsted SH. Small indigenous fish species in Bangladesh: contribution to vitamin A, calcium and iron intakes. J Nutr Am Soc Nutr Sci. 2003:4021S–6S.
37. Ruxton C, Reed S, Simpson M, Millington K. The health benefits of omega-3 polyunsaturated fatty acids: a review of the evidence. J Hum Nutr Diet. 2004;17:449–59. https://www.csuchico.edu/grassfedbeef/research/documents/source/Ruxton%20Omega3%20review.pdf.
38. Shafakatullah N, Shetty S, Orison R, Krishnamoorthy M. Nutritional analysis of freshwater bivalves, Lamellidens spp. from River Tunga, Karnataka, India. Res J Recent Sci. 2013;2:120–3.
39. Shrestha SM, Dixit H, Malla MM, Prasai BR, Shrestha RM, Pandey BR. Endemic goiter in Nepal. J Med Assoc Nepal. 1973;11:1–2.
40. SINA. Statistical Information on Nepalese Agriculture, Ministry of Agriculture Development, Government of Nepal; 2014. p 212. http://www.moad.gov.np/uploads/files/Year%20book%202014.pdf.
41. Soetan KO, Olaiya CO, Oyewole OE. The importance of mineral elements for humans, domestic animals and plants: a review. Afr J Food Sci. 2010;4(5):200–22.
42. Speedy AW. Global production and consumption of animal source foods. J Nutr. 2003;133(11):4048S–53S.
43. Subba BR. Impact of climate change on food value of molluscs in Nepal. Nepal J Biosci. 2012;2:98–108.
44. Sultana S. Flesh estimation of some sis fish. J Life Earth Sci. 2014;7:105–8.
45. Swanson D, Block R, Mousa SA. Omega-3 fatty acids EPA and DHA: health benefits throughout life. Adv Nutr. 2012;3:1–7.
46. Tacon AGJ, Matian M. Fish matters: importance of aquatic foods in human nutrition and global food supply. Rev Fish Sci. 2013;21(1):22–38.
47. Varadharajan D, Soundarapandian P. Proximate composition and mineral contents of freshwater crab Spiralothelphusa hydrodroma (Herbst, 1794) from Parangipettai, South East Coast of India. J Aquac Res Dev. 2014;5:217.
48. Vasantha S, Vijaylakshmi S, Kiran P. Review on impact of changing lifestyles on dietary pattern. Int J Curr Aca Rev. 2015;3–6:135–47.

49. Vilain C, Baran E, Gloria G, Samadee S. Fish and the nutrition of rural cambodians. Asian J Agric Food Sci. 2016;4–1, 26–34 Asian Online J. www.ajouronline.com.
50. WHO. Joint WHO/FAO expert consultation on diet, nutrition and the prevention of chronic diseases: Geneva, Switzerland: report of a joint WHO/FAO expert consultation. Geneva; 2002.
51. Woods B. Links between food price behavior and nutrition in the developing world. Stud Undergrad Res Guelph. 2012;5(2):51–62. https://journal.lib.uoguelph.ca/index.php/surg/article/download/1785/2506.
52. Zaman M, Naser MN, Abdullah ATM, Khan N. Nutrient contents of some popular freshwater and Marine fish species of Bangladesh. Bangladesh J Zool. 2014;42(2):251–9.

CHAPTER 7

Reconnecting people, place, and nature: examining alternative food networks in Newfoundland's fisheries

Chloé Poitevin Des Rivières[1]*, Ratana Chuenpagdee[2] and Charles Mather[2]

[1]*Department of Geography and Environmental Studies, Carleton University, Ottawa, ON, Canada.*

[2]*Department of Geography, Memorial University of Newfoundland, St. John's, NL, Canada.*

ABSTRACT

Background

Fisheries have played a significant role in the economic development of Newfoundland and Labrador and continue to contribute to coastal livelihoods and local food security. However, the access to and availability of locally harvested fish and seafood in the province is notably lacking, particularly since federal and provincial policies have prioritized the development of export markets in the fisheries over local sales. In this paper, we examine market actors and civil society organizations in St. John's (the provincial capital) and Petty Harbour (a nearby fishing community) that have embarked on initiatives to better include fish and seafood in the local food system. The success of these initiatives depends in part on the degree of connectivity between consumers and harvesters, as well as between people, the culture of fishing, and the marine environment.

Results

Reconnections are the foundations of alternative food networks (AFNs), which aim to develop more environmentally and socially sustainable and localized ways of producing, distributing, retailing, and consuming food. This paper explores the social, cultural, and ecological values of the fisheries as means to forging reconnections. The research reveals that values pertaining to traditional food culture, environmental stewardship and conservation, localization, and social cohesion were significant motivating factors for stakeholders in seeking out reconnection in their food system.

Conclusions

While AFNs in Newfoundland's fisheries are currently underdeveloped, reconnections are nonetheless taking place along the fisheries supply chain and can contribute to enhancing the accessibility and availability of locally harvested fish. The study also finds that AFNs must be context appropriate and reflect local values, as there are significant opportunities in building on Newfoundland's strong fishing culture and heritage.

Keywords: Alternative food networks, Fisheries, Values, Reconnection

1. BACKGROUND AND RATIONALE

Alternative food networks (AFNs) have arisen within the food movement as practical and localized responses to growing concerns over the wide-ranging socio-economic, cultural, and environmental impacts of conventional food production. In addressing the shortcomings of conventional food systems, AFNs seek to create shortened and localized supply chains in order to enhance consumer awareness about where their food comes from and develop alternate means of producing, distributing, and retailing food [1]. AFNs were shaped by concerns over agricultural food production; however, they are equally relevant to the diverse and complex challenges faced in fisheries-based food systems. Much like industrial agriculture, commercial fisheries are part of a highly globalized, technology-driven, and resource-intensive food system. The advent of highly efficient catch methods has resulted in the overexploitation and degradation of marine resources, particularly in the light of a rising global demand for fish and seafood [2, 3].

Lengthening supply chains are of equal concern in the fisheries, in which consumers and fish harvesters have become increasingly distanced both socially and geographically [4]. The notion of disconnection highlighted by AFN literature can provide insight into ways to localize and simplify fisheries food systems. While local and sustainable food movements have seldom included fisheries, certain approaches derived from agricultural contexts have been adapted to enhance consumers' access to locally sourced and sustainably caught fish [4, 5]. With the emergence of AFNs that include fish and seafood, there remain significant questions with regard to how these fisheries AFNs operate and are structured, especially given that the institutional, socio-cultural, and economic contexts of fisheries are vastly different from those of agricultural food systems [6].

This research examines ways in which fisheries AFNs may present sustainable and appropriate alternatives to conventional fisheries harvest, marketing, and consumption practices in two locations in Newfoundland: the St. John's metropolitan area and the nearby fishing community of Petty Harbour–Maddox Cove (henceforth Petty Harbour). In the following section, we present the AFN concept of 'reconnection' and assess the extent to which it exists in Newfoundland. Next, we describe the study locations and research methods, followed by the main findings. In this analysis, care ethics and the interactive governance perspective inform an understanding of the ways in which the values and ethical motivations that guide decision-making processes enable reconnections in AFNs. We conclude with a discussion about opportunities and limitations in developing fisheries AFNs in contexts similar to Newfoundland and elsewhere.

2. RECONNECTING THROUGH ALTERNATIVE FOOD NETWORKS

Critical understandings of AFNs evoke a broad definition and appearance of 'alternative foods' to reflect the diverse contexts in which alternatives take place, rather than simply presenting 'alternative' in a dichotomous opposition to conventional food production [1, 7, 8, 9]. Conventional food systems have largely been framed as 'bad', particularly since associated food production processes generate significant ecological issues, including habitat degradation and loss of biodiversity; and socio-cultural and economic issues, such as the homogenization of food cultures and the concentration of power in the hands of a few corporate entities [1, 10]. Despite the many failings of global conventional food systems, it is problematic to assume that alternative food strategies and networks are intrinsically sustainable. The variability in alternative food practices has given considerable weight to the significance of the term 'alternative', particularly since many originally 'alternative' approaches have been absorbed into mainstream and industrial practices [11].

One potential answer to the question of what makes AFNs 'truly' alternative can be found in the concept of 'reconnection'. Reconnection has become a central notion in many leading AFN discourses and actions, given that the connection between the different actors throughout the supply chain is one of the key drivers behind direct marketing schemes and other alternative food initiatives [7, 12]. In accounting for the complex social and natural systems and governing institutions that shape food systems, reconnection as part of AFNs is best viewed as a process that produces varied relations and outcomes [13, 14]. Therefore, reconnection can be economically motivated due to the fact that producers seek to enhance localized market opportunities and add value to their products, although they can also be motivated by a desire to create an ethical and ecologically sustainable food system [14].

These multifaceted rationales for reconnection in food systems are linked to conceptions of moral economies through 'care ethics', which refers to personal concerns and care for ecosystems, culture, local economy, sustainability, and local communities [13]. In this concept, the economy is understood as performative in that it is a part of and produced by social relations, rather than a wholly separate entity [15, 16]. A moral economy is an attempt to rework and challenge the ways in which relations of economic exchange are perceived as enacted by rational actors in the pursuit of profit maximization [15, 17]. Alternative economic possibilities, such as AFNs, include ethical considerations as part of economic relations, thus offering a means to examine interpersonal and economic relations in food systems and fisheries as guided by values and ethics, rather than presumed individualistic self-interest [15, 18]. In this sense, economic choices and preferences are understood as the combined result of behaviours that are learnt through social and economic institutions and transmitted by culture and society by means of personal relationships [17, 19].

In AFNs, food choices are based both on self-interest, including enjoyment, health, and nutrition, and on outward cares and concerns, such as care for others, the local community, and the natural environment [13]. Although these cares are central to the AFN model, consumers are nonetheless constrained by economic factors and are not always able to make food choices based on individual tastes and moral preferences [20]. The concept of reconnection allows for an examination of the values that affect people's food choices, which leads to a better understanding of how AFNs emerge and what motivates people to participate in alternative practices in their food system.

While the motivations for seeking alternatives in fisheries-based food systems bear similarities to those in agriculture, there is a need to understand the differences in the social, biological, and governing systems that create a new and unique set of challenges for this food system. In this light, the idea of 'reflexive localism' is especially important for defining locality in a way that takes context into account and is receptive to change [1]. Reflexive localism is process oriented, in the sense that localization is not an end goal but a continually evolving set of relations between different actors and at different scales. Notions of place making and sense of place are useful in understanding the ways in which people relate to landscapes and places, and how these may impact localization processes. In particular, place is created and recreated through different food practices, cultures, and physical geographies [21].

Landscapes, particularly physical geographic realities such as soil composition, weather, and topography, shape food cultures and localization processes through the biophysical limits imposed on how and what foods can be produced or caught [21]. It is equally important to consider temporal scales, since 'local' foods conceptually include the adoption of a seasonal diet, which can limit the choices and availability of fresh produce throughout the year depending on climate [22]. Both the geographic and temporal scales of fisheries are different from agriculture, and many regions are simply not located in areas that can support commercial fishing [12]. While biological factors, such as seasonal migrations and spawning, play a role in determining the availability of fish and seafood, fisheries management regulations determine when and how much fish can be caught.

3. AFNS AND FISHERIES IN NEWFOUNDLAND

Fisheries in Newfoundland and Labrador hold an immense cultural and historical significance and are highly complex in terms of actors, institutions, and ecology. The province's cultural identity and historical economic development are intrinsically linked with the fisheries, particularly with respect to Atlantic cod [23]. Newfoundland's fishery has gone through dramatic changes and challenges in the last century, specifically with the collapse of the commercial cod fishery in 1992, following nearly 50 years of intensive overfishing [24]. While the subsequent moratorium on the commercial Atlantic cod fishery significantly impacted coastal communities and livelihoods, fishing continues to be an important source of revenue and employment in the province. Snow crab and shrimp have replaced cod as the primary commercial and export species, currently accounting for 83% of capture fisheries' landed value, and have become more lucrative than the cod fishery was prior to the collapse [25, 26]. There remains a restricted commercial quota for Atlantic cod and a limited recreational cod fishery [25, 26].

More recently, the sustainability and durability of the Northern shrimp fishery has been brought into question, as shrimp stocks have been in decline since the 2000s due to increased catches and unfavourable environmental conditions [27]. There is also a push to re-establish the cod fishery, with some evidence pointing to the recovery of Atlantic cod stocks [26]. Other significant challenges to the fishery include a shrinking workforce due to ageing and outmigration, which is exacerbated by the seasonal, economically prohibitive, and often-insecure nature of fishing employment that discourages new entrants [24, 25].

The development of a resilient and sustainable local food system is vital to ensuring food security, particularly in rural communities in Newfoundland, due to the geographic isolation of the island, which presents significant challenges to residents' access to nutritious, affordable, and culturally appropriate foods. In fact, the majority of fresh fruits and vegetables consumed within the province are imported from mainland Canada or internationally, which can lead to food shortages as food distribution networks are vulnerable to interruptions such as natural events that prevent shipping [28, 29]. The access to and availability of fresh, nutritious foods is further challenged by the remoteness of many rural coastal communities, where residents often face lengthy travel times to the nearest full-service grocery store [29].

Although AFNs in the fisheries are developing in many coastal communities in Canada, they have been particularly slow to emerge in Newfoundland and Labrador compared to other coastal provinces. Efforts to introduce fisheries AFNs in Canada mostly come in the form of community-supported fisheries (CSFs), which aim to create a guaranteed local market for fishers and a source of fresh and sustainable fish for consumers. Notable CSFs in Canada include Off the Hook in Halifax, Nova Scotia and Skipper Otto's in Vancouver, British Columbia. While AFNs in Canadian fisheries are still underdeveloped, the number of AFNs in Newfoundland's agricultural systems has increased, with a growing network of farmers' markets and direct marketing schemes for agricultural products. In addition, there is a rich history of self-provisioning practices, including fishing, gardening, hunting, and berry picking, that have contributed significantly to food security [28, 30]. These practices, and other types of informal labour, hold culturally and economically significance for many in Newfoundland, particularly people residing in remote rural areas, such as the many small fishing communities in the province [31]. The seasonal nature of fisheries economies and the strong social ties in these communities, as well as their geographic isolation, create a strong role for informal economic networks as a key means to ensure livelihoods and food security for many residents [28, 31].

Significant barriers exist in the form of provincial and federal legislation, which restrict the development of alternate and informal markets [32, 33, 34]. In particular, both the federal and provincial fisheries governing bodies, the Department of Fisheries and Oceans (DFO) and Department of Fisheries and Aquaculture (DFA), respectively, have neglected the development of local markets for fish and seafood, instead focusing heavily on export-oriented production [25]. Equally, the access to and availability of locally harvested fish was further limited by competing interests in the fisheries, primarily between fishers and fish processors and workers [33]. The direct sales of fish-by-fish harvesters were prohibited in Newfoundland and Labrador until late 2015, justified on the grounds of maintaining the viability of the fish-processing industry and to ensure food safety [33]. Despite strict provincial regulations, a handful of alternative food efforts emerged to increase the availability of locally sourced fish and seafood.

Fisheries regulations and policies are also impacted by the Fish, Food and Allied Workers Union (FFAW-Unifor), an important lobbying group in the province that is active in developing local market opportunities, which the union see as an important means to support inshore fish harvesters. The FFAW-Unifor are also committed to developing external markets for Newfoundland fish and seafood in an effort to support the interests of their diverse membership, which includes large-scale fishing operations and fish plant workers in addition to small-scale inshore fish harvesters.

The geographic focus of the study is on the St. John's metropolitan area and the community of Petty Harbour, located both on the Avalon Peninsula and on the east coast of the island. These cases offer an opportunity to examine emerging alternatives, in terms of how they arise, organize, and operate.

St. John's is the capital of Newfoundland and Labrador and is the province's most densely populated area, with the metropolitan region hosting a population of approximately 200,000 [35]. St. John's serves as the province's economic and administrative centre, with the tourism and the offshore petroleum industries generating significant revenue within the city. The local and sustainable food movement has become increasingly prominent in St. John's, with a number of agriculture-oriented AFNs developing in the region, including direct farm sales, community-supported agriculture (CSA), and farmers markets [36]. In terms of fisheries, there is a growing network of food businesses and other actors looking to include and promote local and sustainable fish, and many of these groups have gained public attention through their efforts [37]. Many food businesses have integrated environmental and ethical values and goals into their mandates [37].

Petty Harbour is an important fishing and tourism hub located 15 km south of St. John's. Fish harvesters in this area have a unique long-standing commitment to sustainability and stewardship practices and are actively involved in local fisheries governance [38]. An example of this mentality is the establishment of the Petty Harbour Fisherman's Co-operative, which operates as a fishing enterprise that is socially just and sustainable while reinvesting in the local community. Moreover, efforts to reconnect people with the local environment and their fishing heritage have been strengthened by the creation of the Petty Harbour Mini-Aquarium, a non-profit organization aimed at raising awareness about local marine habitats and animals, and the Island Rooms, a fisheries and Newfoundland culture education program for children and youth.

4. METHODS

The study employed semi-structured interviews that covered a range of topics relating to the fisheries and food systems sustainability and governance in Newfoundland. This method allows the possibility to ask probing and follow-up questions and enables respondents to provide perspectives, experiences, and reflections in their own words [39, 40]. Selected topics were drawn from the interactive governance framework, focusing particularly on the interactions within and between the different components of the food system and the fisheries [34, 41]. Interactive governance frames values as the foundation of decision-making processes, and stresses that understanding how people come to make choices is essential to understanding and improving governance processes [41].

The interview questions used in the study aimed at discerning the underlying values that govern the decision-making processes of actors participating in alternative food initiatives in Newfoundland's fisheries. These values were drawn from a selection of topics previously identified in fisheries governance literature by Song et al. [34]. In their study, 24 value types were identified and categorized into four value orientations, relating to perspectives on fisheries governance and decision-making processes. These value orientations make explicit the 'ethics of care' performed in fisheries, as well as AFNs. They include 'better world' values (e.g. altruistic, common good), 'good life' (personal well-being), 'personal virtues' (inner personal qualities), and 'outward aspirations' (relationships with others and objects) [34]. In addition, the interviews comprised questions in which participants outlined their roles and relationships in the fishery and/or food system, as well as their involvement in enhancing alternative markets, spaces, and knowledge for fish and seafood.

Respondents were selected using purposive sampling method in order to cover a range of actors engaged in alternative food practices. A total of 13 interviews were completed in the St. John's and Petty Harbour areas, each lasting between 15 and 45 min. Participants included restaurant owners, chefs, retailers, fish harvesters, and local food promotion and conservation education organizations. Interview data were analysed thematically using NVivo 9 software (QSR International), guided partly by the categories outlined by Song et al. [34].

The interviews were complemented by an analysis of secondary data, including academic and grey literature, in particular news articles, podcasts, and documentaries, in order to provide a social, cultural, economic, and historical context to the interview data. Federal and provincial policy documents were examined to assess the roles of governing actors and institutions in Newfoundland's fisheries and identify potential barriers and opportunities for emerging AFNs.

5. RESULTS

The identification of values underpinning alternative food practices aids in understanding the types of reconnections taking place, and the underlying motivations that lead people to seek these connections. Three primary value types were identified in the interviews: food system localization, social cohesion, and education and conservation.

5.1 Food system localization

The interview respondents interpreted 'local' in different ways and associated different benefits and values to the prospect of localizing food systems. However, these values pertained predominantly to developing market opportunities. The scale of 'local' in terms of Newfoundland's fisheries is considered larger than that of land-based food production. For instance, fish and seafood products originating from the whole island were referred to as 'local', while 'local' for agricultural products was generally seen as encompassing only the Avalon Peninsula. As well, fish originating from Atlantic Canada was viewed as relatively local, as they come from the same waters.

Restaurants and retailers interviewed for the study indicated reasons for choosing locally harvested fish and seafood around varied notions of quality. Most replied that local fish was preferable as the product 'hasn't travelled', thus perceived as fresh and possessing a small overall environmental footprint. Better taste, nutrition, and higher product quality were also mentioned. For instance, one restaurant owner viewed Newfoundland waters to be more 'pure and clean', resulting in better tasting fish and seafood. Certain restaurant owners also expressed a preference for purchasing fish and seafood from small-scale fishing vessels as these were viewed as engaging in more sustainable harvest methods and better product handling practices. Another motive for purchasing locally caught fish that many respondents noted was the desire to support the local economy, particularly rural fishing communities.

Despite a preference for local products, interview respondents expressed varying degrees of difficulty accessing locally harvested fish and seafood, either for their business or for personal consumption. The seasonality of fish and seafood was one reason that affected local availability of certain species, due to both biological limits and fisheries management regimes. The restaurant owners and chefs that have made local sourcing a part of their business plan change their menus according to seasonal variability, and turn to in-house methods of preservation for local fish, meats, and produce to overcome limited product availability, especially in the winter. Some respondents did note, however, that for some species, the fishing seasons enforced by DFO are too restrictive and do not align with the province's tourism season (which generally runs from June until September). This mismatch applies particularly to halibut and lobster, which means that restaurant owners either have to resort to frozen product, in the case of halibut, or imports from the mainland, as with lobster.

The export-oriented fisheries policy in Newfoundland was another reason frequently listed as a cause of limited access and availability to locally harvested fish. Nonetheless, export was perceived as a necessity, with one fisher based out of Petty Harbour, stressing that export was necessary to make a living. The fishers interviewed noted that the species with the highest local demand is cod, although quotas and seasons are very restricted, resulting in low profitability. The main species harvested by the Petty Harbour Fisherman's Co-op, particularly crab and capelin, are of high value on the export market, while respondents feel that the local demand simply is not there.

The one species that all of the restaurants and retailers interviewed reported no issues accessing locally was cod. Rather, it is the access to other species, such as whelks, squid, crab, and octopus, which are caught locally and destined for foreign markets or for use as bait, that pose a significant challenge to local consumption. The restaurants in St. John's were limited only to what was available through their suppliers, since they could not purchase fish directly from harvesters as per provincial regulations at the time of the interviews. Product availability was further restricted by minimum processing requirements, which are implemented by DFA to support the fish-processing industry, which dictate in what form fish and seafood can be sold. In fact, many restaurants found it difficult to access whole, unprocessed fish. In the words of one restaurant owner and chef: 'A lot of our work is in the sourcing of the product than [the] preparations. Half the work is actually finding it'.

5.2 Social cohesion

The most important value identified throughout the interviews was that of social cohesion, which refers to the processes of social connection, interpersonal relationships, and community values [34]. One respondent noted that the considerable disconnection between consumers and fish harvesters in Newfoundland has impeded the flow of information about how fish gets from ocean to plate. The motivations of respondents to seek personal relationships are focused on increasing the knowledge about the product itself and the processes of harvesting fish and seafood. There is limited, if any, information available to retailers, chefs, and consumers about where, how, and by whom fish is caught, thus impeding traceability. An opportunity for creating social connections, albeit at a distance, is the seafood traceability project headed by the FFAW-Unifor on the west coast of the island. This initiative, adapted from Ecotrust Canada's ThisFish project, allows consumers to connect with fish harvesters on an online platform by means of a traceable tag on fish and seafood products. The group of restaurants in St. John's seeking to access the traceable fish see this approach as an opportunity to engage consumers with their food and the fishery and feel that, to a limited degree, the project may help in creating social connections between fish harvesters, restaurants, and consumers.

With an increasing network of alternatives in the agricultural sector in Newfoundland, restaurateurs and consumers are becoming able to build personal relationships with local food producers. Restaurants in St. John's are able to purchase directly from both local farmers and licensed hunters, and owners expressed a desire to connect in the same ways with fish harvesters: 'We can develop relationships with all kinds of farmers, all kinds of producers of everything, and I have a personal relationship with them, but I can't have a personal relationship with my fisherman,' said a restaurant owner.

Another restaurant owner echoed the importance of personal relationships in accessing local products and stressed that proximity was key in facilitating these networks. The restaurant owner relayed their experience living and working in a rural fishing community in Newfoundland and noted that the proximity to an active fishery allowed them to develop networks and personal relationships with harvesters, which subsequently enhanced their access to locally caught fish. Many respondents viewed this type of relationship with fish harvesters as being very difficult to build from within the urban St. John's area.

While formalized food networks for fish and seafood have, up until very recently, been difficult to form between harvesters, retailers, and consumers, there are nonetheless established informal food networks, particularly in rural communities. One respondent in Petty Harbour noted that she had no trouble accessing local fish in the community due to personal relationships with fish harvesters: 'when the commercial fishery starts up, then that's not a problem for me personally because I have a brother in law who's a commercial fisherman. But if you're not, if you're outside the harbour, you're going to have some problems…'. In Petty Harbour, an informal network did seem to take place with regard to cod. One fisher explained: 'Fishermen do it, fishermen take chances and do it for a few extra dollars. That's the way it is in all the communities'. Employees of the Petty Harbour Mini-Aquarium observed an informal trade in the community, and their location beside the co-operative has positioned the organization as a link between local consumers and tourists that are interested in finding fresh, local seafood.

5.3 Education and conservation of culture and nature

Given that the fisheries in Newfoundland are fundamentally intertwined with place, culture, history, and community, it is no surprise that alternative initiatives are seeking to integrate traditional values and practices. Community actors and restaurants have stepped into try to educate people about food traditions, skills, and the natural environment, in order to create a demand for and interest in locally harvested fish and seafood.

A number of restaurants in St. John's have developed menus that feature not only local products, but also traditional Newfoundland dishes. As well, many have sought to diversify consumer tastes for seafood to include a wider variety of species. 'Fish' is synonymous with cod in Newfoundland, and consumer demand for other species that are caught locally is generally low. While the inadequate access to a diversity of fish species creates barriers, another issue lies in the lack of food skills; people often have limited capabilities and knowledge in preparing fish and seafood other than cod, which restricts what they will purchase. A seafood retailer in St. John's pointed out that this issue did affect what he sold to local consumers, noting that people living in rural coastal areas have more experience and knowledge with regard to preparing a larger variety of fish and seafood.

Seeking to increase food skills among consumers can also help to increase food security and solve food-related health issues in the province. Food First NL seeks to emphasize the importance that traditional food skills, such as berry picking, preserving, gardening, and fishing, have played in the past and the ways they can contribute to ensuring food security and self-sufficiency. Equally, Food First NL advocates for the development of food skills and increased access to and availability of local foods, including fish, as means to help mitigate the high prevalence of diet-related health problems in Newfoundland.

The need to build self-sufficiency in Newfoundland's food system was another concerned raised by interview respondents. Young people, in particular, lack knowledge about fishing heritage. A fisher in Petty Harbour asserted that, when he was growing up, fishermen were seen as the heroes in the community, and 'it was easy to fall into the fishery trap'. The Island Rooms program looks at using food and fisheries as a way to engage youth in the food system and with nature by developing food skills through fishing workshops and by teaching other forms of self-provisioning. With the workforce declining in Newfoundland's fisheries, it is hoped that the project will introduce young people to a possible career in fishing.

A number of interview respondents stressed that there is a disconnection between different food systems actors and the natural environment that must be addressed in Newfoundland, which can be tied to a lack of knowledge about the ecological impacts of food production and fisheries. The Island Rooms and the Mini-Aquarium are possible avenues to reconnect people to the marine environment through education initiatives related to ecosystem conservation. The Petty Harbour Mini-Aquarium in particular provides an opportunity for people to connect with the local marine environment by learning about, seeing, and handling different aquatic animals, including species that are commercially harvested in Newfoundland such as cod, lobster, and flatfish. By bringing the 'ocean up to eye level', the Mini-Aquarium not only contributes to public education, but also provides fish harvesters with a new means to understand, experience, and appreciate the marine ecosystem on which their livelihoods depend. For instance, the members of the Petty Harbour Fisherman's co-op had never seen a codfish swim and were unaware of some of the characteristics and natural behaviours of the species they fish. In turn, they were also able to share their knowledge and experience of the marine environment with the employees of the aquarium, and there are plans to create a program that would allow visitors to the aquarium to engage with the local fishers.

6. DISCUSSION

6.1 Developing skills, tastes, and knowledge

The access to locally sourced fish in Newfoundland is a significant challenge, though there are many potential endeavours that, together, may help in developing fisheries AFNs in the province. Opportunities to develop these alternatives lie in strengthening relations between food systems actors, as well as the connections to nature, place, and Newfoundland culture. The re-emergence of traditional Newfoundland cuisine provides a unique prospect to reconnect with place, culture, and fisheries. Thus, consumers are able to gain an appreciation for and a better knowledge of food that is available locally, which in turn can help create a demand for locally source seafood. The interviews illustrated the integral role that fisheries play in place-making and localization processes, as respondents emphasized the significance of fish within food culture, heritage, and identity in Newfoundland. Cod in particular is a reflection of food preferences and tastes for Newfoundlanders, as well as a manifestation of sense of place, culture, and identity that is established by family, social norms and traditions, economic realities, and landscapes [13, 19, 21]. The efforts of both market and civil society actors help to situate food and fish in particular places and ecologies in Newfoundland by reconnecting individuals with the local marine environment, their fishing heritage, and the local food system.

Market actors play a role in enhancing reconnections in the local food system by engaging with local food cultures and attempting to improve product availability. More specifically, the ways in which restaurateurs in St. John's craft menus that are reflective of seasonality and locally available products can make consumers more aware about the fish and seafood being harvested in Newfoundland. Through this practice, restaurants help to re-instil a pride of place and culture locally and with visitors. Despite the key role that restaurants and chefs can play in promoting locally sourced food and sustainable food systems, their adherence to 'local' is often oriented towards the characteristic of the food itself rather that the processes and conditions through which food is produced [42]. In doing so, they are engaging with weaker conceptions of 'alterity' that neglect broader socio-economic issues. To some extent, this is true of the restaurateurs interviewed, as quality and taste were mentioned as important reasons for preferring locally harvested fish. Nevertheless, most chefs and restaurant owners also expressed a desire to engage in the local food system as a means to support fish harvesters and rural communities, as well as to celebrate local food cultures and traditions.

Encouraging the diversification of the fish species that are consumed locally is a fundamental part of developing alternatives in the fisheries and enhancing local economies. However, as previously noted, food skills impact food choices by limiting the foods that individuals are able to prepare and can prevent people from accessing healthy foods and participating in food cultures [43]. In order for fisheries-based AFNs to take root and be sustained in Newfoundland, consumers not only need to be interested in and able to access local fish, but also must have the capacities to prepare it.

The education initiatives put forward by the Island Rooms, Food First NL, and the Mini-Aquarium, in addition to the efforts of St. John's restaurateurs to develop tastes for locally sourced fish, are a means of diversifying local consumer food preferences and enhancing food skills. O'Hara and Stagl [19] assert that food preferences are dynamic and continuously adapting, and it is through education and learning opportunities that aim to increase food knowledge and skills that consumer preferences and values can be altered. This reality is evidenced through the endeavours of the Island Rooms and Food First NL, specifically in their attempts to revitalize self-provisioning skills in order to generate a local food system that is truly sustainable. More specifically, one of the main aims of the Island Rooms is to enable young people to engage in subsistence fishing throughout the province in hopes of creating a more culturally appropriate and secure food system. Subsistence food production and fishing enhance food security in Newfoundland, particularly in rural communities, since people at all income levels are able to access food outside formal markets [28]. Equally, self-provisioning skills allow people to reduce dependency on conventional food supply chains, which lessens the impacts of unforeseen shocks to food supply in Newfoundland, including weather and transportation issues [28].

6.2 Challenges to developing fisheries AFNs

Despite the considerable opportunities and efforts to develop AFNs in Newfoundland's fisheries, the complex political economic structures of the fisheries and food system generate significant challenges. Specifically, fisheries policies at the provincial and federal levels remain highly problematic, even with the recent change in direct sale regulations. Andrée et al. [9] found that in regions where export-oriented food production policies are implemented, the development of AFNs is arduous, since governments prioritize the growth of international markets over stimulating local opportunities. This is the case in Newfoundland's fisheries, where governments at both federal and provincial levels have sought to develop external markets to the detriment of the local one.

The implications of the new regulations for direct fish sales on local markets remain unclear, as many details have yet to be worked out, such as the structure of reporting and monitoring systems. The benefits of this regulatory change for fish harvesters and buyers is equally uncertain, given that the majority of fish harvesters still rely on export markets for most of their catch due to the small local market. This may lead to a more 'hybridized' version of AFNs, in which food producers continue to depend on some conventional aspects of the food system in order for their business to subsist, while waiting for the local demand to expand [9, 44]. As with the fisheries in Newfoundland, this hybridity can often be the result of policy environments that encourage and incentivize all producers, including those engaging in alternative practices and niche markets, to participate in the conventional supply chain [9]. Direct marketing strategies therefore offer a means to enhance the strength, or 'alternativeness' of AFNs in these environments [45].

Prior to the shift in provincial regulation, the social connections between harvesters and consumers were maintained through informal food networks that have enabled the direct distribution of fish and seafood. Self-provisioning practices, including the food fishery, are important alternatives to conventional market practices, as well as meaningful facets of Newfoundland heritage. However, as noted, these informal food networks in the fisheries are geographically and socially limited to those near active fisheries and to those who know fish harvesters. AFN literature is limited in its ability to address these informal practices, despite the fact that many of these are, by current definitions, AFNs. Conceptually, AFNs are positioned as a response to conventional food production practices, while informal networks such as those taking place in Newfoundland's fisheries existed before conventional practices were developed. People in many Newfoundland communities have historically bought fish directly from harvesters on the wharf. As one fish harvester put it: 'I mean it was done for year and years and years before all this [provincial legislation] came in'.

AFNs that operate as part of informal economies, such as bartering and self-provisioning practices, are culturally significant to Newfoundlanders and will likely remain an important means to access fish, particularly in rural and remote coastal communities. However, these informal practices are not accessible to all, and AFNs that operate within the formal economy also need to be developed. In many instances, formalized alternative food economies are required to enable consumer–producer connections, which can take place in many different ways, such as through direct marketing, local retailers, and restaurants. Newfoundland food systems actors can look elsewhere in Canada, where direct fish sales have legally been allowed for longer periods of time, to draw inspiration in developing fisheries AFNs. Community-supported fisheries (CSFs) are cited as a means for people in Newfoundland, particularly in urban areas, to access fresh fish and seafood while enhancing small-scale fisheries and coastal community livelihoods [46]. The CSF model has the capacity to be adapted to a variety of contexts and to meet diverse socio-economic and ecological needs, as long as ethical and environmental concerns are integrated and enacted adequately [6, 18]. Community and economy are intertwined and integral to the CSF model, thus opening up possibilities for alternative means to managing fisheries resources in a way that builds towards a just and sustainable food system [6, 18].

In the light of the legalized direct sales of fish and the increased opportunity to develop strategies such as CSFs, careful consideration must be given to which groups are targeted by AFNs and how these strategies can be made to include a diverse demographic. Many have criticized cases of the practical implementation of AFNs as catering to higher-income and more privileged groups, based on the fact that strategies such as CSA and farmers' markets can prove costly to the consumer [11]. In these cases, central values to AFNs such as social justice, democracy, and inclusivity are neglected, especially as only certain groups have access and are able to participate in alternatives [11]. Many AFNs operate under the assumption that consumers are willing to pay a premium to access sustainably produced and local foods [20]. Conversely, profitability remains a challenge for many alternative food strategies in agricultural contexts, particularly with the goal of attaining a fair price for consumers while ensuring producers' livelihoods [20, 47]. These challenges are likely to pose limitations on fisheries AFNs in balancing the need to create an inclusive and affordable model while guaranteeing fish harvesters an adequate income.

Many of the efforts towards creating fisheries AFNs in St. John's tend to cater to higher-income populations, particularly due to the fact that most of the restaurants that are most active in supporting local fish are higher end and quite expensive. Equally, the reliance on market-driven mechanisms to create more sustainable, localized, and just food systems places the responsibility squarely on the shoulders of consumers, who are expected to vote with their dollar [6, 48]. These strategies restrict all but those with the economic means from participating in their local food system and having the ability to make food choices based on their values and ethics. These more exclusive AFN strategies risk mirroring the problematic and unequal political economic structures that underpin the conventional food system [6, 11, 20]. In order for systemic changes to take place in the food system, the access to and availability of locally harvested fish needs to be tied to broader food security and food justice goals [6]. The efforts of civil society organizations, such as Food First NL and the Island Rooms, to highlight the importance of local fish and food in building towards community food security in Newfoundland offer a means to better engage with issues of accessibility and inclusivity in AFNs.

7. CONCLUSIONS

The question of how alternatives can be developed in Newfoundland's fisheries is best answered by looking at the reconnections taking place within the food system, especially in terms of re-engaging consumers with their food through culture and traditions. The emphasis on relationships and reconnections in AFNs allows for the recognition of the multiple actors involved in building alternative markets and how they relate to one another. This study found that, while fisheries hold important socio-cultural values in Newfoundland, they are primarily valued by government policy for their economic contributions, which has created a significant barrier to building AFNs. Conversely, recognizing these alternate values is key to strengthening the local access and availability of fish, especially those that relate to and build on Newfoundland food cultures and traditions. In fact, the values identified demonstrate that choosing locally harvested fish and seafood is not simply related to personal ideals such as quality, health, and nutrition, but to 'care ethics' related to outward aspirations such as environmental sustainability and enhancing coastal communities.

Furthermore, there is a need for improved access to local fish and seafood in diverse contexts and for many different people in Newfoundland. Along with AFNs, informal food practices enhance the food system in Newfoundland and can provide opportunities to develop alternative practices in the fisheries. More work needs to be done on informal networks in the fisheries, particularly with regard to how these networks and traditional practices in rural coastal communities may be impacted by the new possibility for legal direct sales.

This research contributes to an understanding of how AFNs can emerge and develop in unfavourable policy environments that impede the development of localized markets for fish and seafood. Moreover, while export-oriented fisheries policies are barriers to developing AFNs, this study finds that reconnections are already taking place in the broader food system, and specifically within the fisheries, that aim to create better access to and availability of locally caught fish and seafood in Newfoundland. These reconnections offer a starting point on which to build fisheries AFNs by strengthening existing food systems and economic relations that include ethical and moral values. In other words, there is significant value in 'starting where you are', as it draws attention to the existing community assets and efforts in building towards sustainable and just food systems [49, 50]. These alternate ways of valuing fisheries are reflective of the diverse possibilities in Newfoundland that already exist and that can help to strengthen and create AFNs in the fisheries. In summation, there must be careful consideration of the types of AFNs that are best suited to support community livelihoods, cultural ties, and ecological sustainability in order to create a just and secure food systems.

ACKNOWLEDGEMENTS

This research is part of the Master thesis of the first author and contributes to Too Big to Ignore: Global Partnership for Small-Scale Fisheries Research. The authors would like to express their gratitude to all those who participated in the interviews. We are grateful to Brennan Lowery for his helpful feedback. The authors also thank the reviewers for their useful comments on the manuscript.

REFERENCES

1. Harris EM. Eat local? Constructions of place in alternative food politics. Geogr Compass. 2010;4(4):355–69. doi:10.1111/j.1749-8198.2009.00298.x.
2. Ponte S. Greener than thou: the political economy of fish ecolabeling and its local manifestations in South Africa. World Dev. 2008;36(1):159–75. doi:10.1016/j.worlddev.2007.02.014.
3. Chuenpagdee R, Morgan LE, Maxwell SM, Norse EA, Pauly D. Assessing collateral impacts of fishing methods in US waters. Front Ecol Environ. 2003;1(10):517–24.
4. Lowitt K, Nagy M, Nelson C, Bavington D. Where's the fish? Altern J. 2013;39(4):1–6 **(Food and Drink)**.
5. Loring PA, Gerlach SC, Harrison HL. Seafood as local food: food security and locally caught seafood on Alaska's Kenai Peninsula. J Agric Food Syst Community Dev. 2013;3(3):13–30.
6. Olson J, Clay PM, Pinto da Silva P. Putting the seafood in sustainable food systems. Mar Policy. 2014;43:104–11. doi:10.1016/j.marpol.2013.05.001.
7. Winter M. Embeddedness, the new food economy and defensive localism. J Rural Stud. 2003;19(1):23–32. doi:10.1016/S0743-0167(02)00053-0.
8. Holloway L, Kneafsey M, Cox R, Venn L, Dowler E, Helena T. Beyond the 'alternative'-'conventional' divide? Thinking differently about food production-consumption relationships. In: Maye D, Holloway L, Kneafsey M, editors. Alternative food geographies: representation and practice. Bingley: Emerald; 2008. p. 77–93.
9. Andrée P, Dibden J, Higgins V, Cocklin C. Competitive productivism and Australia's emerging "alternative" agri-food networks: producing for farmers' markets in Victoria and beyond. Aust Geogr. 2015;2010(41):307–22. doi:10.1080/00049182.2010.498038.
10. Howard PH. Concentration and power in the food system. London: Bloomsbury; 2016.
11. Guthman J. Bringing good food to others: investigating the subjects of alternative food practice. Cult Geogr. 2008;15:431–47. doi:10.1177/1474474008094315.
12. Campbell LM, Boucquey N, Stoll J, Coppola H, Smith MD. From vegetable box to seafood cooler: applying the community-supported agriculture model to fisheries. Soc Nat Resour. 2014;27(1):88–106. doi:10.1080/08941920.2013.842276.
13. Dowler E, Kneafsey M, Cox R, Holloway L. "Doing food differently": reconnecting biological and social relationships through care for food. Editor Board Sociol Rev. 2009;57(s2):200–21.
14. Kneafsey M, Cox R, Holloway L, Dowler E, Venn L, Tuomainen H. Reconnecting consumers, producers and food: exploring alternatives. Oxford: Berg Publishers; 2008.
15. Lind C. Rumours of a moral economy. Halifax: Fernwood Publishing; 2010.

16. Gibson-Graham JK. Diverse economies: performative practices for "other worlds". Prog Hum Geogr. 2008;32(5):613–32. doi:10.1177/0309132508090821.
17. Sayer A. Moral economy as critique. New Polit Econ. 2007;12(2):261–70. doi:10.1080/13563460701303008.
18. Snyder R, St. Martin K. A fishery for the future: the midcoast fishermen's association and the work of economic being-in-common. In: Roelvink G, St. Martin K, Gibson-Graham JK, editors. Making other worlds possible: performing diverse economies. Minneapolis: University of Minnesota Press; 2015. p. 26–52.
19. O'Hara SU, Stagl S. Endogenous preferences and sustainable development. J Socio Econ. 2002;31:511–27.
20. Allen P. Realizing justice in local food systems. Camb J Reg Econ Soc. 2010;3(2):295–308. doi:10.1093/cjres/rsq015.
21. DeLind LB. Of bodies, place, and culture: re-situating local food. J Agric Environ Ethics. 2006;19(2):121–46. doi:10.1007/s10806-005-1803-z.
22. Parkins W, Craig G. Culture and the politics of alternative food networks. Food Cult Soc An Int J Multidiscip Res. 2009;12:77–103. doi:10.2752/155280109X368679.
23. Kurlansky M. Cod: a biography of the fish that changed the world. Toronto: Random House Canada; 1997.
24. Schrank WE. The Newfoundland fishery: ten years after the moratorium. Mar Policy. 2005;29(5):407–20. doi:10.1016/j.marpol.2004.06.005.
25. Department of Fisheries and Aquaculture (DFA). Annual Report 2013-14. 2014. http://www.fishaq.gov.nl.ca/publications/pdf/annual_report_2013_14.pdf.
26. Mather C. From cod to shellfish and back again? The new resource geography and Newfoundland's fish economy. Appl Geogr. 2013;45:402–9. doi:10.1016/j.apgeog.2013.06.009.
27. Foley P. The political economy of marine stewardship council certification: processors and access in Newfoundland and Labrador's inshore shrimp industry. Agrar Change. 2012;12(2):436–57.
28. Lowitt K. Examining fisheries contributions to community food security: findings from a household seafood consumption survey on the west coast of Newfoundland. J Hunger Environ Nutr. 2013;8(2):221–41. doi:10.1080/19320248.2013.786668.
29. Food First NL. Everybody eats: a discussion paper on food security in Newfoundland and Labrador. 2015. http://static1.squarespace.com/static/54d9128be4b0de7874 ec9a82/t/56572239e4b0b807738c5ad0/1448550969272/Everybody+Eats_NL+Discussion+Paper+2015.pdf.
30. Teitelbaum S, Beckley T. Harvested, hunted and home grown: the prevalence of self-provisioning in rural Canada. J Rural Community Dev. 2006;1:114–30.
31. Felt LF, Sinclair PR. "Everyone does it": unpaid work in a rural peripheral region. Work Employ Soc. 1992;6(1):43–64. doi:10.1177/095001709261003.
32. Murphy I, Neis B. Navigating the legislative requirements for fisheries—tourism initiatives in Newfoundland and Labrador. 2012. http://www.curra.ca/documents/TCR_Fisheries Tourism_Regulations_Report.pdf.
33. Eric Dunne Consulting Initiatives. Report of the review of the regulations and policy for direct fish sales in Newfoundland. 2010: p. 1–51. http://www.fishaq.gov. nl.ca/publications/pdf/Direct_Fish_Sales_NL_2010.pdf.
34. Song AM, Chuenpagdee R. A principle-based analysis of multilevel policy areas on inshore fisheries in Newfoundland and Labrador, Canada. In: Jentoft S, Chuenpagdee R, editors. Interactive governance for small-scale fisheries. MARE Publication Series. Vol. 5. Cham: Springer; 2015. p. 435–65. doi:10.1007/978-3-319-17034-3.
35. Department of Finance- Newfoundland & Labrador Statistics Agency. Population estimates. 2016. http://www.stats.gov.nl.ca/statistics/population/PDF/Population_Estimates_CDCMA.pdf. Accessed 26 Jan 2017.
36. Root Cellars Rock! Local Links. 2014. http://rootcellarsrock.ca/local-links/. Accessed 14 April 2014.

37. Gollner AL. Jeremy Charles: Canada's real top chef. Globe Mail. 2014. http://www.theglobeandmail.com/life/jeremy-charles-canadas-real-top-chef/article20495580/. Accessed 1 May 2015.
38. Protected Areas Association of Newfoundland and Labrador. Ancient rights: The protected fishing area of Petty Harbour-Maddox Cove. 1996. http://collections.mun.ca/PDFs/ich_avalon/AncientRightsBooklet.pdf.
39. Cope M. Interviewing. In: Warf B, editor. Encyclopedia of human geography. Thousand Oaks: Sage Publications; 2006. p. 262–63. doi: 10.4135/9781412952422.n163.
40. Mason J. Semi-structured interview. In: Lewis-Beck MS, Bryman A, Liao TM, editors. The SAGE encyclopedia of social science research methods. Vol. 3; Thousand Oaks: Sage Publications; 2004. p. 1020–22. doi:10.4135/9781412950589.
41. Kooiman J, Jentoft S. Meta-governance: values, norms and principles, and the making of hard choices. Public Adm. 2009;87(4):818–36. doi:10.1111/j.1467-9299.2009.01780.x.
42. Inwood SM, Sharp JS, Moore RH, Stinner DH. Restaurants, chefs and local foods: insights drawn from application of a diffusion of innovation framework. Agric Hum Values. 2008;26(3):177–91. doi:10.1007/s10460-008-9165-6.
43. Stead M, Caraher M, Wrieden W, Longbottom P, Valentine K, Anderson A. Confident, fearful and hopeless cooks: findings from the development of a food-skills initiative. Br Food J. 2004;106(4):274–87. doi:10.1108/00070700410529546.
44. Watts DCH, Ilbery B, Maye D. Making reconnections in agro-food geography: alternative systems of food provision. Prog Hum Geogr. 2005;29(1):22–40. doi:10.1191/0309132505ph526oa.
45. Mount P. Growing local food: scale and local food systems governance. Agric Hum Values. 2011;29(1):107–21. doi:10.1007/s10460-011-9331-0.
46. Neis B, Ommer R. Moving forward: building economically, socially and ecologically resilient fisheries and coastal communities. 2014. http://www.curra.ca/documents/Revised_CURRA.
47. Galt RE. The moral economy is a double-edged sword: explaining farmers' earnings and self-exploitation in community-supported agriculture. Econ Geogr. 2013;89(4):341–65.
48. Konefal J. Environmental movements, market-based approaches, and neoliberalization: a case study of the sustainable seafood movement. Organ Environ. 2012;26(3):336–52. doi:10.1177/1086026612467982.
49. Gibson-Graham JK. Postcapitalist politics. Minneapolis: University of Minnesota Press; 2006. doi:10.1017/CBO9781107415324.004.
50. Harris E. Neoliberal subjectivities or a politics of the possible? Reading for difference in alternative food networks. Area. 2009;41:55–63. doi:10.1111/j.1475-4762.2008.00848.x.

CHAPTER 8

The humble sardine (small pelagics): fish as food or fodder

Moenieba Isaacs*

Institute for Poverty, Land and Agrarian Studies (PLAAS), University of the Western Cape (UWC), Cape Town, South Africa

ABSTRACT

Background

The group of small pelagic fish is the largest species group landed globally. A significant proportion of this nutrient-rich food is processed and lost to livestock feed, fish feed, fish oil, pet food and omega-rich vitamins. The nutritional importance of small pelagics as an easily digestible protein source, rich in essential lipids with fatty acids (EPA/DHA), essential amino acids, minerals and vitamins, is well known and documented. Small pelagics contain all the elements of a healthy and nutritionally optimal food source for humans and are an important contributor to the food and nutritional security of many poor, low-income households in developing countries.

Findings

Large-scale and small-scale fisheries play an important role in contributing to food security and nutrition. Yet, all of the anchovy landings from large-scale fisheries are reduced to animal feed, fish oil and pet food in South Africa. The size of the species, labour costs and lack of incentives by the state are some of the challenges to redirecting anchovy for human consumption. This trend is also now prevalent in Tanzania, where most (84%) of the dagaa fished is reduced to fishmeal in Kenya, mainly to feed chickens. The main challenges are post-harvest handling and sanitation.

Conclusions

The redirecting of small pelagics to human consumption will depend on the role of the state in Tanzania and South Africa in investing in post-harvest processing. The role small-scale fisheries play in providing fish for food security needs to be understood in the context of economic viability and of how data are reported in this sector as compared to large-scale fisheries.

Keywords: South Africa, Tanzania Food, security, Nutrition Consumption, Small-scale, Large-scale, Small pelagics

1. BACKGROUND

Long-term food insecurity leads to undernourishment, in which the body receives insufficient nutrients, either macro- or micronutrients. Protein is an example of a macronutrient, and amino acids, vitamins and minerals are examples of micronutrients. Undernourishment has been studied extensively in underweight children stunted by growth and wasted around the waist. In an attempt to stress the importance of nutrition, pregnant women and breastfeeding mothers are also important focus groups being studied extensively. In 2013, UNICEF reported the alarming malnutrition statistic that over one-third of all stunted (low height for age, caused by long-term insufficient nutrient intake) children and one-third of all wasted (low weight for height, a strong predictor of mortality among children under five) children lived in Africa [1]. About one-third of all deaths among under-5-year children in Africa are caused by malnutrition. In addition, food and nutrition security is still one of the major constraints facing child survival in many African countries today. Furthermore, micronutrient deficiency is a major problem worldwide, as two billion people are reported to be deficient in the micronutrients crucial for human development [2, 3]. In many countries in Africa, diet consists predominantly of starch, while fish is fairly readily available in many riparian and coastal communities and could play an important role in enriching diets deficient in vitamins and minerals.

Fish as food does not represent a significant percentage of the calorie intake for many poor people, although it is a healthy and affordable food source, rich in high-quality protein and essential micronutrients and vitamins. Adding fish to the rice, maize, bread and noodles, etc., consumed by many poor, would contribute significantly to providing vitally needed nutrients.

2. INTERNATIONAL RECOGNITION OF FISH AS NUTRIENT SOURCE

The International Conference on Sustainable Contribution of Fisheries to Food Supply held in Kyoto, Japan, in 1995 identified fish (marine, inland or aquaculture) as playing a significant role in food security. Delegate states approved a Declaration and Action Plan to enhance the contribution of fisheries to human food supply [4,5].

To address the right to food for many poor and malnourished people, there is an increased emphasis in international organisations (Committee on Food Security, Food and Agriculture Organisation and many donor countries) on the importance of the nutritional contribution of fish as food to be considered in the diet of the poor. International instruments and commitments, such as the *Voluntary Guidelines on Responsible Governance of Tenure of Land, Fisheries and Forests in the context of National Food Security (Tenure Guidelines)* [6] and the *Voluntary Guidelines on the Progressive Realization of the Right to Adequate Food in the Context of National Food Security (Right to Food Guidelines)*[7], endorse the important role of small-scale fisheries in contributing to food security. More specifically, the UN's Special Rapporteur on the Right to Food makes "the explicit link between right to food and rights of those who produce it, to fair access to resources such as fish and water" [8].

3. INTERNATIONAL RECOGNITION OF SMALL FISHERIES

The often ignored, underestimated, undervalued and overlooked small-scale fisheries finally gained traction when the *Voluntary Guidelines for Securing Sustainable Small-Scale Fisheries in the Context of Food Security and Poverty (Small-Scale Fisheries Guidelines)* [9] by the Committee on Fisheries of the Food and Agriculture Organization of the United Nations (FAO) was adopted in 2014. The Committee on Food Security (CFS/FAO) Higher Level Panel of Experts (HLPE) [3] report on *Sustainable Fisheries and Aquaculture for Food Security and Nutrition* reconfirmed the importance of small-scale fisheries in contributing to food security and nutrition.

Internationally the debates on fish as food, livelihood, income, trade, feed and forage have all been gaining traction internationally with specific reference to small-scale fisheries and the right to food campaigns by civil society. The nutritional importance of small pelagics, which include sardines, mackerels, anchovies, etc., is widely known, and many small-scale fishing communities consume the small fish. Since fish is one of the most traded commodities in the world, the nutritional importance of protein and micronutrients in developing countries is often compromised for sale on the local, regional and international markets. Feeding into the debate fish as a livelihood and income to provide indirect food security, the findings of the FAO/NORAD report in 2013 indicated that local fishers rarely benefit from "fish for whom" debates concerning the international demand for small pelagics competing with the fish as food [10].

4. CURRENT PREDOMINANT USE OF SMALL PELAGICS WORLDWIDE

More small pelagic fish are landed globally than any other type of fish. ("Landed" refers to the portion of the catch retained and not discarded at sea.) Large-scale trawling for pelagic species, with notable exceptions, is most often linked to supplying fish feed for industrial aquaculture [11]. Forage species, into which small pelagic fish fall, are among the most populous in the world, and demand for products derived from forage fish is increasing. Some would argue that forage species are at the heart of the human food system.

Small pelagics are used to fuel the growing global demand for animal protein as global incomes rise and diets change. Large-scale industries process forage fish using what is called "reduction", which involves cooking, grinding and chemically separating the fat from its protein and micronutrients. The fat and protein form key ingredients in the chemical inputs into animal feed for aquaculture, industrial livestock, chicken farming, a growing pet food sector and bait. The micronutrients such as nitrogen, phosphorus and potassium are converted into crop and vegetable fertiliser. This oil- and protein-rich meal becomes animal feed. The fats are used to make human nutritional supplements such as fish oil and omega-3 fatty acid tablets. Small pelagics are as important for human consumption as they are for the ecosystem, and the balance is important [12-16].

5. DEFINING SMALL PELAGICS

Small pelagics are small forage fish that live in the surface and near-surface waters over the continental shelf of most of the coast. Forage fish are small schooling fish that feed on plankton and occupy a vital place in marine foodwebs. Generally, these species eat herbivorous (phytoplankton) or carnivorous (zooplankton) plankton and are eaten by larger predators that occur higher up in the foodweb. Forage fish therefore play a fundamental role in marine ecosystems by converting energy from lower trophic levels into food for larger fish, marine mammals and seabirds—essentially all predators at higher trophic levels in the marine environment [17]. This paper focuses on small pelagic species in two case study sites in South Africa (marine—ocean fisheries) and Tanzania (inland—freshwater lake fisheries).

6. THE SOUTH AFRICAN WEST COAST SMALL PELAGIC FISHERIES

The South African, small pelagics (anchovy *Engraulis encrasicolus* and sardine *Sardinops sagax*), together with associated bycatch for red-eye round herring (*Eriumeus whiteheadii*) and cape horse mackerel (*Trachurus trachurus*), form a capital-intensive industrial fishery based on the west coast of South Africa in the southern Benguela region (Fig.1). Most of the small pelagic fishery infrastructure (the fishing fleet and major canning and fishmeal factories) is situated on the west coast in the greater Saldanha Bay, St Helena Bay and Velddrif areas. Since the 1940s, canning and fishmeal have been part of this sector's operations, and 6 canning and processing companies and 15 packing entities have been operational. Coastal communities have settled in and around this area, largely based on employment derived from the fishing industry [17,18].

Sardine catches have dominated to the extent that 60–90% of the pelagic industry in South Africa consisted of pilchard (sardine) and anchovy (see Fig. 2). The ratio between pilchard and anchovy landings has, however, fluctuated over the decades. Two collapses of the sardine fishery have been reported: in the early 1960s and 1997 due to intensive overfishing and in the early 2000s due to the regime shift of sardine to the southern Benguela coast [21, 22]. The significant shifts in winds in the El Niño years of 1983 and 1997 resulted in increased upwelling variability, causing a sudden upsurge in abundance of small pelagic in the southern Benguela.

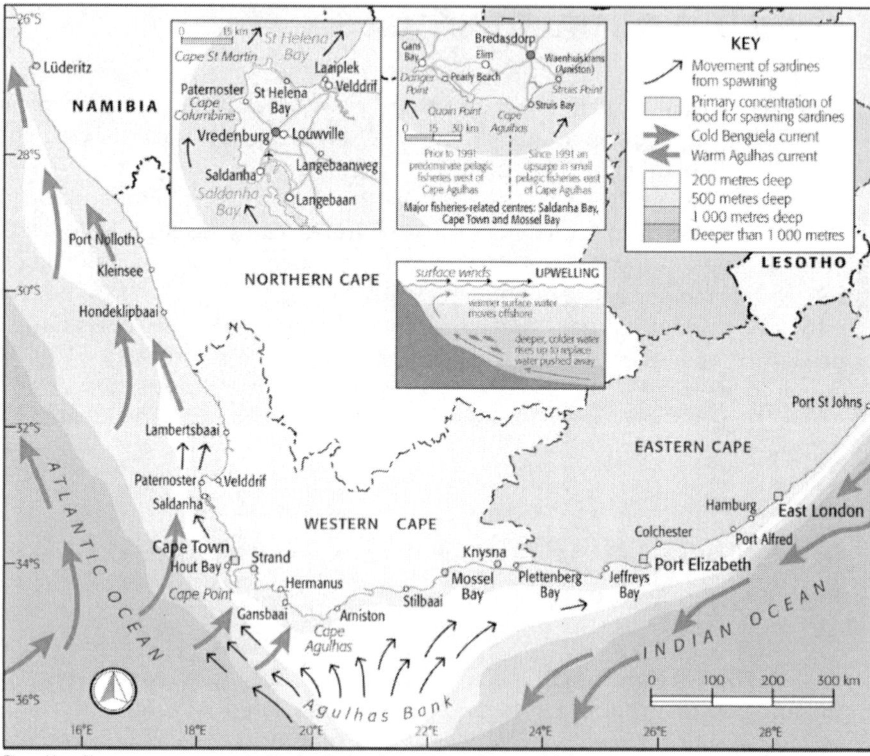

Figure. 1 Map of location, spawning grounds of Agulhas Bank, surface winds upwelling, spatial shifts

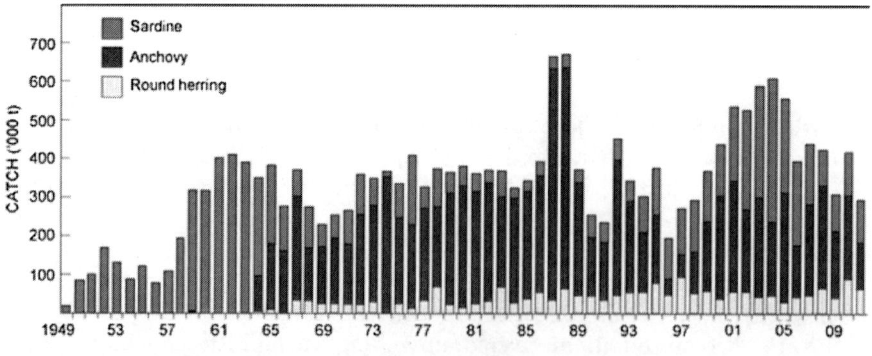

Figure. 2 Annual catches of sardine, anchovy and round herring taken by South African pelagics fishery from 1949 to 2011 [58: 46]

Views on what caused these environmental changes are contested, but El Niño, climate change, global warming and overfishing were frequently mentioned as factors that impacted on the productivity of the small pelagic resources, fishing possibilities and the distribution of the resource [23].

7. GOVERNANCE OF SOUTH AFRICAN SMALL PELAGIC FISHERIES

South Africa's fisheries are a right-based system of allocating Individual Transferable Quotas based on scientific inputs through Operational Management Procedures (OMPs) that sets annual TAC for anchovy and sardines. The Ecosystem Approach to Fisheries has governed the fisheries since 2002 [19, 20] supported by a consultative and advisory form of comanagement via the Scientific Working Group and Resource Management Working Groups, on which fisheries department scientists, University scientists, government officials, NGOs (WWF and industry associations) are represented.

8. THE LAKE VICTORIA SMALL PELAGIC FISHERIES

Lake Victoria is the largest fresh water body in Africa, and Kenya, Uganda and Tanzania share the management responsibilities (see Fig. 3). Commercially important species are Nile perch, dagaa and tilapia. The Nile perch and Nile tilapia were introduced to Lake Victoria in the 1950s and experienced explosive population growth in the 1970s, while negatively impacting the 300 indigenous species almost to extinction. This study specifically focuses on the small pelagic dagaa silver cyprinid (*Rastrineobola agrentea*), a species in the family of Cyprinidae, found in Lake Victoria [24]. The silver cyprinid is known by various local names such as *dagaa* in Tanzania, *mukene* in Uganda and *omena in* Kenya.

This fishery is a labour intensive, artisanal and small scale in nature using traditional methods of fishing in dug-out canoes, purse-seine and bright lanterns as fishing activity mainly takes place at night. When the fish is landed, it is sold to women who spread it out in the sun to dry for a period of 2 days. Drying is often done on sand, rocks and nets and some instances on drying racks [24, 25].

The Dagaa stocks have steadily increased, along with the catches, since the mid-1980s (see Fig. 4). Since 2005 it has become the largest fishery in the lake by weight and there are no signs of overexploitation.

As the stock size, based on hydroacoustic surveys, still appears to be increasing, this would indicate that there is still room for expansion of this fishery [26].

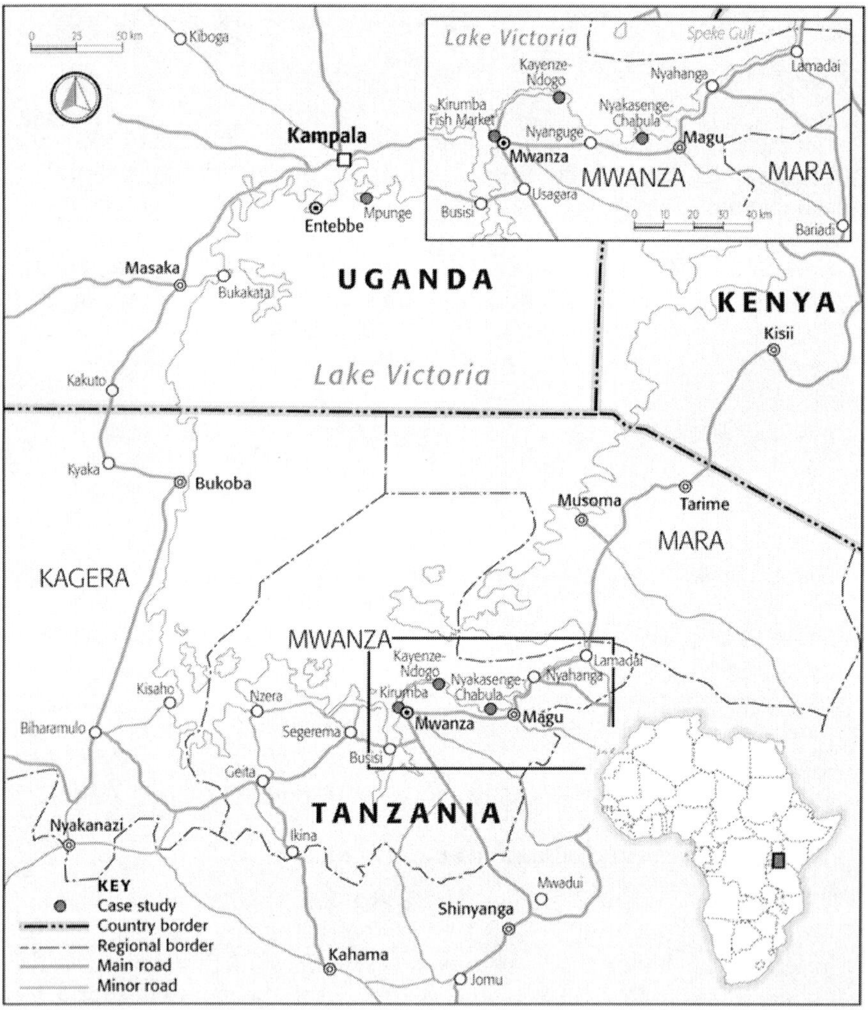

Figure. 3 Map of Lake Victoria, field sites in the Magu district, Mwanza Tanzania

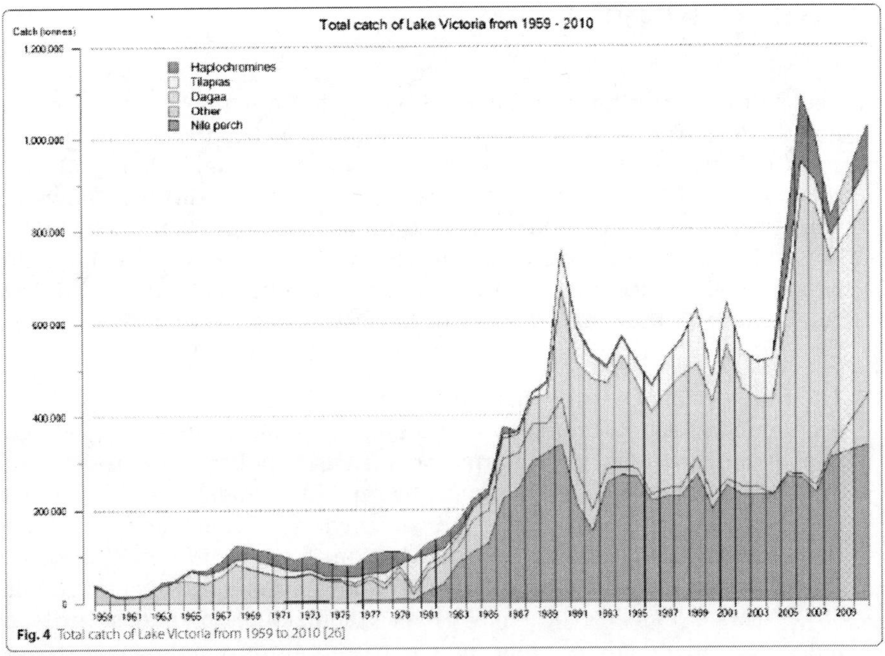

Figure. 4 Total catch of Lake Victoria from 1959 to 2010 [26]

9. GOVERNANCE OF DAGAA FISHERIES ON LAKE VICTORIA

The dagaa fishery is situated in traditional communal tenure systems that link the management of land, lake (water) and fish species. An open-access fisheries management system is still prevalent on Lake Victoria for the dagaa fisheries, with a comanagement system established with the creation of Beach Management Units (BMUs). The role of the BMU is mainly self-policing, but it is not actually part of the decision-making structures. Onyango 2011 [27] labels the comanagement introduced by the World Bank as a top-down system that lacks legitimacy and compliance at local level around the lake. More or less at the same time a regional management institution, known as the Lake Victoria Fisheries Organisation (LVFO), was established, and all states (Kenya, Tanzania and Uganda) that share the resource formed part of this structure to manage and develop the utilisation of the fisheries, as well as to coordinate research and management regulations [26, 27].

10. METHODS

The methodology for this paper is gleaned from literature review and fieldwork in Tanzania and South Africa on unpacking the contribution of small pelagics to fish as food or income. Data collection in South Africa for this paper is gleaned from key informant interviews and interviews with workers, unions and industry associations between 2012 and 2015. In Tanzania interviews were conducted with Beach Village Committee members, fisheries officials, cooperatives in the fish markets and fisheries researchers in the Magu district of Kayenze Ndogo, Nyakasenge Chabula, Kayenze and Kirumba fish market in October 2015 (Fig.3).

The frame of analysis for this paper is mainly gleaned from the social and economic indicators originally developed by Thomson in 1980 (see Fig. 5) to show how national and international fisheries policies tend to favour the large-scale sector through fuel subsidies and investments, while small-scale fisheries remain undervalued, data on them are often not reported, and they are often ignored in their contribution to food and nutrition security [28]. The key indicators for this paper aim to examine the extent to which large- versus small-scale small pelagics contribute to food security and nutrition, and are landed value, human consumption, employment, discards and fishmeal reduction. The main argument for this paper is based on the following key questions: What is the nutritional importance of small pelagics especially to the poor and vulnerable in the developing countries? To what extent are small and large-scale fisheries contributing to food security and nutrition? What percentage of small pelagics (marine and inland) is reduced to animal feed, aquaculture feed, fish oil, etc., in South Africa and Tanzania, and what percentage is directed for human consumption?

10.1 Contextualising food and nutrition security

10.1.1 Situating food security in context of the food system

The notions of food security and the nutritional needs of the poor need to be understood within the food system; therefore, access to food includes the availability, affordability and consumption practices of food. Linking food security to the food system needs to address the issue of what the economic, social and environmental drivers are that affect the food security of the poor [29]. At the centre of these concerns is the demand for high-quality fish at low prices [30], and a better understanding of consumption and purchasing practices among the global rich [31]. How does the demand for high-quality fish impact on the livelihoods and income of many small-scale fisheries around the world? As one of the most traded commodities in the world, the trade of fish is integrated into national economies and to the global economy through trade and investment rules and privatisation and is supported by technological advances. On the other hand, localisation is a process that reverses the trend of globalisation by discriminating in favour of the local [32,33]. Researchers [3,10, 34-37] warn us that the food quality standards imposed by international trade institutions and importing countries are excluding small-scale fishers from benefitting from the high-value fish trade. Hence, the many small-scale fisher organisations are promoting direct local trade instead of international trade.

Advantage ✓	Large scale company owned	Small scale artisanal
Number of fishermen employed	Around 2 million	Over 30 million ✓
Marine fish caught for human consumption	Around 40 million tonnes annually ✓	Around 40 million tonnes annually ✓
Capital cost of each job on fishing vessel	$$$$$$$$$ $$$$$$$$ $$$$$$$$ $30 000 to $300 000	$ $100 to $5 000 ✓
Bycatch discarded at sea	Around 5 million tonnes annually	Around 2 million tonnes annually ✓
Marine fish caught for industrial reduction to meat and oil, etc.	Around 15 million tonnes annually	Almost none ✓
Fuel oil consumption	30 to 40 million tonnes annually	3 to 15 million tonnes annually
Fish landed per tonne of fuel consumption	= 1 to 2 tonnes	= 3 to 15 tonnes ✓
Fishermen employed for each $1 million invested in fishing vessels	3 to 30	200 to 10 000 ✓

Figure. 5 Advantages for small-scale fisheries when comparing large-scale fisheries operations [updated from 3, 28, 53, 64]

Malnutrition can be defined as chronic food consumption deficiencies, or the excess consumption of nutritionally poor food, or the consumption of mainly industrialised, processed food, or an abnormal physiological condition caused by deficiencies, excesses or imbalances in energy, protein and/or other nutrients [2:4]. Malnutrition can include being under- or overweight, due to the correct nutrients not being consumed [39]. Malnutrition is not only caused as a result of a shortage of food, but is a result of a lack of nutritionally adequate food [38]. South Africa, a middle-income country, faces this "double burden" of malnutrition: both undernourishment and obesity [40]. The double burden is related to South Africa's extreme levels of inequality and the country's large poor population and is exacerbated by rapid change in the South African food system. Many poor people are increasingly distanced from self-sufficiency as they abandon local dietary foods and increase their dependence on processed and industrialised foods [41].Families eat the same very limited foods day in and day out. Highly processed starchy foods are prioritised for energy and take up substantial space on the plate. The animal proteins many families consume tend to be of very poor quality, for example, processed meat and chicken feet and heads with increased levels of salt and fat. "Women are carrying their poverty on their bodies" with the increased intake in starchy, sugary and processed cheap foods [42,26].

10.1.2 Value of small pelagics in nutrition

Fish provides an excellent solution for food insecurity and malnutrition, especially small pelagics that are rich in macro- and micronutrients. Fish is a particularly good source of protein for children, which is vital for their growth, and their underdeveloped digestive systems often cannot process the more starchy staple foods [43]. Fish in the human diet makes a significant contribution to meeting health needs, as it provides protein that contains all essential amino acids, lipids with essential omega-3 fatty acids and other vitamins and minerals [3]. It is crucial that fish protein of sustainably harvested species like small pelagics is directed towards human consumption [41]. Small pelagics provide easily digestible protein, and the bioavailability (the rate of absorption and availability of a substance to the human body) of the protein in fish is 5–15% higher than the protein from plant sources [3,43,44].

The nutritional benefits of small pelagic fish go beyond the importance of the role it plays as a source of protein. The lipid composition of fish is different from that of other proteins, and the long-chain, polyunsaturated fatty acids (LCPUFAs) have a number of potential health benefits [3]. The omega-3 docosahexaenoic (DHA) is also believed to be important during pregnancy, breastfeeding and infancy for neurodevelopment [3, 45]. This omega-3 fatty acid, an important building block for neural and brain development in young children, is also found in various vegetable oils, but in these it is a form of fatty acid (alphalinolenic acid—ALA) that needs to be converted into DHA to be effective. The human body is, in fact, inefficient at converting the ALA into the necessary DHA. Therefore, fish is a far more reliable source of DHA. For brain development in children to be optimal, it is believed that an intake of only 150 mg of small pelagic fish daily is necessary [3].

The omega-3 oils found in fish have also been found to reduce unhealthy cholesterol and triglycerides (both of which are causes of heart disease) as well as increase healthy cholesterol [3,46,47,48,49].

Fish is a source of lysine and methionine, which are essential micronutrients in the form of amino acids [3, 43]. Further micronutrients found in fish include Vitamins A, D, B1 and B2, as well as iron, phosphorous and calcium [3, 43]. Vitamin A is present in fish as retinol and an hydroretinol, which is more easily preserved through the cooking process and more readily absorbed by humans than the form of Vitamin A found in vegetables [50]. The calcium in fish, digested when the bones and organs are eaten, is even slightly more effective than milk as a source of calcium [50]. Fish is a rich source of iron and also aids the bioavailability of iron of other foods in a meal, such as rice [50]. The deaths of 800,000 children globally each year are attributed to zinc deficiency [3]. Small pelagic fish contain all of these micronutrients in significant amounts.

Over 30% of people in the world are reported to be iron deficient, including up to 50% of pregnant women and children in developing countries, with approximately two billion being anaemic [3, 38]. Lack of iron impairs cognitive development, growth and immune function, very often resulting in suboptimal performance at school and in adulthood [38]. Fish added to the traditional blend of staple diets will increase the utilisation of micronutrients, indirectly boosting the immune system. Vulnerable populations such as young children, the aged and immune-compromised persons living with HIV and AIDS would benefit significantly from introducing small pelagics into their diet [5]. More importantly, fish as food is separated from the consumption of other foods produced on the land.

10.2 Findings

The consumption of fish contributes approximately 17% to the global intake of protein. In some countries in Africa, particularly West Africa, this can be significantly higher—up to 72% in Sierra Leone and 55% in Ghana and the Gambia [3]. The fish frequently consumed by local populations in Africa is the small pelagic fish, often regarded as a lower-value fish in terms of international trade.

10.2.1 Difficulties of small pelagics for human consumption

Small pelagics with their high fat content tend to spoil easily, which results in many post-harvest losses during handling, sanitation and drying. When small pelagics are placed directly on the soil, they become contaminated and become rancid, and placing fish on mats, concrete or drying racks improves the quality of the product [50]. In small pelagics, rancidity is caused by the oxidation of fish lipids, which leaves a bad odour and bitter taste. Smoking and salting small pelagics are not very popular in most parts of Africa. Salt is very expensive, and it changes the product, because, as with smoking, the species is easily charred [50] (blackened). Canning is one way of preserving the fish for human consumption, but is expensive, and canning is mostly confined to South Africa and Namibia.

10.2.2 Advantages of small-scale fisheries in small pelagic fishing

Large-scale pelagic trawls straddle temperate and tropical waters to harvest large shoals of small pelagic species within the Economic Exclusive Zones (extending 200 nautical miles of ocean space) many countries. Such operations are undertaken on a perennial basis. Along with the target pelagic species, significant numbers of larger-sized fish and marine mammals are also harvested, since the former are prey for the latter. Small-scale fishers would have caught these larger fish using more selective methods during the appropriate seasons. This again results in competition with local small-scale fishers, who are also often totally marginalised. In many quota-managed fisheries, these larger species are considered overexploited and hence denied to small-scale fishers, but the "incidental" catch by trawlers is often overlooked or condoned [43]. In these debates on fish as food, small-scale fisheries play the role of being more affordable, available and accessible to many poor populations and important to their livelihood [3]. The direct and indirect contribution to food security from small-scale fisheries outweighs the contribution of large-scale fisheries. The series of socio-economic indicators developed by Thomson in 1980 has been updated several times over the past 36years [28,51,52,53,54,55] to compare the contribution of large- versus small-scale fisheries, to address global estimates of employment, catches, bycatch and discards, human consumption and employment created through investments.

10.3 Case studies

10.3.1 The case of sardine and anchovy in South Africa

The economic boom for purse-seining started off along the underdeveloped west coast of South Africa in the late 1930s and significantly increased with the post-Second World War demand for canned fish, technological advances and high unemployment rates [56]. This sector provided increased employment opportunities for many locals in the area [57]. From the 1950s to the 1980s, during the closed seasons between August and December, workers engaged in alternative livelihoods, i.e. fishing for west coast rock lobster and handline as a form of subsistence [57].

These fish are targeted by around 100 purse-seiners (dragnet fishing boats) that catch up to 400 tonnes a single haul. Anchovy and most of the round herring caught are reduced to fishmeal and fish oil in industries along the west coast. Sardines (also known as pilchards) are caught on the west and south coast of South Africa and canned or frozen locally for human consumption, pet food and bait. Established in the 1940s, the South African small pelagic industry, of which sardine is one of the main species fished, is the largest by landed volume and second only to demersal hake in terms of value [15, 58, 59, 60]. The employment in the sector is around 4500, but because operation is seasonal, it is difficult to differentiate employment that is permanent, temporary or seasonal.

The number of operational fishing vessels in the sector has declined by nearly one-third in the past decade, mainly as a result of most independent vessel owners leaving the industry. Also, factories have bought new and bigger boats following the assurance of being granted medium- and long-term rights. Further, owners have replaced old wooden vessels with larger, more fuel-efficient steel or fibreglass vessels with refrigeration systems. The maximum number of vessels allowed in the sector is about 100 [20]. The number of vessels operating at any given time is largely influenced by the annual quota size; for example, only 71 vessels were declared operational for the 2011 season, based on permits issued. The use of fewer, more efficient, mechanised vessels has resulted in job losses for skippers and crew [60]. Smaller operators off the Gansbaai coast have been pushed out by larger companies, because when there is an oversupply, large companies prioritise their own vessels over contracted vessels [18, 60]. Increased competition for fish and crew employment to land an increased quantity and quality of catch has lead to the closing of fishing areas, high grading and dumping [60].

10.3.2 Use of sardine in South Africa

Canned sardine (pilchard) is one of the most important food items in the diets of South Africans, especially of the poor. The most common form is canned pilchards which come in mostly three recipes: in tomato sauce, in chilli sauce or in brine. Ease of storage and transportation and consistent quality count in the favour of canned sardines or pilchards. Another advantage is that canned sardines come in small tins of 155, 215 or 400g, all priced below one US dollar equivalent in South Africa. They can be eaten with bread, rice, potatoes, *pap* (maize meal) or even on their own. Canned pilchards can be found in all corners of South Africa, right into rural areas.

The increasing expansion of South African retail chains into Africa is increasing the distribution and reach of food items such as canned sardines in the rest of Africa. Sardines are rich in micronutrients, vitamins and proteins (e.g. vitamins D and B12, protein and calcium). They are known to be one of the most concentrated sources of the omega-3 fatty acids, which lower triglycerides and cholesterol levels. There is no doubt therefore that sardine is vital for protein food security for the majority of South Africans, and increasingly Southern Africans, especially the poor [3]. It is no surprise that the sardine catches are caught at almost full capacity of the total allowable catch (TAC),see Fig.6.

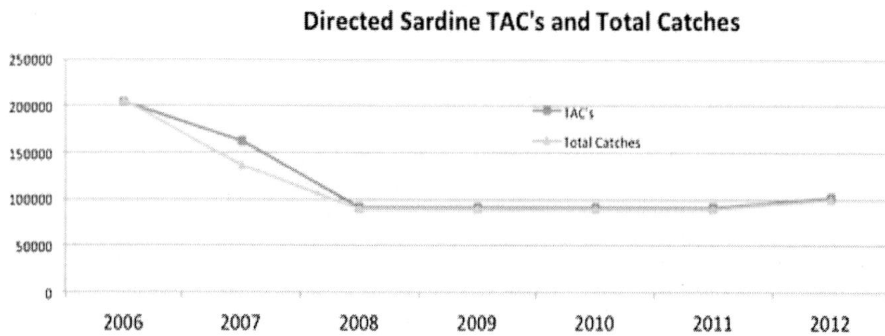

Figure. 6 Directed sardine total catches [61]

10.3.3 Use and underuse of anchovy in South Africa

Since the 1970s, South Africa's average annual anchovy landings have exceeded 300,000 tonnes. Some of the products produced earlier in South Africa were anchovy salted, dried or cured in oil, and anchovy paste. With the decrease in the average size of the anchovies (less than 12 cm) in the 1980s, there was an increase in the manual labour required to prepare each fish individually, with a high risk of oxidation and rancidity, and the industry resorted to reducing far more anchovy for fishmeal [61, 62, 63]. Figure 7 shows the annual catches of anchovy since 2006, and for a number of years less than 50% of the TAC have been caught [61].

The undercatching of anchovy is a concern within the Department of Agriculture, Forestry and Fisheries in South Africa, as there is a need for food security and job creation in this sector. To mitigate this, in 2012–2013 Ukloba Fishing experimented with a new process of salting, curing and filleting anchovy for human consumption. The outcomes of this experiment were interesting, because filleting anchovy is labour intensive and therefore has job-creation potential. However, the company decided not to continue as the profitability would be low due to the labour costs [61]. In the end, mechanisation of the anchovy product for animal feed and fish oil seems to be the choice of many fishing companies in this sector.

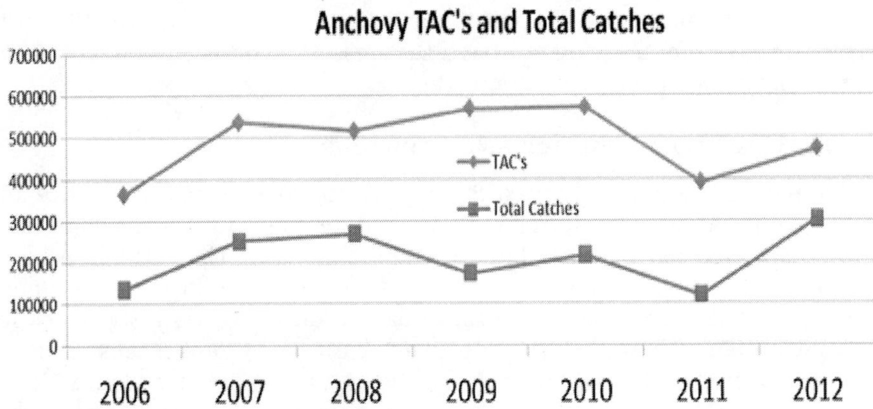

Figure. 7 Anchovy total catches [61]

What is the role of small-scale fisheries in the product development of anchovy fillets for the higher end market [61, 62, 63]? Small-scale fishers do not target small pelagics due to the vessel size required and distance of the fishing from shore. When sardine shoals are swimming within the small-scale fishing zone, fishers will target them, but will use them for bait rather than consumption. Even small-scale fisher communities are far more likely to obtain their sardines from a tin than fresh.

10.3.4 The case of the dagaa fisheries in Lake Victoria, Tanzania

Lake Victoria is Africa's largest lake, covering an area of 68,800 square km, and the lake is shared between Tanzania 51%, Uganda 43% and Kenya 6%. The maximum depth of the lake is 84 m, and it has a shoreline of 3450 km. The lake supports 30 million people living in the basin and is a major source of food, livelihoods and employment. Nile perch and Nile tilapia were introduced in the lake in the 1950s, which negatively impacted over 300 indigenous species to near extinction [64, 65].

10.3.5 Dagaa fishing: methods and challenges

The dagaa is a low-value species that is largely consumed around the lake and inland rural areas and animal feed (fishmeal). Trading dagaa on the global market is major challenge, as the post-harvest handling does not meet the minimum hygiene standards. Dagaa fishing is largely subsistence in nature due to the importance of fish consumption for many poor households around the lake, and in Eastern and Southern Africa, and has been a livelihood and income source for many small-scale fishers since the early 1970s [24]. The fish is caught at night with bright lanterns around the dug-out canoe by artisanal and small-scale fishers using traditional methods. When the fish is landed, it is sold to women who spread it out in the sun to dry for a period of 2 days. Drying is often done on sand, rocks and nets and some instances on drying racks.

10.3.6 Loss of dagaa to human consumption

The nature of the post-harvest handling of the fish, no cold chain (ice and refrigeration) and few drying racks mean that dagaa fishing does not meet the minimum hygiene standards to export globally. The Kirumba market sells mainly to local and regional markets, and currently most—almost 80%—of the dagaa sold at the Kirumba market in Mwanza is reduced for animal feed. The product is packed with sand and shells and for transportation in huge quantities to agents for the animal feed industries or to individual poultry farmers, mainly in the region, for example to Kenya, Burundi, Rwanda, Congo and within Tanzania [24]. There are new markets in the Far East such as Malaysia and Cambodia that import dagaa from Kirumba market [24, 25] in small quantities for their aquaculture production [24,25].

10.3.7 Potential value of dagaa fishing

In 2007, the estimated harvest level for the three riparian countries of Tanzania, Uganda and Kenya reached approximately 600,000 tonnes annually [65]. Since 2005, dagaa has become the largest fishing focus in the lake by weight, and there are no signs of overexploitation. Scientists and government officials agree that there is room for expansion in this arena of fishing [25, 26]. The development goals for the dagaa fisheries in Lake Victoria are to provide food security, employment and wealth creation. In line with these goals is the formalisation of the value chain to create more wealth and employment.

10.3.8 Nutritional value of dagaa

The nutritional value of dagaa is similar to that of marine small pelagic fish, in that they contain protein and micronutrients essential in the diet of growing children, pregnant and breastfeeding women and undernourished and malnourished communities. However, the handling and processing of the fish is key in ensuring the nutrition value is sustained [24].

11. DISCUSSION

11.1 Benefits of small-scale fisheries

Globally, the landed value of large-scale fisheries is around 50 million tonnes annually, while for small-scale fisheries it is 40 million tonnes. About 8% of the catch is dumped before landing, and global discards are around 7 million tonnes annually [3]. Small-scale fisheries generate less wastage in the form of discards. There are many artisanal fishers, for example, in Mozambique that collect discards from large-scale vessels [3, 66]. In the large-scale fisheries sector, huge quantities of small pelagic fish are reduced to fishmeal and fish oil. About 27 million tonnes of fish (34 per cent of the world fisheries catch) are processed into fishmeal, oil and pet food, etc., annually [2, 7, 39, 67]. In the case study of South Africa, 100% of the anchovy and red-eye round herring are reduced to fishmeal, fish oil and pet food.

More of the small pelagic fish, even that reduced to fishmeal, could be used for human consumption. In Southeast Asia, specifically Thailand, most of the fishmeal is converted to products for human consumption. In the case of the dagaa fisheries, tonnes are reduced to fishmeal for chicken feed, and increasing amounts are exported to the new Far East markets for aquaculture production. This is of great concern as an important protein and micronutrients are reduced to protein for animal feed for foreign direct income, wealth creation and job creation, while millions of poor people around Lake Victoria are at risk from malnourishment.

Most of the fish—90–95% [68]—caught by small-scale fisheries is destined for human consumption, while only 80% of the large-scale fisheries' landings goes to human consumption [3]. Although large-scale fisheries produce four times the catch, the same amount in tonnes as for small-scale fisheries goes for human consumption. Mills et al. [68: 8] state that "an estimated 94% (13 million tonnes) of small-scale inland production is consumed within the country of origin". Ultimately, both large-scale and small-scale fisheries make an important contribution to food security. The dagaa fish are consumed by the poor rural and urban people of the riparian states around the Lake Victoria, and they are clearly an important protein and micronutrient source to many. The sardine canning industry plays a significant role in providing nutritious meals to many urban and rural poor in South Africa. The trade of canned fish from South Africa also extends to Lesotho, Swaziland, Namibia, Zimbabwe and Malawi.

In terms of employment, however, over 90% of the fishers and fish workers are employed in small-scale fisheries. Employment in small-scale fisheries is several times higher per tonne than in large-scale fisheries [69]. An estimated 35 million commercial fishers are engaged in harvesting operations in the developing and developed countries [3].

Kurien's work on food security and trade makes the distinction between Bluefin tuna as fish for "luxury consumption" and anchovies as fish for "nutritional consumption" [43: 5]. The trade link to luxury consumption in Europe, Japan and the USA [43] results in the increase of income to small-scale fishers. In Tanzania, this rationale results in the high-value species Nile perch (fillet of white flesh fish) being exported to Europe and USA, while the bones and heads remain for locals to eat. This trend of increasing the amount of dagaa for animal feed will make the species more desirable for foreign income and compromise the protein and micronutrients available to the local poor. The food and nutrition needs of the poor in developing countries are often compromised by the need for income from trade and foreign investment, especially with the higher-value species [10]. The poor, who often harvest the fish, are not able to consume it. Protecting nutritionally important species for local consumption, rather than reducing it to animal feed, would be a first step in addressing the malnutrition of many vulnerable poor in Africa and especially Southern Africa. In addition, the small pelagics sector shows that the focus on efficiency, profit and the mechanisation of vessels and factories have failed dismally in creating employment, because these imperatives cause the shedding of jobs or zero growth in new jobs.

11.2 Relook at what economic viability means for small-scale fisheries

The next step is for a better understanding of the concept of economic viability for small-scale fisheries is in their contribution to food security and fish consumption. In analysing economic viability, Schuhbauer and Sumaila expanded the socio-economic indicators of the Thomson scale to include: the degree of economic dependence on fishing; the number of jobs per employment type; the fair distribution of benefits, subsidies, costs and cost structure and fish consumption; and governance indicators [55].

In Tanzania, the case of dagaa shows that the nutritional importance of the fish to food security is compromised by the safety and post-harvest handling of the fish, which causes almost 80% of the landed volume to be reduced to fishmeal. The sustainability of the resource is also threatened by the increasing use of motorised vessels and the decreasing use of dug-out canoes and sailboats. In South Africa, almost 100% of anchovy is reduced to animal meal. More investment in post-harvest handling, safety and improved hygienic standards, would allow a greater direction of dagaa towards human consumption. In South Africa, with the targeting of larger anchovy (10–12 cm), workers will need up-skilling in the process of deheading and cleaning the anchovy for human consumption. This processing will be labour intensive, which brings opportunities for job creation. It will also bring new economic opportunities for product development in small-scale fisheries. At the same time, there is potential for addressing malnutrition [62].

Despite these opportunities for job creation and food security, it seems that the investment from the industry in further developing this sector is conditional upon government support and economic incentives. The small-scale sector could possibly benefit from product development and incentives from the state.

11.3 Small-scale fisheries data—separate but equal

Following on the need to define economic viability for small-scale fisheries, Pauly and Zeller reconstructed fisheries data and found many gaps in the data of the FAO especially when it comes to small-scale fisheries [70]. The data of small-scale fisheries and also subsistence fisheries are often not reported or are underestimated, ignored and rarely included in the national catch statistics [3]. This is especially important for small-scale fisheries and food security in particular inland fisheries and in rural household diets [3]. Pauly and Zeller call for a policy change that would be easy for "the FAO to coordinate and implement with all countries around the world [which] is to request countries to submit their annual catch statistics separately for large-scale and small-scale fisheries, which would be an excellent contribution towards the implementation of the *Voluntary Guidelines for Securing Sustainable Small-scale Fisheries in the Context of Food Security and Poverty Eradication*" [70: 5].

11.4 Implementation and use of the soft laws

The implementation and use of the soft laws (Tenure Guidelines, Right to Food Guidelines, Small-scale fisheries Guidelines [6, 7, 9]) should go further than only raising the profile and importance of small-scale fisheries rights to food and nutrition security, and protection of their livelihoods, by ensuring that these rights are protected in national policies and that the rights are realised. These rights and the livelihoods, lives, shelter and food of many small-scale fishers are also threatened and infringed on a daily basis by access to resources by coastal development, tourism, mining and large-scale acquisitions.

12. CONCLUSION

Small pelagic fish (marine and inland) play a critical role in the nutrition of many vulnerable and poor communities, especially in Africa. Mechanisation, overfishing and fishmeal production for cultured fish and animal feed all compromise the key position of small pelagics as food for the many malnourished and vulnerable coastal and inland populations.

The often ignored, undervalued, underestimated role and position of small-scale fisheries is gaining traction with support from the FAO voluntary guidelines (right to food, tenure and small-scale fisheries). Academics, conservation NGOs, donors and funders are all taking a keen interest in this sector. Space for action has been created for small-scale fishery civil society movements, both internationally and locally, to carefully align these soft law tools to the nature and structure of small-scale fisheries. This can be seen in the rights system, the management system (Territorial User Rights System) and the creation of marine protected areas in small-scale fishing zones. To fully realise the critical role small-scale fisheries play in food security and nutrition, they will need their governments' support in the development of these sector legislation investments.

Large-scale and small-scale fisheries play an important role in providing food security, as shown in the case of canning in South Africa and dagaa in Tanzania for human consumption. In cases like South Africa, where canning is providing high-quality protein, further investment is needed in the research and development of the role of small-scale fisheries in the use of the anchovy species for human consumption. Supporting investments in local markets and the local economy, and creating local value chains will be important in providing indirect food security through income and livelihood. The case of dagaa in Tanzania shows the need for post-harvest investments to increasing the quality of the fish for human consumption. The complex ecosystem and economic roles of small pelagic fish necessitate a holistic view of their dynamics [16, 17, 71].

ACKNOWLEDGEMENTS

Dr. Paul Onyango is my research partner in Tanzania, and he was responsible for organising the fieldtrips and field interviews in Mwanza, Lake Victoria, Tanzania. This paper is part of the TBTI fish as food research cluster. Funding was provided by National Research Foundation (RSA) (Grant No. 86909).

REFERENCES

1. UNICEF. Improving child nutrition. The achievable imperative for global progress. United Nations Children's Fund. 2013. http://www.data.unicef.org/corecode/uploads/document6/uploaded_pdfs/corecode/NutritionReport_April2013_Final_29.pdf. Accessed 28 March 2016.
2. Allison E. Aquaculture, fisheries, poverty & food security. Working paper 2011-65. Penang: The World Fish Center. 2011.
3. HLPE. Sustainable fisheries and aquaculture for food security and nutrition. A report by the High Level Panel of Experts on Food Security and Nutrition of the Committee on World Food Security. Food and Agriculture Organization of the United Nations. Rome: High Level Panel of Experts. 2014.
4. Thompson B, Amoroso L. Combating micronutrient deficiencies: food based approaches. Food and Agriculture Organization of the United Nations. 2011. ISBN-13: 978 92 5 106546 4.
5. Scheiber HN, Paik J. Regions, institutions and law of the sea. Leiden: Martinus Nijhoff Publishers; 2013.
6. FAO. Voluntary Guidelines on the Responsible Governance of Tenure of Land, Fisheries and Forests in the context of National Food Security. Committee on World Food Security. Rome: Food and Agriculture Organization of the United Nations. 2012.
7. FAO. Voluntary guidelines to support the progressive realization of the right to adequate food in the context of national food security. Rome: Food and Agriculture Organization of the United Nations; 2005.
8. De Schutter O. Note to the Secretary-General from the Special Rapporteur on the right to food. New York: United Nations, Sixty-seventh session General Assembly; 2012.
9. FAO. Voluntary guidelines for securing sustainable small-scale fisheries in the context of food security and poverty. Rome: Food and Agriculture Organization of the United Nations; 2014.
10. NORAD, FAO. A value-chain analysis of international fish trade and food security with an impact assessment of the small-scale sector: Summary report. Norwegian Agency for Development Cooperation and Food and Agriculture Organization of the United Nations. 2013.
11. Naylor R, Burke M. Aquaculture and ocean resources: raising tigers of the sea. Annu Rev Environ Resour. 2005;30:185–218.
12. Schleifer D. The fish at the heart of the food system. 2016: [Internet] 2016. http://limn.it/the-fish-at-the-heart-of-the-food-system/. Accessed 20 March 2016.
13. Tidwell J, Allan G. Fish as food: aquaculture's contribution. EMBO Rep. 2001;2(11):958–63. doi:10.1093/embo-reports/kve236.
14. Alder J, et al. Forage fish: from ecosystems to markets. Annu Rev Environ Resour. 2008;33:153–66. doi:10.1146/annurev.environ.33.020807.143204.
15. Tacon A, Metian M. Fishing for feed or fishing for food: increasing global competition for small pelagic forage fish. Ambio J Hum Environ. 2009;38(6):294–302. doi:10.1579/08-A-574.1.
16. Pikitch E, Boersma PD, Boyd IL, Conover DO, Cury P, et al. Little fish, big impact: managing a crucial link in ocean food webs. Washington, DC: Lenfest Ocean Program; 2012.
17. Coetzee J, Badenhorst A. Status and management of the South African small pelagic fishery—August 2013. Cape Town: Oceana Fishing Group;2013.
18. Hara M, Rogerson J, de Goede J, Raakjær J. Fragmented participation in management of the fishery for small pelagic fish in South Africa—inclusion of small-rights holders is a complex matter. Afr J Marine Sci. 2014;. doi:10.2989/1814232X.2014.930708.
19. Shannon LJ, Cochrane KL, Moloney CL, Fréon P. Ecosystem approach to fisheries management in the southern Benguela: a workshop overview Afr J Marine Sci. 2004;26(1):1–8. doi:10.2989/18142320409504046

20. RSA—Republic of South Africa. General Policy on the Allocation and Management of Long-term Commercial Rights: 2005. Pretoria: Minister of Environmental Affairs and Tourism
21. Fairweather T, Hara M, van der Lingen C, Raakjær J, Shannon L, Louw G, Degnbol P, Crawford R. A knowledge base for management of the capital-intensive fishery for small pelagic fish off South Africa. Afr J Marine Sci. 2006;28(3–4):645–60
22. Coetzee JC, van der Lingen CD, Hutchings L, Fairweather TP. Has the fishery contributed to a major shift in the distribution of South African sardine? ICES J Marine Sci. 2008;65:1676–88.
23. Isaacs M. Climate variability, overfishing and transformation in the small pelagics sector in South Africa. In: Guillotreau P, Bundy A, Perry I, editors. Global change in marine systems: societal and governing responses (forthcoming). Routledge Studies in Environment, Culture, and Society. Taylor Francis Group; 2017
24. Legros D, Luomba J. Dagaa value chain analysis and proposal for trade development. IOC-Smart fish programme technical document.2011.
25. Luomba J, Onyango P. Dagaa fishery, the unknown wealth of Lake Victoria. Visible possibilities: The economics of sustainable fisheries, aquaculture and seafood trade. In: International Institute of Fisheries Economics and Trade (IIFET) 2012 Tanzania Proceedings. Dar res Salaam, Tanzania. 2012.
26. Kolding J, Medard M, Mkumbo O, Zwieten PAM. Status, trends and management of the Lake Victoria Fisheries. In: Welcomme RL, Valbo- Jørgensen J, Halls AS, editors. Inland fisheries evolution and management— case studies from four continents. FAO Fisheries and Aquaculture Technical Paper 579; 2014.
27. Onyango P. Governance of tenure in the Lake Victoria fisheries, Tanzania. Land Land Tenure J. 2013; 1
28. Thomson D. Conflict within the fishing industry. ICLARM Newslett. 1980;1980(3):3–4.
29. Ericksen PJ. Conceptualising food systems for global environmental change research. Glob Environ Change. 2008;18(1):234–45. doi:10.1016/j. gloenvch.2007.09.002.
30. Freidberg SE. Culture, conventions and colonial constructs of rurality in south–north horticultural trades. J Rural Stud. 2003;19(1):97–109. doi:10.1016/S0743-0167(02)00037-2.
31. Kneafsey M, Dowler E, Lambie-Mumford H, Inman A, Collier R. Consumers and food security: Uncertain or empowered? J Rural Stud. 2013;29:101– 12. doi:10.1016/ j.jrurstud.2012.05.005.
32. Hines C. Localization: a global manifesto. London: Earthscan; 2000.
33. Hinrichs CC, Sterling VA. The practice and politics of food system localization. J Rural Stud. 2003;19(1):33–45. doi:10.1016/S07430167(02)00040-2.
34. Henson SJ, Brouder AM, Mitullah W. Food safety requirements and food exports from developing countries: the case of fish exports from Kenya to the European Union. Am J Agric Econ. 2000;82(5):1159–69.
35. Henson SJ, Mitullah, W. Kenyan exports of Nile Perch: impact of food safety standards on an export-oriented supply chain. World Bank Policy Research Working Paper, 3349. Washington, DC: World Bank. 2004.
36. Mansfield B. Fish, factory trawlers, and imitation crab: the nature of quality in the seafood industry. J Rural Stud. 2003;19(1):9–21. doi:10.1016/ S0743-0167(2)00036-0.
37. Gibbon P, Ponte S. Trading down: Africa, value chains and the global economy. Philadelphia: Temple University Press; 2005.
38. Kambewa E, Ingenbleek P, Tilburg A. Improving income positions of primary producers in international marketing channels: the Lake Victoria–EU Nile Perch case. J Macromarketing. 2008;28(1):53–67. doi:10.1177/0276146707311305.
39. FAO. The right to food. United Nations General Assembly. New York: Food and Agriculture Organization of the United Nations; 2012.
40. Monteiro CA, Moura EC, Conde WL, Popkin BM. Socioeconomic status and obesity in adult populations of developing countries a review. Bull World Health Organ. 2004;82(12):940–6. doi:10.1590/ S0042-96862004001200011.

41. Isaacs M. The governability of small-scale fisheries food system in South Africa—The case of snoek and West Coast rock lobster in Ocean View. In: Jentoft S, Cheunpagdee R, editors Interactive governance for small-scale fisheries: global reflections; 2005, p. 245–262. doi: 10.1007/978-3-319-17034-3_13
42. PACSA. Food Price Barometer. Annual Report. Pietermaritzburg Agency for Community Social Action 2015.
43. Kurien J. Fish trade for the people: toward understanding the relationship between international fish trade and food security. Report of the Study on the impact of international trade in fishery products on food security. Food and Agriculture Organization of the United Nations and the Royal Norwegian Ministry of Foreign Affairs; 2004.
44. Bell J, Kronen M, Vunisea A, Nash W, Keeble G, Demmke A. Planning the use of fish for food security in the Pacific. Marine Policy. 2009;33(1):64–76. doi:10.1016/j.marpol.2008.04.002.
45. Mozaffarian D, Rimm EB. Fish intake, contaminants, and human health: evaluating the risks and the benefits. J Am Med Assoc. 2006;296(15):1885–99. doi:10.1001/jama.296.15.1885.
46. Seedat A. Know your fats and oils. The Heart and Stroke Foundation article. 2014. http://www.heartfoundation.co.za/topical-articles/ know-your-fats-and-oils.
47. Mozaffarian D, Lemaitre R, Kuller L, Burke G, Tracy R, Siscovick D. Cardiac benefits of fish consumption may depend on the type of fish meal consumed: the Cardiovascular Health Study. Circulation. 2003;107(10):1372– 7. doi:10.1161/.CIR.0000055315.79177.16.
48. Sidhu K. Health benefits and potential risks related to consumption of fish or fish oil. Regul Toxicol Pharmacol. 2003;38(3):336–44. doi:10.1016/j. yrtph.2003.07.002.
49. FAO/World Bank/WorldFish Center. Small-scale capture fisheries: a global overview with emphasis on developing countries. A Preliminary Report of the Big Numbers Project. Food and Agriculture Organization of the United Nations, World Bank, WorldFish Centre 2008.
50. Thilsted S, Roos N, Hassan N. The role of small indigenous fish species in food and nutrition security in Bangladesh. Naga ICLARM Q. 1997;20(3–4):13–5.
51. Kumulo-Johnson CA, Ndimele PE. A review of post-harvest losses in some African Countries. J Fish Aquatic Sci. 2011;6(4):365–78. doi:10.3923/ jfas2011365-378.
52. Berkes F, Mahon R, McConney P, Pollnac R, Pomeroy R. Managing smallscale fisheries. Alternative directions and methods. Ottawa: International Development Research Centre (IDRC); 2001
53. Pauly D. Major trends in small-scale marine fisheries, with emphasis on developing countries, and some implications for the social sciences. MAST. 2006;4:7–22.
54. Kolding J, Béné C, Bavinck M. Small-scale fisheries—importance, vulnerability and deficient knowledge. In: Garcia S, Rice J, Charles A, editors. Governance for marine fisheries and biodiversity. London: Wiley; 2004. doi:10.1002/978118392607.ch22.
55. Schuhbauer A, Sumaila R. Economic viability and small-scale fisheries: a review. Ecol Econ. 2016;124:69–75. doi:10.1016/j.ecolecon.2016.01.018.
56. Crawford RJM. Food, fishing and seabirds in the Benguela upwelling system. J Ornithol. 2007;148(Suppl 2):S253–60. doi:10.1007/ s10336-007-0228-z.
57. Jarre A, Ragaller SM, Hutchings L. Long-term, ecosystem-scale changes in the southern Benguela marine pelagic social-ecological system: interaction of natural and human drivers. Ecol Soc. 2013;18(4):55. doi:10.5751/ ES-05917-180455.
58. DAFF. Status of the South African Marine Fishery Resources 2012. Republic of South Africa. Department Agriculture, Forestry and Fisheries. 2012.
59. DAFF. Status of the South African Marine Fishery Resources 2014. Republic of South Africa. Department of Agriculture, Forestry and Fisheries. 2014.
60. Hara M. Efficacy of rights-based management of small pelagic fish within an ecosystems approach to fisheries in South Africa. Afr J Marine Sci. 2013;35:315–22. doi:10.1080/03768350903181423.

61. De Goede J. Current fisheries management measures. Presentation at the anchovy for human consumption workshop, 12–13 March 2013 Marine Resources Management. Department of Agriculture, Forestry and Fisheries (DAFF). 2013.
62. Backteman K. Anchovy from fishmeal to fish for people. 2010.[Internet] 2016. http://www.efdinitiative.org/our-work/policy-interactions/anchovyfishmeal- food-people. Accessed 15 April 2016.
63. Japp D. Feasibility of Utilising South African Anchovy for Human Consumption. Capfish. Presentation at the Anchovy for Human Consumption Workshop, 12–13 March 2013.
64. World Bank and Food and Agriculture Organization of the United Nations. The Sunken Billions: the economic justification for fisheries reform. Washington DC: World Bank, Agriculture and Rural Development Department, Sustainable Development Network. 2009, p 130.
65. LVFO. Regional Status Report on Lake Victoria Bi-ennial Frame Surveys between 2000 and 2010. Kenya, Tanzania and Uganda. Supported by IFMP EDF Project Number 8 ACP/ROR/029 and GEF/LVEMP Phase II Project, Jinja, Uganda. Lake Victoria Fisheries Organisation. 2010.
66. Bene C, Macfadyen G, Allison E. Increasing the contribution of small-scale fisheries to poverty alleviation and food security. FAO fisheries technical paper 481. Rome: Food and Agriculture Organization of the United Nations. 2007.
67. Bostock J, McAndrew B, Richards R, Jauncey K, Telfer T, Lorenzen K, Little D, Ross L, Handisyde N, Gatward I, Corner R. Aquaculture: global status and trends. Philos Trans R Soc. 2010;365:2897–912. doi:10.1098/ rstb.2010.0170.
68. Mills DJ, Westlund L, De Graaf G, Willmann R, Kura Y, Kelleher K. Underreported and undervalued: small-scale fisheries in the developing world. In: Andrew NL, Pomeroy R, editors. Small-scale fisheries management: frameworks and approaches for the developing world. Wallingford: CABI; 2011.
69. Béné C, Hersoug B, Allison EH. "Not by rent alone": analysing the pro-poorfunctions of small-scale fisheries in developing countries. Dev Policy Rev. 2010;28:325–58. doi:10.1111/j.1467-7679.2010.00486.x.
70. Pauly D, Zeller D. Catch reconstructions reveal that global marine fisheries catches are higher than reported and declining. Nat Commun. 2016;. doi:10.1038/ncomms10244.
71. Checkley DM, Alheit J, Oozeki Y, Roy C. Climate change and small pelagic fish. Cambridge: Cambridge University Press; 2009.

CHAPTER 9

Growth Rate Assessment of Alcyonacean Sarcophyton glaucum from Northern Hurghada, Red Sea, Egypt

Tarek A. A. Mohammed[1*], Abdel-Hamid A. Ali[2], Mohamed M. El-Komi[3], Mohammed A. H. Ezz El-Arab[2], Fayez A. M. Shoukr[4]

[1]*Marine Invertebrates Department, National Institute of Oceanography & Fisheries (NIOF), Hurghada Branch, Hurghada, Egypt*

[2]*Hydrobiology Department, National Institute of Oceanography & Fisheries (NIOF), Suez Branch, Suez, Egypt*

[3]*Hydrobiology Department, National Institute of Oceanography & Fisheries (NIOF), Alexandria, Egypt*

[4]*Zoology Department, Faculty of Science, Tanta University, Tanta, Egypt*

ABSTRACT

Eighteen specimens of the soft coral Sarcophyton glaucum have been collected and tagged to determine the growth rate using the weight variations technique and to investigate the reproductive season histologically. Six colonies were totally weighed and 12 fragments (produced from three colonies) were marked and weighed. Some samples showed sever shrinkage that lead to death. The maximum semiannual growth rate was recorded in summer (30.05 gm/6 m). The average monthly growth reached 3.95 gm/month. Many factors affect the growth rate such as water currents, turbidity, sedimentation, temperature and reproduction process. The studied species is a dioecious species (unisex), where the Autozooid polyps are the reproductive organs while siphonozooids are sterile. Female gonads were observed to develop at late July before male gonads which began at late September. Larvae releasing started at late January and the gonads are completely spawned in February. Fertilization occurs internally to form planulae larvae which librated completely in February. The oogenesis exhibited prolonged about 23 months, which gave rise to oocytes of different developmental stages in female colonies at any given time, and a spermatogenic cycle of only 10 - 12 months.

Keywords: Sarcophyton glaucum, Growth Rate, Reproduction, Hurghada-Red Sea

1. INTRODUCTION

The growth rate of family Alcyoniidae has been avoided by most researchers due to the softness of the coral animal and difficulties in the determination of growth. Very few investigations were related to the reproduction of soft corals in the Red Sea and not concerned with their growth rates; this action makes a gap in their biology, especially with regard to the sexual reproduction. So, the present study investigates the growth rate of Sarcophyton glaucum and its effect on the reproduction process. On the other hand, high growth rates of asexual reproduction were found in the Red Sea Xenia macrospiculata by Benayahu and Loya [1]. The radial growth was determined in both Sinularia and Sarcophyton which reached about 0.5 cm · yr^{-1}. Sarcophyton compensate for their low rates of reproduction by having low mortality and living to a great age [2]. Some soft corals display opportunistic life history features such as fast growth rates, high fecundity and asexual reproduction [3].

Some studies have focused on planulating species of the family Xeniidae [1] [4]. Among members of the family Alcyoniidae several modes of sexual reproduction have been recorded. The temperate species Alcyonium digitatum releases gametes and A. hibernicum has internal parthenogenesis of planular development; although alcyoniids are very common on coral reefs, little is known about their sexual reproduction and gonadal development. Yamazato [5] present circumstantial evidence for gamete spawning in Lobophytum crassum, and an external mode of planular brooding was described in the alcyoniid Parerythropodiumfulvumfulvum [6]. Moreover, recently spawned eggs were collected from several alcyoniids in the Great Barrier Reef, Australia [7]. Sarcophyton glaucum is one of the most conspicuous alcyoniids on the Red Sea reefs and at numerous sites across the Indo-West Pacific Ocean [8] [9].

Benayahu and Loya [10] examined various aspects of sexual reproduction in S. glaucum and presented the annual gonadal development, mode of reproduction, and its annual duration with their synchronous. However, this species illustrated a relatively high survival rate especially for the produced fragments [11]. On the other hand, there are many factors could affect the growth as temperature, sedimentation, turbidity and water motion [12] as well as the light intensity, salinity and feeding regimes which affect the survival rate [13]. Chavanich [14] pointed out that the elevated temperature appeared to be a more powerful causative factor than salinity on the survivorship especially during the extremely low tides (at June). In addition, the Antrophogenic factors such as catchment uses, unsustainable and destructive fishing may cause coral reefs degradation [15].

Cultured corals are known to be highly prized for three main markets: a) the bioprospecting of natural products for the pharmacological and biomedical applications [16] [17] ; b) the marine aquarium trade [18] ; and c) coral reef restoration efforts [19] [20]. These cultured corals were influenced by numerous factors, such as water movement [21], temperature [13], and light [22] [23]. While the ex situ culture of corals involves higher production costs, it has the advantage of maximizing survival and growth rates through the optimization of culture conditions [24]. The aims of the current study were: i) to assess and determine the growth rate using the weight variations technique underwater for the most common soft coral S. glaucum; ii) to determine the survival rate of the whole colonies and the fragmented pieces and their growth rates; and iii) to investigate the reproductive season histologically.

2. MATERIALS AND METHODS

2.1. Area of Study

The studied area in front of the National Institute of Oceanography and Fisheries (NIOF formally mentioned Marine Biological Station, MBS) was chosen due to its naturally protected area from most of the anthropogenic, human activities and the diving and/or the fishing processes. It is also not affect by any anthropogenic impacts or any human activities but affected by sedimentation processes due to the water movements and the north coming currents. The studied site is located at 5 km north Hurghada City (Figure 1) at latitudes of 27°17'13"N and longitudes of 33°46'43"E, where it is characterized by high sedimentation rates [25] - [27].

2.2. Methods

Some physical factors, as temperature (°C), salinity (‰), PH and dissolved oxygen (mg · l^{-1}) were measured as averages for one year at each site and were measured seasonal mean at the studied site using Hydrolab instrument (Model Surveyor 4, 1997).

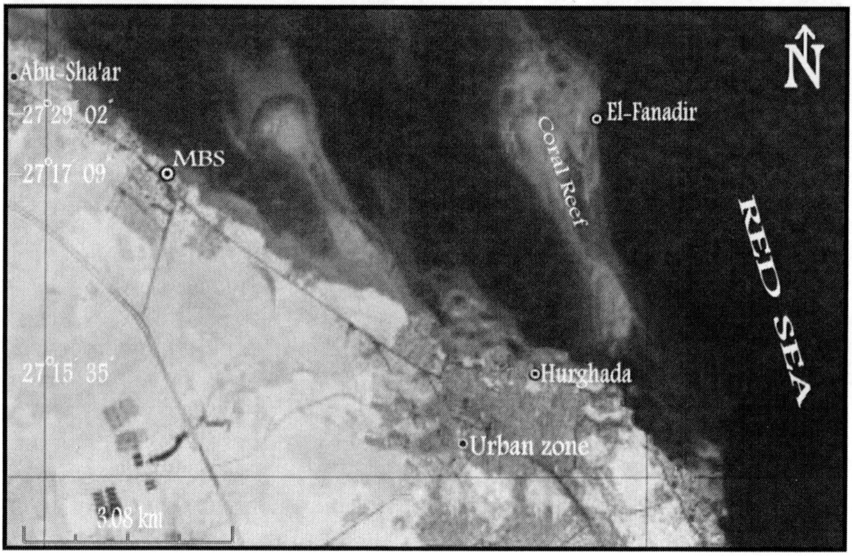

Figure 1. The location map of National Institute of Ocean- ography and Fisheries (Marine Biological Station MBS).

2.3. Growth Rate and Reproduction Determination

Nine colonies were chosen for studying the growth rate. Six colonies were totally weighing according to weighing technique of Bake [28] ; the other three colonies were fragmented into 12 fragments (see Figure 2) according to the fragmentation technique of English [29] which were applied using a pair of stainless steel scissors for the semiannual measurements of the increase in growth rate. The present method modify Back's method using weighing technique which was used for the whole six colonies by underwater weighing scale in grams and milligrams. This technique was applied due to the irregular shape of the coral colonies and its great sensitivity for losing water which cause shrinkage and death of the colony [30].

The fragments were dipped in fine sediments immediately after cutting; this helps to seal the exposed area to prevent the loss of the animal fluid, and adds negative buoyancy to the cut area of fragment [30]. Before weighing process, the particulate matter and sediment are cleaned carefully to avoid error in measurements (balance standard error is ±0.1 during weighing) and avoid any damage for the coral. The coral fragments were weighed separately before planting, to avoid water loss. Sample was placed in a tumbler filled with known weight of water, then by the calculating the difference between the weight of the tumbler with water and the weight of the tumbler containing water and sample. So, we determined the sample weight without water loss.

After the weighing process, coral fragments are fixed on plastic plates using a thin wire (or pike tooth) by piercing through the part, then marked and tagged for the following stages. Then the coral samples were placed in the sea at protected area from turbulence at 5 m depth, so the site should be chosen primarily for low turbulence and care must be taken to protect the corals during the attachment phase. After two weeks the thin wire removed and the fragments became attached well with the substrate and then samples weighing again. The growth rates for the coral fragments and the whole colonies were determined and measured semiannually (twice per year; once each six months due to the colonies sensitivity and to avoid the colonies shrinkage and death) at the warm season (summer) and the cold season (winter).

Some the expanded and relaxed polyps were collected immersed rapidly in 10% formalin solution for the histological study. Some polyps were fixed in Bouin's fluid, and then were decalcified within a solution of formic acid and sodium citrate for about 15 - 20 minutes according to Rinkevich and Loya [31] for the following steps of histology and examinations [25]. The gained planulae larvae were examined microscopically and amplified up to ×400. The sectioned tissues were examined under the stereo binocular microscope and then photographed using microscopic camera. The gonads size was measured by eyepiece micrometer. Camera Lucida was used to draw the sections.

3. RESULTS

3.1. Growth Rate

The preliminary weights of the whole six colonies and the twelve fragments were determined before planting and reweight again throughout a year for determination of the growth rate. However, the maximum growth in the whole or the naturally grown colony reached 30.05 gm/6 month in summer and the least growth reached 20.15 gm/6 month, while in winter the range of growth was varied between 21.04 - 28.36 gm/6 month (Figure 3). On the other hand, the transplanted fragments grow with sub-equal values or slightly lower. They recorded a range of 19.20 - 29.70 gm/6 months in summer and a range of 18.50 - 29.30 gm/6 months (Figure 3).

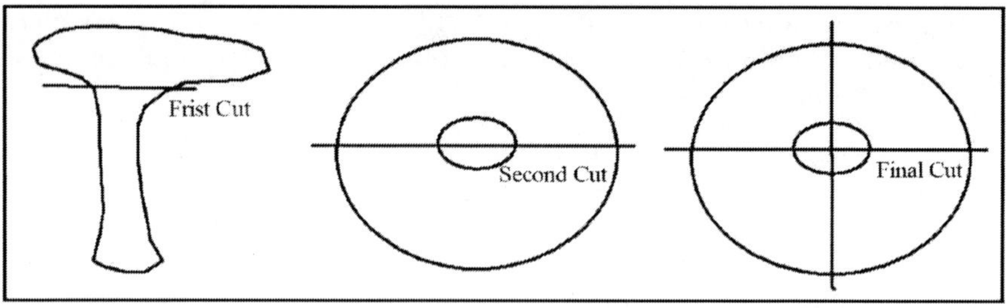

Figure 2. The fragmentation method of the selected colonies of S. glaucum.

The average semiannual growth of the naturally growing colonies recorded the highest value (25.875 ± 1.4058 gm/6 m) in summer; while in winter they illustrated a slightly low value (23.533 ± 1.2018 gm/6 months). While the coral fragments illustrated lower average than the natural grown colonies but appeared as sub-equal values in both summer and winter. The average growth recorded 23.483 ± 0.9171 gm/6 m and 23.083 ± 1.001 gm/6 m in summer and winter respectively (Table 1). Moreover, the average annual growth rates of the natural growing colonies recorded a relatively higher rate, 49.408 ± 2.4105 gm/y, than the fragmented colonies which reached 46.567 ± 1.9094 gm/y (Table 1 and Figure 4).

One-way ANOVA indicated that, there is no significant difference between the recorded data of the growth rate of the coral fragments during the first six months (during summer) and second six months (during winter) at P-value = 0.205 > 0.05 (Table 2). Moreover, the survival rate of the used colonies (n = 12 fragments and n = 6 small colonies) were approximately 100%.

3.2. Gonads Formation, Position and Sex Determination

The colony contains numerous dimorphism polyps (the autozooids and the siphonozooids) (see Figure 5). The autozooids (reproductive organs-carriers) appeared during the protrusion and extension outside the colony ground (Figure 6(a) and Figure 6(c)). It reached to 8 - 9 mm length and 0.8 to 0.9 mm base diameter. The siphonozooids appeared as minute dots scattered between the autozooids (Figure 6(c)). It is obviously noticed that, the water lose of the colony occurred through the siphonozooids. The autozooids have a cylindrical shape body opened with a mouth opening which is surrounded by few numbers of tiny tentacles (always 8 tentacles). The whole colony (stalk, disc and polyps, in the surface or in the coenenchyme) has calcium carbonate spicules (Figure 6(b)), which help in identifying the species and give the colony its internal skeleton.

The autozooids is a tube differentiated morphologically into three main regions: the main body (anthocodial stem), the oral disc and the tentacles. The autozooids are communicating by means of transverse tubes through the coenenchyme of the colony; these tubes are composed mainly of the continuation of the gastro vascular cells, which help in the asexual reproduction and developing the new polyps.

Gonads:

The gonads, either male (in the male colonies) or female gonads (in the female colonies), were found and formed in the second half of the autozooid length on the mesenteries and never extending upward to the oral disc or to the base of tentacles, while the siphonozooids have no gonads. The gonads start their development from the cells of the endoderm of the mesenteries (the interstitial cells) it begins by minute swollen where it divide into a series of independent pieces and form a row of globular structure until it ripe and swell from the mesenteries to be differentiated into one large eggs or sacs of spemaries. The size of gonads varied with the size of the colony size, the male gonads in mature stage ranged in general between 500 and 600 μm, while the female gonads diameter in mature stage ranged between 400 and 450 μm.

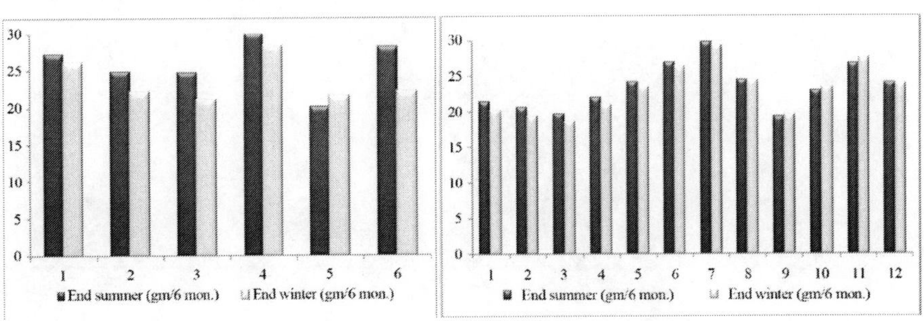

Figure 3. The six natural growing colonies (left) and the twelve fragments (right).

The eight septa or the mesenteries of the autozooid that are attaching to the endoderm layer of the Actinopharynx (stomodaeum) and extending to the base of the autozooid are consisting of three layers; two outer endodermic layers and an inner mesogloea layer. There are primordial generative cells developed in the endoderm of the ventral and lateral mesenteries in the mesenteric chambers and may migrate to the mesogloea during the early development of the genital cells.

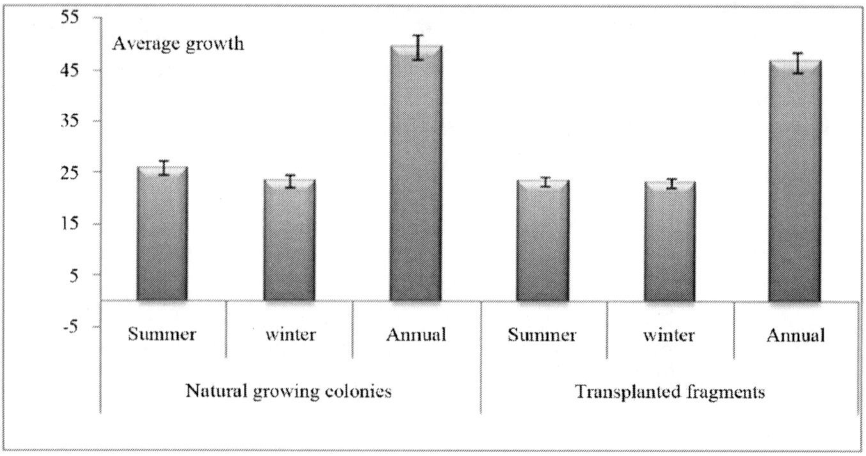

Figure 4. The average semiannual and annual growth of the natural and fragmented colonies.

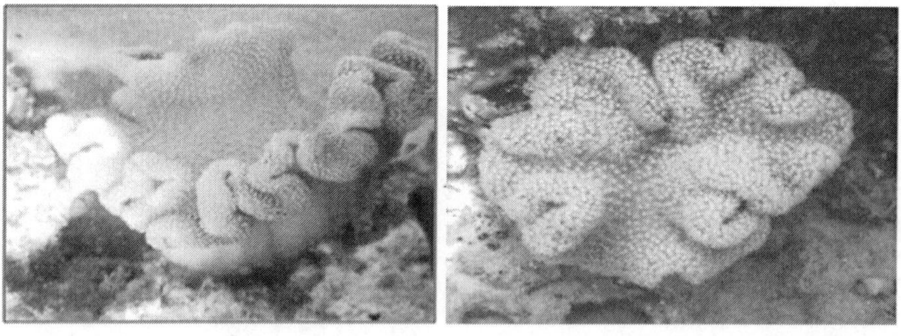

Figure 5. Fragmented colony (left) and the natural grown colony (right).

Table 1. The average and range of the whole and the fragmented colonies through a year.

Categories	Type of Growth	Summer	Winter	Annual
Average growth	Natural growing colonies (whole colonies)	25.875 ± 1.4058	23.533 ± 1.2018	49.408 ± 2.4105
Range		20.15 - 30.05	21.04 - 28.36	41.71 - 58.41
Average growth	Transplanted fragments	23.483 ± 0.9171	23.083 ± 1.001	46.567 ± 1.9094
Range		19.2 - 29.7	18.5 - 29.3	38.2 - 59

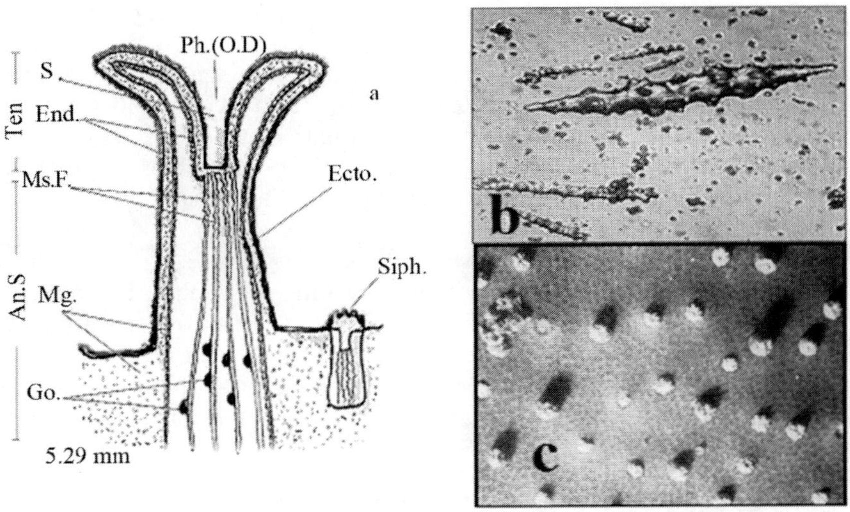

Figure 6. The general anatomy of the autozooid (a), sclerite (b) and partofcolonysurface (c). {An. S: Anthocodial Stem; Ecto: Ectoderm; End: Endoderm; Go: Gonads; Mg: Mesogloea; Ms.F: Mesenteric Filament; O. D: Oral Disc; Ph: Pharynx; S: Siphonoglyph; Siph: Siphonozooid; Te: Tentacle.}.

Table 2. One-way ANOVA analysis of the coral fragments growth rate per 6 months.

	Sum of Squares	D.F.	Mean Square	F-Test	Sig.
Between Groups	110.240	10	11.024	14.11	0.205
Within Groups	0.781	1	0.781		
Total	111.022	11			

Sexes of S. glaucum couldn't be distinguished by external characteristics, but by the microscopic examining for the transverse sections of the monthly collected autozooids and the siphonozooids polyps of marked colonies. The examined colonies illustrated that, the male gonads (spermatogenesis) or the female gonads (oogenesis) were formed and found only in the autozooids; while the siphonozooids are sterile and not include neither male nor female gonads. S. glaucum is a dioecious species (unisex) where the investigated colonies were separated and containing either the male gonads only or the female gonads only. The gonads of both sexes develop in the autozooids on the four lateral and the two ventral mesenteries (mesenteries at the side of the siphonoglyph). Gonads start their development initially from the primordial generative cells in the mesentery then be covered by the mesogolea and gastroderm is of the mesenteries (Figure 7) to be clustered and increased in size forming either male gonads (in the male colony) or female gonads (in the female colony). Gonads were observed extending at the posterior third part of the polyp length at the sub-stomodaeal region of the autozooids near the polyp base. The gonads mostly were not observed in the anterior part of the polyp.

The Oogenesis:

The mature female gonads were found swelling containing the young ova and detected during the months October-November 2006 covered by a common layer of endodermal cells of the mesenteries. Different stages of the young ova that are still attaching to the mesenteries (mature gonads) in addition to little of the pre-mature ova were detected through the period from December to the following February 2007.

The ova were separated from the mesenteric filament forming the oocytes in October 2006 (Figure 7(c)). The free mature oocytes reached about 400 - 450 μm (Figure 7(d)) and were recorded in the coelenteron at the sub-stomodaeal region, then the mature oocytes were moved to the anterior part of the autozooid during December to the next January 2007 where it apparently becomes ready for fertilization (Figure 8(a)). The autozooids becomes empty in the end of February to May 2008 (Figure 8(b)), but the mesenteries simply started to be swollen in July perhaps for the new reproductive cycle which will repeated again. Investigation of the autozooids began in October 2006 and was continued to July 2008, where the maturation of the oocyte took in probability 20 -23 months.

Spermatogenesis:

The immature male gonads (spermatogonia) started slightly swelling into the mesenteries including the primordial generative cells and are covered by both of the mesogloea and the endodermal layer of the mesenteries; they were detected at the sub-stomodaeal region during the February 2007. Spermatocytes were formed within spherical cysts (spermaries) from April to August 2007 filling up all cavities of the posterior portion of the autozooids of another colony. The young spermaries are attaching to the mesenteric filaments at the sub-stomod- aeal region and are densely packed with the spermatocytes and including a central cavity with spherical and not flagellated cells (Figure 8(c) and Figure 8(d)).

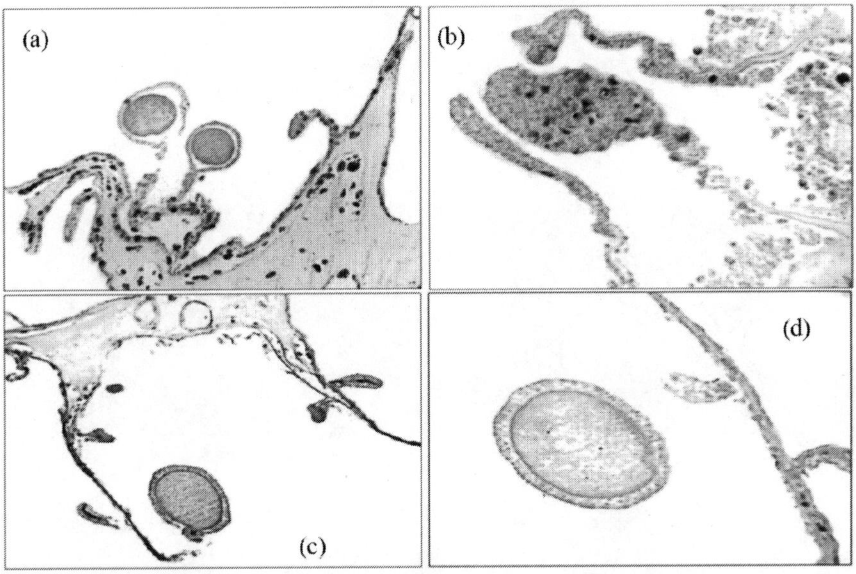

Figure 7. The ova at different maturity stages, some still attached to the mesentery by stalks (a); the mature female gonad extended posteriorin the autozooid ×400 (b); the pre-mature ova still attaching to the mesenteries at the sub-stomodaeum region (c); the detached mature ova in the coelenteron's cavity ×200 (d).

Further development by successive division and increasing in number takes place then the spermatocytes developed into the spermatozoa, where the spherical cells became flagellated, the flagella are extending to the center of the spermaries. The spermary that including the pre-mature spermatozoa reached to 300 - 400 μm in size and a small cavity appeared in the center. Mature spermatozoa were completed their formation during the last days of November to December 2007, where the spermary ranged between 500 and 600 μm (Figure 8(e) and Figure 8(f)). It was difficult to detect the shedding process of the sperms. The build-up of the spermatogenesis generally takes less than 12 months.

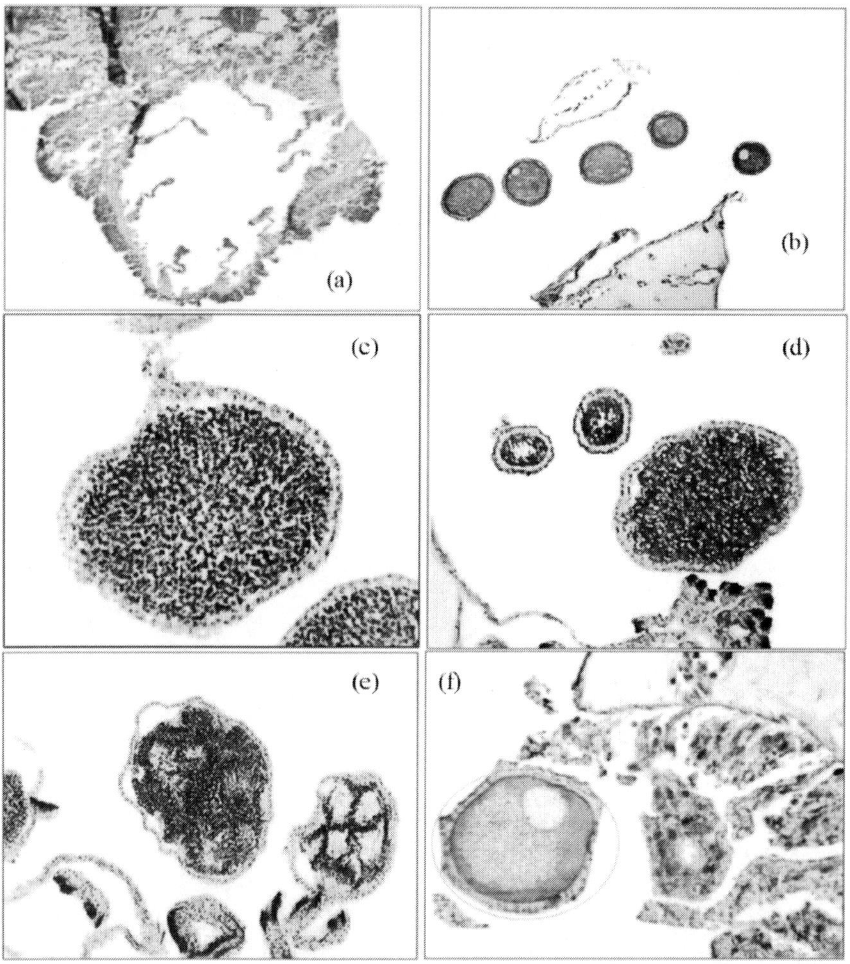

Figure 8. Free mature oocytes at the sub-stomodaeum region ×100 (a); the simple swelling at the periphery of the mesenteries ×100 (b); spermary attaching to the mesenteries and completely filled with the spermatocytes ×400 (c); the different developmental stages of spermaries ×200 (d); the free spermaries in coelenteric cavity of autozooid (e); the planulae larvae incorporated within the autozooid ×200 (f).

3.3. Reproduction and Spawning Season

Spawning Season & Planulae Larvae:

S. glaucum is a brooder species, where, fertilization and development takes place internally; this fact was concluded when the planulae larvae shed in the laboratory aquarium in the last days of January 2008 (Figure 9(a) and Figure 9(b)), where, it was noticed that the autozooids expelled brown minute tissue. After the investigation, the authors found the water containing the larvae of the species, which are counted to reach to about 30 larvae approximately belonging to a piece of S. glaucum measured 30 cm^2. According to the timing of the shedding of the planulae larvae that occurred in the last days of January and beginning of February 2008, so we can concluded that, fertilization took place at the end months of 2007, and it was observed that the female gonad develops before those of the male, where the female gonad begins its development in the last days of October 2006, while the male gonads begins its development in the last days of February 2007.

Number of the collected larvae reaches to about 30 larvae approximately belonging to about 40 autozooids only of a collected colony. The larvae are rounded ciliated forms, having negative buoyancy and observed to swim in a rotary manner containing very small tentacles (eight tentacles) at the oral disc surrounding it. These cilia are very fin help the larvae in swimming, which reached to about 300 micron. The posterior end of the larvae contains a disc-like structure which will form the colony substrate with its stalk.

4. DISCUSSION

4.1. Growth Rate

S. glaucum could not constitute a fixed shape during the current study, this is may be due to its irregular shape and the rapid changes in its shape during touch or slight disturbance, in addition of its great sensitivity for losing water that cause shrinkage and death of the colony in some cases. S. glaucum is one of the most common and frequent alcyonarian soft coral species which were chosen to measure and determine their growth rate, using a modification of Bake [28] method which was used for measuring growth rate of hard corals. Bak [28] studied the growth of Montastrea annularis and Madracis mirabilis as hard coral using a new technique of buoyancy of colony weight. Mitchell [32] applied new method for measuring the growth rate of two gorgonians corals Leptogorgia hebes and Leptogorgia virgulata through the study of the annual periodicity, Marschal [33] used another new technique to determine the age of the red coral Corallium rubrum colonies based on staining the organic matrix found in the axial calcareous skeleton. Holbrook and Schmitt [34] studied the growth rate of the sea anemone Heteractis magnica and related it to the presence of the mutualist or absence. At the present study, the normal weighing technique was used for S. glaucum with a modification of Bak's method due to its irregular shape and the rapid changes in its shape during touch or slight disturbance, in addition of its great sensitivity for losing water that cause shrinkage and death of the colony. Generally, the growth rate of family alcyoniidae was avoided due to the difficulties in the growth determination and due to the softness of the coral animal, and most of the growth rate investigations are applied on the hard corals. High potential for asexual propagation, fast growth rates characterize the nephtheid Dendronephthya hemprichi from the red sea [35]. The late authors studied the growth rate of Sinularia and Sarcophyton on the mid- and the outer-shelf reefs of the Great Barrier Reef, using tagging colonies over 3.5 years.

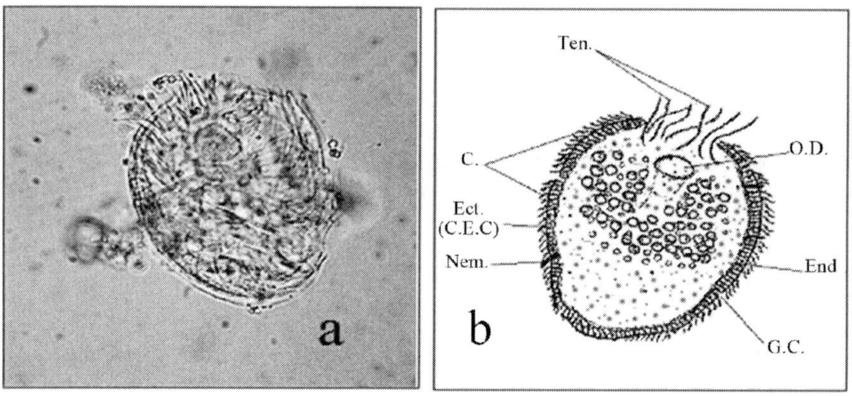

Figure 9. Planulae larvae of Sarcophyton glaucum (a); Photomicrograph of the planulae larvae ×200 (b). [C: Cilia; C.E.C: Columnar Epithelial Cells; Ect: Ectoderm; End: Endoderm; G.C: Glandular Cell; Nem: Nematocyst; O.D: Oral Disc; Ten: Tentacles].

Many physical and chemical factors affect the growth of the corals such as temperature, salinity, dissolved oxygen, sedimentation, turbidity and water motion [25] [36] [37] (Mc Cain et al., 1984; Kotb, 1996; Mohammed, 2003). Additionally, Vine [38] reported that the sea water temperatures in shallow areas of the Red Sea were close to the optimum range for coral growth (25°C - 29°C). So, temperature is considering one of the most important factors that influence the growth rate of corals and their distribution, particularly in the Red Sea [39]. Where, the growth rate of S. glaucum during the present study is promoted and fast in summer (warm season), while it is almost crippled in winter. The growth may be detained in winter due to the devotion of a large part of metabolism to reproduction rather than to growth [12]. Where, the animal exploits its energy in the formation of gonads. The most of the year is being exploited especially the winter and the autumn seasons and little of the summer season for the formation of the gametes stages [40]. While during summer seasons, the highest growth rate may be referred to the end of gonad formation and be ready to release gametes, where, the average growth in the present study is ranged between 23.48 ± 0.92 gm/6 months (Spring & Summer) and 23.08 ± 1 gm/6 months (Autumn & Winter) for the fragmented colonies; while the average growth rate of the naturally grown colonies are higher than the fragmented ones and recorded 25.875 ± 1.4058 gm/6 m in summer and 23.533 ± 1.2018 gm/6 months in winter with an annual average growth reached 49.408 ± 2.4105 gm/y and 46.57 ± 1.9094 gm/y for the naturally grown and the fragmented colonies respectively [40].

The growth rate of S. glaucum is considered higher than some other soft corals which previously studied by Putri and Fahrudin [15] who found that, S. glaucum grow faster than Nepthia sp. with as survival rate of 100%. According to Harriot and Fisk [41], coral transplantation declared successful if survival rate of transplanted coral fragments reaches more than 50%. Moreover, according to Rocha [22], the present study illustrated that, there is no significant difference in the growth rate of the transplanted fragments with survival approximately 100% [13] [15].

Benayahu [42] mentioned that, Sarcophyton glaucum increased a few millimeters per year in colony diameter from the northern Red Sea. He illustrated that, the nutrients affect the growth rate of alcyoniids and has a broad physiological tolerance. During the present study there, no effects for the nutrients on the growth rate of S. glaucum (may be in their optimal or normal range). Cornish [43] reported that, colonies of large sizes grew with higher rates than the small sizes for the alcyoniid corals. The present study illustrated that, the colonies of medium sizes appeared to grow faster than the smaller ones where the average annual growth rates of the natural growing colonies recorded a relatively higher rate (49.408 ± 2.4105 gm/y) than the fragmented colonies (46.567 ± 1.9094 gm/y).

4.2. The Sexual Reproduction of S. glaucum

Three different forms of sexual reproduction are common in the soft corals; broadcasting of eggs and sperm, internal brooding of larvae, and external brooding of larvae; where, the internal brooding occurs when a small number of eggs has fertilized and develop into larvae inside the females. Days to weeks later, the larvae are released when they are almost ready for metamorphosis [44]. The frequency of brooding versus broadcasting spawning in Octocoral varies with taxonomic order. Brooding is common in the Order alcyonacea [45] - [47]. Sarcophyton glaucum is a member of the order alcyonacea. On the other hand, Coll [48] consider broadcast spawning to be the most common sexual reproductive strategy for alcyonacean.

Benayahu and Schleyer [49] examined the species of Anthelia glauca and reported that the months January to March is the peak of the breeding season, the brood of species is found within the pharyngeal cavity. Slattery [40] described the mass spawning behavior of the soft coral Sinularia polydactyla; the gametogenic cycle of female colonies lasted 12 months while male colonies produced viable sperm within 9 months, and the species exhibited a split spawn between March and June. Moreover, the reproductive cycle of a female S. polydactyla colony requires one year for the production of mature eggs. In contrast, spermatogenesis within male colonies of S. polydactyla lasted less than one year. Gohar [4] stated that the gonads of the Xeniidae are developing on the six ventral and lateral mesenteries but not the dorsal pair. The author elucidated that the hermaphroditism is defining in Heteroxeniafuscescens, H. ghardaqensis and H. elizabethae, but not in H. capensis, also all other species of Xenia, Anthelia and Sympodium were found to be unisexual. All the species of the genera Xenia and Heteroxenia in the Red Sea, with the exception of Xenia garciae, are brooding species. Benayahu [50] described the reproductive patterns and developmental pathway of 21 xeniid species in the Red Sea. Gonochorism is the commonest sexual mode but simultaneous hermaphroditism was recorded in four species and brooding of planulae was observed in 15 species. Dahan and Benayahu [51] studied the sexual reproduction in the a zooxanthellateoctocoral Dendronephthya hemprichi and stated that the sperm and egg release were observed in the laboratory at night on all dates, from October 1989 to April 1991; the soft coral Dendronephthya hemprichi is a gonochoric broadcast spawner, reproducing all year round. Sexual reproduction of the Alcyonacean octocoral Lobophytum pauciflorum should be stated that was studied by Fan [52] through histological examinations of gonad development on monthly samples. Lobophytum pauciflorum is a gonochoric broadcast spawner; mature eggs were 400 - 870 μm in diameter. Oogenesis and spermatogenesis took about 12 months. Spawning occurred from July-September during late summer to early autumn.

The present work elucidated that S. glaucum is a brooder species, where fertilization takes place internally during December 2006/January 2007. It was observed that the female gonad develops before those of the male, where the female gonad began its development in the last days of October 2006, while the male gonads began its development in the last days of February 2007 up to December. However, the oogenesis exhibited prolonged about 23 months, which gave rise to oocytes of different developmental stages in female colonies at any given time, and a spermatogenic cycle of only 10 - 12 months. These features are shared with S. elegans on the GBR [53] who pointed out that, the oogenesis took 19 - 24 months, with a new cycle commencing every year, and spermatogenesis took 10 - 12 months as well as Lobophyton crassum in Okinawa [5], and Sinularia polydactyla in Guam [40].

The present study reported and revealed that the mature oocyte of S. glaucum with a diameter of 400 - 450 µm was recorded in December 2006, and the oocyte growth proceeded in probability through about 20 - 23 months. In contrast, Benayahu and Loya [10] stated that the mature oocyte with a diameter of 500 - 650 µm was recorded in July, and the oocyte growth proceeded through 22 - 23 months, while Schleyer [54] stated that the gametogenesis in male and female colonies of S. glaucum takes 9 - 10 and 16 - 18 months respectively and spawning occurs annually in March. Additionally, the present investigation reported that the reproductive cycle the male S. glaucum colony attained the maturity and the free mature spermaries with a diameter of 500 - 600 µm have been appeared in December 2006 and their build-up takes less than 12 months, while Benayahu and Loya [10] reported that the young spermaries are found during August-September and their build up generally takes 10 - 12 months and approximately reach to 400 µm prior to spawning (Table 3).

The reproductive cycle of S. glaucum for production of the mature eggs is characterizing to be prolonged, there is an agreement to what was stated that brooders of all taxa soft corals or others typically have prolonged reproductive periods (gametogenesis and planula release) [55], this may confident our result about the manner of S. glaucum, where during the present study, it is noticed that the development of their eggs occurred internally inside the female polyps to be fertilized and developed into larvae; whereas, the larvae were collected at the laboratory aquarium from the autozooids of the female colonies, where, some coral species releases planula larvae after internal fertilization or release gametes, during colonies transportation to the laboratory [56], therefore, the colony transportation may be considered as a stressful condition that cause the coral respond to release of their sexual products [57]. On the other hand, Benayahu and Loya [10] at the Northern Gulf of Eilat and Schleyer et al.

Table 3. A comparison between the present study and the other studies in gonads development of Sarcophyton glaucum and other species.

	Oogenesis	Spermatogenesis	Mature oocyte	References
S. glaucum	In probability 20 - 23 months	<12 months	December-January	Present Study (2008)
S. glaucum	22 - 23 months	10 - 12 months	July	Benayahu & Loya (1986)
S. glaucum	16 - 18 months	9 - 10 months	March	Schleyer *et al.* (2004)
Sinularia polydactyla	12 months	9 months	March-June	Slattery *et al.* (1999)
Lobophytum pauciflorum	12 months	12 months	July-September	Fan *et al.* (2005)

[54] in KwaZulu-Natal, South Africa indicated that, the soft coral Sarcophyton glaucum is a broadcasting species with externally developing larvae; where they pointed out that, the gametes were externally fertilized forming larvae. The present study pointed out that, there is a conflict between the present work and their results in the point of the formation of the larvae; where, the fertilization occurred internally producing the mobile ciliated larval stage. These results may illustrates the reproductive manner of the soft coral S. glaucum, but the difference may be due to some environmental factors which could affect their physiological responses as in most of the marine invertebrates, specifically gametogenesis formation, via coordination of intrinsic hormonal signals [58]. Moreover, the temperature is an important factor may affect the reproduction as well as the growth and coral distribution. On the other hand, the geographic distribution of reef corals and other coral reef organisms may control the timing of reproduction and reproductive behavior, as well as the seasonal changes in the abiotic features that may affect the reproduction manner; it was clearly mentioned that the seasonal changes of the water may have an impact on reproduction in Heteroxenia fuscescens [59].

In both sexes, gonads develop along the ventral and lateral mesenteries, often in the basal region of the polyp [60] [61]. The transverse sections of the autozooids and the siphonozooids polyps of the examined colonies were investigated to record and reveal that either the autozooids not the siphonozooids are including the stages of the oogenesis or the female sexual component (oocytes) mostly in the coelenteron (at the sub-stomodaeal region) of the most collected autozooids. Additionally, the autozooids are including the stages of the spermatogenesis or the male sexual component (spermaries) in the coelenteron (at the sub-stomodaeal region) of few collected mature autozooids that were collected from the middle of the colony disc. The siphonozooids of the examined colonies are sterile not include neither male nor female components. S. glaucum according to the later outcome is a dioecious species its colonies are separated into male and female colonies [10] [54]. The oogonia or primary oocytes first appear within the mesogloea of the septa [62] [63]. The common observation is that Scleractinian gametes develop within the mesogleal linings (lamellae) of the septa (mesenteries) which partition the coelenteron [57]. Germ cells originate in the gastrodermis and are initially surrounded by a layer of mesogloea with overlying gastrodermis [64] - [66]. Clusters of primordial germs cells are often attached to mesenteries (or mesentarial filaments in Pennatulacean) by a short pedicel of "stalk" [60] [67] [68]. The present study assures the entire previous conclusion, where the primary reproductive female components and the primary reproductive male components have been recorded attaching to the periphery of the mesenteric filament.

Hwang and Song [69] elucidated that the planulae of the soft coral Dendronephthya gigantea is ciliated planulae that had negative buoyancy after planulation, and showed rapid metamorphosis into a primary polyp stage within 2 days. Ben-David-Zaslow [59] recorded that the planulae of the Alcyonarian soft coral Heteroxenia fuscescens released during summer and is longer, with almost zero percent deformation. Fully formed Scleractinian larvae are elongate flagellated forms with ectoderm, mesoglea, and endoderm surrounding a central coenlenteron [57]. Swimming planulae of Scleractinian have common behavior patterns Most planulae are observed to swim in a rotary clockwise and counter-clock-wise manner around the anterior-posterior (aboral-oral) axis [70]. During the present study, the recorded planulae larvae are moderately agreed to what was stated by Fadlallah [57] and Hwang and Song [69], but is conflicted with Ben-David-Zaslow [59], where, our observed larvae are rounded ciliated forms, having negative buoyancy and watched to swim in a rotary manner around the anterior- posterior containing very small tentacles (eight tentacles) at the oral disc surrounding it, and the cilia are very fin help in the larval swimming.

5. CONCLUSIONS

It was concluded that:

- Temperature fluctuation between summer and winter influences the growth rate, so growth is faster in summer and slower in winter.
- High sedimentation rates and water currents are some factors affecting the growth rate especially in winter.
- Fragment size is the effective factor for fragments growth which causes the decrease of fragments growth than the naturally grown colonies.
- Reproduction is another factor affecting and reducing the growth in winter due to the devotion of a large part of metabolism to reproduction rather than to growth.
- Autozooid polyps of S. glaucum are the reproductive organs while siphonozooids are sterile.
- The oogenesis exhibited prolonged about 23 months, which gave rise to oocytes of different developmental stages in female colonies at any given time, and a spermatogenic cycle of only 10 - 12 months.
- Fertilization occurs internally and form larvae which librated at late February to liberate planulae larvae librated during February.

REFERENCES

1. Benayahu, Y. and Loya, Y. (1984) Life History Studies of the Red Sea Soft Coral Xenia macrospiculata Gohar, 1940. I. Annual Dynamics of Gonadal Development. Biological Bulletin, 166, 32-43. http://dx.doi.org/10.2307/1541428
2. Fabricius, K.E. (1995) Slow Population Turnover in the Soft Coral Genera Sinularia and Sarcophyton on Mid-Shelf and Outer-Shelf Reefs of the Great Barrier Reef. Marine Ecology Progress Series, 126, 145-152. http://dx.doi.org/10.3354/meps126145
3. Bednarz, V.N., Cardini, U., Hoytema, N., Al-Rshaidat, M.M.D. and Wild, C. (2015) Seasonal Variation in Dinitrogen Fixation and Oxygen Fluxes Associated with Two Dominant Zooxanthellate Soft Corals from the Northern Red Sea. Marine Ecology Progress Series, 519, 141-152. http://dx.doi.org/10.3354/meps11091
4. Gohar, H.A.F. (1940) Studies on the Xeniidae of the Red Sea. Publication of Marine Biological Station of Ghardaqa, Red Sea, Egypt, 25-118.
5. Yamazato, K.M., Sato, M. and Yamashiro, H. (1981) Reproductive Biology of an Alcyonacean Coral Lobophytumcrassum Marenzeller. Proceedings of the Fourth International Coral Reef Symposium, Manila, 2, 671-678.
6. Benayahu, Y. and Loya, Y. (1983) Surface Brooding in the Red Sea Soft Coral Pareryhtropodium fulvumfulvum (Forskal 1775). Biological Bulletin, 165, 353-369. http://dx.doi.org/10.2307/1541201
7. Bowden, B., Coli, J., Tapiolas, D. and Wilus, A.R. (1985) Some Chemical Aspects of Spawning in Alcyonacean Corals. Proceedings of the Fifth International Coral Reef Congress, Tahiti, 4, 325-329.
8. Tanaka, J., Yoshida, T. and Benayahu, Y. (2005) Chemical Diversity of Sarcophton Soft Corals in Okinawa. Galaxea (JCRS), 7, 1-9. http://dx.doi.org/10.3755/jcrs.2005.1
9. Nguyen, X.C., Tran, A.T., Phan, V.K., Chau, V.M., Eun, M.C. and Young, H.K. (2008) New Cembranoidditerpenes from the Vietnamese Soft Coral Sarcophyton mililatensis Stimulate Osteoblastic Differentiation in MC3T3-E1 Cells. Chemical and Pharmaceutical Bulletin, 56, 988-992. http://dx.doi.org/10.1248/cpb.56.988
10. Benayahu, Y. and Loya, Y. (1986) Sexual Reproduction of a Soft Coral: Synchronous and Brief Annual Spawning of S. glaucum (Quoy and Gaimard, 1833). Biological Bulletin, 70, 32-42. http://dx.doi.org/10.2307/1541378
11. Olivotto, I., Planas, M., Simoes, N., Holt, G.J., Avella, M.A. and Calado, R. (2011) Advances in Breeding and Rearing Marine Ornamentals. Journal of the World Aquaculture Society, 42, 135-166. http://dx.doi.org/10.1111/j.1749-7345.2011.00453.x
12. Mohammed, T.A.-A., Kotb, M.M.A., Ghobashy, A.-F.A. and El-Deek, M.S. (2007) Reproduction and Growth Rate of Two Scleractinian Coral Species in the Northern Red Sea, Egypt. Journal of Aquatic Research, 33, 70-86.
13. Sella, I. and Benayahu, Y. (2010) Rearing Cuttings of the Soft Coral Sarcophytonglaucum (Octocorallia, Alcyonacea): towards Mass Production in a Closed Seawater System. Aquaculture Research, 41, 1748-1758. http://dx.doi.org/10.1111/j.1365-2109.2009.02475.x
14. Chavanich, S., Viyakarn, V., Loyjiw, T., Pattaratamrong, P. and Chankong, A. (2009) Mass Bleaching of Soft Coral, Sarcophyton spp. in Thailand and the Role of Temperature and Salinity Stress. ICES Journal of Marine Science, 66, 1515-1519. http://dx.doi.org/10.1093/icesjms/fsp048
15. Putri, L.S.E. and Fahrudin, F. (2014) Corals Transplantation Using Rack and Substrate Method at Badul Island. Ujung Kulon, Indonesia. Proceedings of the International Conference on Advances in Bio-Informatics, Bio-Technology and Environmental Engineering-ABBE, 103, 75-78.
16. Blunt, J.W., Copp, B.R., Keyzers, R.A., Munro, M.H.G. and Prinsep, M.R. (2013) Marine Natural Products. Natural Product Reports, 30, 237-323. http://dx.doi.org/ 10.1039/C2NP20112G

17. Leal, M.C., Calado, R., Sheridan, C., Alimonti, A. and Osinga, R. (2013) Coral Aquaculture to Support Drug Discovery. Trends in Biotechnology, 31, 555-561. http://dx.doi.org/10.1016/j.tibtech.2013.06.004
18. Rhyne, A.L., Tlusty, M.F. and Kaufman, L. (2012) Long-Term Trends of Coral Imports into the United States Indicate Future Opportunities for Ecosystem and Societal Benefits. Conservation Letters, 5, 478-485. http://dx.doi.org/10.1111/j.1755-263X.2012.00265.x
19. Rinkevich, B. (2005) Conservation of Coral Reefs through Active Restoration Measures: Recent Approaches and Last Decade Progress. Environmental Science and Technology, 39, 4333-4342. http://dx.doi.org/10.1021/es0482583
20. Shafir, S., Van Rijn, J. and Rinkevich, B. (2006) Steps in the Construction of Underwater Coral Nursery, an Essential Component in Reef Restoration Acts. Marine Biology, 149, 679-687. http://dx.doi.org/10.1007/s00227-005-0236-6
21. Riegl, B., Heine, C. and Branch, G.M. (1996) Function of Funnel-Shaped Coral Growth in a High-Sedimentation Environment. Marine Ecology Progress Series, 145, 87-93. http://dx.doi.org/10.3354/meps145087
22. Rocha, R.J.M., Calado, R., Cartaxana, P., Furtado, J. and Serôdio, J. (2013) Photobiology and Growth of Leather Coral Sarcophyton cf. glaucum Fragments Stocked under Low Light in a Recirculated System. Aquaculture, 414-415, 235-242. http://dx.doi.org/10.1016/j.aquaculture.2013.08.018
23. Schutter, M., van der Ven, R.M., Janse, M., Verreth, J.A.J., Wijffels, R.H. and Osinga, R. (2012) Light Intensity, Photoperiod Duration, Daily Light flux and Coral Growth of Galaxea fascicularis in an Aquarium Setting: A Matter of Photons? Journal of the Marine Biological Association of the United Kingdom, 92, 703-712. http://dx.doi.org/10.1017/S0025315411000920
24. Forsman, Z., Rinkevich, B. and Hunter, C. (2006) Investigating Fragment Size for Culturing Reef-Building Corals (Porites lobata and P. compressa) in ex Situ Nurseries. Aquaculture, 261, 89-97. http://dx.doi.org/10.1016/j.aquaculture.2006.06.040
25. Mohammed, T.A.A. (2003) Study of Growth and Reproduction of Some Corals at Hurghada Region with Reference to the Effect of Some Pollutants in the Area. Zool. Dept. Faculty of Science, Suez Canal University, Ismaïlia, 204 p.
26. Ammar, M.S.A. and Mahmoud, M.A. (2004) Distribution and Space Monopolization by Corals in the Middle Reef at Hurghada Research Station, Red Sea, Egypt. The Journal of Egyptian Academic Society for Environmental Development (D-Environmental Studies), 5, 69-83.
27. Mohammed, T.A.A. and Mohamed, M.A. (2005) Some Ecological Factors Affecting Coral Reef Assemblages off Hurghada, Red Sea, Egypt. Egyptian Journal of Aquatic Research, 31, 133-145.
28. Bak, R.P.M. (1973) Coral Weight Increment in Situ. A New Method to Determine Coral Growth. Marine Biology, 20, 45-49. http://dx.doi.org/10.1007/BF00387673
29. English, S., Wilkinson, C. and Baker, V. (1997) Survey Manual of Tropical Marine Resources. 2nd Edition, Australian Institute Resources, Townsville, 385 p.
30. Ellis, S.C. and Sharron, L. (1999) The Culture of Soft Corals (Order: Alcyonacea) for the Marine Aquarium Trade. Publication No. 137, Center for Tropical and Subtropical Aquaculture, Waimanalo, Hawaii, 73 p.
31. Rinkevich, B. and Loya, Y. (1979) The Reproduction of the Red Sea Coral Stylophora pistillata I. Gonads and Planulae. Marine Ecology Progress Series, 1, 133-144. http://dx.doi.org/10.3354/meps001133
32. Mitchell, N.D., Dardeau, M.R. and Schroeder, W.W. (1993) Colony Morphology, Age Structure, and Relative Growth of Two Gorgonian Corals, Leptogorgia hebes (Verrill) and Leptogorgia virgulata (Lamarck), from the Northern Gulf of Mexico. Coral Reefs, 12, 65-70. http://dx.doi.org/10.1007/BF00302103
33. Marschal, C., Garrabou, J., Harmelin, J.G. and Pichon, M. (2004) A New Method for Measuring Growth and Age in the Precious Red Coral Corallium rubrum (L.). Coral Reefs, 23, 423-432. http://dx.doi.org/10.1007/s00338-004-0398-6

34. Holbrook, S.J. and Schmitt, R.J. (2005) Growth, Reproduction and Survival of a Tropical Sea Anemone (Actiniaria): Benefits of Hosting Anemone Fish. Coral Reefs, 24, 67-73. http://dx.doi.org/10.1007/s00338-004-0432-8
35. Fabricius, K., Benayahu, Y. and Genin, A. (1995) Herbivory in Asymbiotic Soft Corals. Science, 268, 90-92. http://dx.doi.org/10.1126/science.268.5207.90
36. McCain, J.C., Tarr, A.B., Carpenter, K.E. and Coles, S.L. (1984) Marine Ecology of Saudi Arabia—A Survey of Coral Reefs and Reef Fishes in the Northern Area, Arabian Gulf, Saudi Arabia. In: Bittiker, W. and Krupp, F., Eds., Fauna of Saudi Arabia, Vol. 6, National Commission for Wildlife Conservation and Development, Riyadh, 102-126.
37. Kotb, M.M.A. (1996) Ecological and Biological Studies on the Coral Reefs at Southern Sinai Coasts, Red Sea, Egypt. Ph.D. Thesis, Faculty of Science, Suez Canal University, Ismaïlia, 174 p.
38. Vine, P. (1986) Red Sea Invertebrates. Immel Publishing, Ely House, 37 Dover St. London WIX 3 RB.
39. Sheppard, C.R. and Sheppard, A.L.S. (1991) Corals and Coral Communities of Arabia. In: Buttiker, W. and Krupp, F., Eds., Fauna of Saudi Arabia Vol. 12, Natural History Museum, Basle, 3-170.
40. Slattery, M., Hines, G.A., Starmer, J. and Paul, V.J. (1999) Chemical Signals in Gametogenesis, Spawning and Larval Settlement and Defense of the Soft Coral Sinularia polydactyla. Coral Reefs, 18, 75-84. http://dx.doi.org/10.1007/s003380050158
41. Harriot, V.J and Fisk, D.A. (1988) Coral Transplantation as Reef Management Option. Proceedings of the 6th International Coral Reef Symposium, 2, 375-379.
42. Benayahu, Y. (1982) Population Dynamics of Soft Corals (Octocorallia, Alcyonacea) at the Coral Reefs of the Gulf of Eilat. Ph.D. Thesis, Tel Aviv University, Tel Aviv. (English summary)
43. Cornish, A.S. and DiDonato, E.M. (2004) Resurvey of a Reef Flat in American Samoa after 85 Years Reveals Devastation to a Soft Coral (Alcyonacea) Community. Marine Pollution Bulletin, 48, 768-777. http://dx.doi.org/10.1016/j.marpolbul.2003.11.004
44. Australian Institute of Marine Science (2001) Reproduction and Propagation of Soft Corals and Sea Fans. http://www.aims.gov.au/pages/softcorals-seafans/soft-corals06-016018.html
45. Alino, P.M. and Coll, J.C. (1989) Observations of the Synchronized Mass Spawning and Post Settlement Activity of Octocorals on the Great Barrier Reef, Australia: Biological Aspects. Bulletin of Marine Science, 45, 697-707.
46. Brazeau, D.A. and Lasker, H.R. (1990) Sexual Reproduction and External Brooding by the Caribbean Gorgonian Briareum asbestinum. Marine Biology, 104, 465-474. http://dx.doi.org/10.1007/BF01314351
47. Coma, R., Zabala, M. and Gili, J.-M. (1995) Sexual Reproductive Effort in the Mediterranean Gorgonian Paramuricea clavata. Marine Ecology Progress Series, 117, 185-192. http://dx.doi.org/10.3354/meps117185 [Citation Time(s):1]
48. Coll, J.C., Leone, P.A., Bowden, B.F., Carroll, A.R., König, G.M., et al. (1995) Chemical Aspects of Mass Spawning in Corals. II. (-)-Epi-Thunbergol, the Sperm Attractant in the Eggs of the Soft Coral Lobophytum crassum (Cnidaria: Octocorallia). Marine Biology, 123, 123-143. http://dx.doi.org/10.1007/BF00350332
49. Benayahu, Y. and Schleyer, M.H. (1998) Reproduction in Anthelia glauca (Octocorallia: Xeniidae). Marine Biology, 131, 433-442. http://dx.doi.org/10.1007/s002270050336
50. Benayahu, Y. (1991) Reproduction and Developmental Pathways of Red Sea Xeniidae (Octocorallia, Alcyonacea). In: Williams, R.B., Cornelius, P.F.S., Hughes, R.G. and Robson, E.A., Eds., Coelenterate Biology: Recent Research on Cnidaria and Ctenophora, Springer, The Netherlands, 125-130. http://dx.doi.org/10.1007/978-94-011-3240-4_17 [Citation Time(s):1]
51. Dahan, M. and Benayahu, Y. (1997) Clonal Propagation by the Azooxanthellate Octocoral Dendronephthya hemprichi. Coral Reefs, 16, 5-12. http://dx.doi.org/10. 1007/ s003380050053
52. Fan, T.Y., Chou, Y.H. and Dai, C.F. (2005) Sexual Reproduction of the Alcyonacean Coral Lobophytum pauciflorum in Southern Taiwan. Bulletin of Marine Science, 76, 143-154.

53. Hellström, M., Kavanagh, K.D. and Benzie, J.A.H. (2010) Multiple Spawning Events and Sexual Reproduction in the Octocoral Sarcophyton elegans (Cnidaria: Alcyonacea) on Lizard Island, Great Barrier Reef. Marine Biology, 157, 383-392. http://dx.doi.org/10.1007/s00227-009-1325-8
54. Schleyer, M.H., Kruger, A. and Benayahu, Y. (2004) Reproduction and the Unusual Condition of Hermaphroditism in Sarcophyton glaucum (Octocorallia, Alcyoniidae) in KwaZulu-Natal, South Africa. Hydrobiologia, 530/531, 399-409.
55. Fadlallah, Y.H. and Pearse, J.S. (1982) Sexual Reproduction in Solitary Corals: Synchronous Gametogenesis and Broadcast Spawning in Paracyathus stearnsii. Marine Biology, 71, 233-239. http://dx.doi.org/10.1007/BF00397040
56. Richmond, R. (1981) Energetic Considerations in the Dispersal of Pocillopora damieornis (Linnaeus) Planulae. Proceedings of the 4th International Coral Reef Symptom, 2, 153-156.
57. Fadlallah, Y.H. (1983) Population Dynamics and Life History of a Solitary Coral Balanophyllia elegans, from Central California. Oecologia, 58, 200-207. http://dx.doi.org/10.1007/BF00399217
58. Giese, A.C. and Pearse, J.S. (1974) Introduction: General Principles. In: Giese, A.C. and Pearse, J.S. Eds., Acoelomate and Pseudocoelomate Metazoans, Academic Press, New York, 1-49. http://dx.doi.org/10.1016/B978-0-12-282501-9.50006-5
59. Ben-David-Zaslow, R., Henning, G., Hofmann, D.K. and Benayahu, Y. (1999) Reproduction in Red Sea Soft Coral Heteroxenia fuscescens, Seasonality and Long-Term Record (1991 to 1997). Marine Biology, 133, 553-559. http://dx.doi.org/10.1007/s002270050495
60. Achituv, Y. and Benayahu, Y. (1990) Polyp Dimorphism and Functional, Sequential Hermaphroditism in the Soft Coral Heteroxenia fuscescens (Octocorallia). Marine Ecology Progress Series, 64, 263-269. http://dx.doi.org/10.3354/meps064263
61. Kruger, A., Schleyer, M.H. and Benayahu, Y. (1998) Reproduction in Anthelia glauca (Octocorallia: Xeniidae). I. Gametogenesis and Larval Brooding. Marine Biology, 131, 423-432.
62. Szmant-Froelich, A., Yevich, P. and Pilson, M.E. (1980) Gametogenesis and Early Development of the Temperate Coral Astrangia danae (Anthozoa: Scleractinia). Biological Bulletin, 158, 257-269. http://dx.doi.org/10.2307/1540935
63. Fadlallah, Y.H. and Pearse, J.S. (1982) Sexual Reproduction in Solitary Corals: Overlapping Oogenic and Brooding Cycle, and Benthic Planulas in Balanophyllia elegans. Marine Biology, 71, 223-231. http://dx.doi.org/10.1007/BF00397039
64. Chia, F.S. and Crawford, B.J. (1973) Some Observations on Gametogenesis, Larval Development and Substratum Selection of the Sea Pen, Ptilosarcus guerneyi. Marine Biology, 23, 73-82. http://dx.doi.org/10.1007/BF00394113
65. Fautin, D.G. and Mariscal, R.N. (1991) Cnidaria: Anthozoa. In: Harrison, F.W. and Westfall, J.A., Eds., Microscopic Anatomy of Invertebrates. Vol. 2: Placozoa, Porifera, Cnidaria, and Ctenophora, Wiley-Liss, New York, 267-358.
66. Gutierrez-Rodriguez, C. and Lasker, H.R. (2004) Reproductive Biology, Development, and Planula Behavior in the Caribbean Gorgonian Pseudopterogorgia elisabethae. Invertebrate Biology, 123, 54-67. http://dx.doi.org/10.1111/j.1744-7410.2004. tb00141.x
67. Farrant, P.A. (1986) Gonadal Development and the Planulae of the Temperate Australian Soft Coral Capnella gaboensis. Marine Biology, 92, 381-392. http://dx.doi.org/10.1007/BF00392679
68. Tyler, P.A., Bronsdon, S.K., Young, C.M. and Rice, A.L. (1995) Ecology and Gametogenic Biology of the Genus Umbellula (Pennatulacea) in the North Atlantic Ocean. Internationale Revue der gesamten Hydrobiologie und Hydrographie, 80, 187-199. http://dx.doi.org/10.1002/iroh.19950800207

69. Hwang, S.J. and Song, J. (2007) Reproductive Biology and Larval Development of the Temperate Soft Coral Dendronephthya gigantea (Alcyonacea: Nephtheidae). Marine Biology, 152, 273-284. http://dx.doi.org/10.1007/s00227-007-0679-z
70. Lewis, J.B. (1974) The Settlement Behaviour of Planulae Larvae of the Hermatypic Coral Favia fragum (Esper). Journal of Experimental Marine Biology and Ecology, 15, 165-172. http://dx.doi.org/10.1016/0022-0981(74)90042-2

CHAPTER 10

Fisheries, food, and health in the USA: the importance of aligning fisheries and health policies

David Clifford Love, Patricia Pinto da Silva, Julia Olson, Jillian Parry Fry, Patricia Mary Clay

ABSTRACT

Background

Food availability, access, and utilization are the three pillars of food security and need to be aligned in order to support a healthy population. United States (US) fisheries policy plays an important role in seafood *availability*. US health policy impacts *access* and *utilization* of seafood in various ways; however, health policies are often disconnected from fisheries policy. Aligning fisheries and health policies is imperative to improve food security. We address two questions with our work: (1) how would US federal fisheries policy be different if our fisheries were managed with beneficial health outcomes for Americans as clear objectives; and (2) how would US health policy be different if one of its goals was to support sustainable domestic fisheries and aquaculture?

Results

We report how fisheries policies and health policies are additive, synergistic, or antagonistic with regard to seafood, and provide illustrative examples of collaboration between health and fisheries communities at different levels of the food system (federal and state policies, corporate partnerships, and civil society). We also develop a list of topics for future research, and opportunities to align and integrate fisheries and health policies.

Conclusions

Managing fisheries to promote optimal nutrition and efficient food production likely requires a different approach to fisheries management—new outcomes will need to be monitored, new approaches found, and fisheries, aquaculture, and health policies better integrated. Health policies rarely consider the source of fish, their connections with US fisheries systems, and global distribution of seafood. Change can begin where the most promising opportunities exist, such as institutional food procurement, Farm to School programs, social marketing campaigns, and private sector start-ups. Continued development in fisheries and health policies, however, will need to occur at multiple levels of federal policy, and across the different domains and dimensions of the food system (e.g., social, political, biophysical, economic).

Keywords: Aquaculture, Fisheries, Food, Health, NOAA Policy, USDA

1. BACKGROUND

Fish and other types of aquatic animal and plant species are an integral part of human economies, cultures, and health. Here, we refer to those species using the term "seafood" to include freshwater and marine species, principally those caught in commercial fisheries or grown through aquaculture.[1] Seafood supports the livelihoods of 10–12% of the world population [1]. Seafood is considered part of a healthy diet, providing nearly 3 billion people with 20% of their dietary protein needs and micronutrients such as iodine, potassium, selenium, B vitamins, and vitamin D [1, 2]. The omega-3 polyunsaturated fatty acids (PUFAs) in seafood support childhood development and reduce risks from heart disease [3, 4]. Underscoring the importance of seafood, the United States (US) and other national dietary recommendations call for increasing seafood consumption, particularly fish high in omega-3 PUFAs and low in mercury [5, 6]. Yet at the same time, fisheries experts and health experts are raising questions about whether national dietary recommendations are in fact achievable, equitable, and sustainable, given what some see as a diminished seafood supply from the world's oceans and general movement of seafood products from low- and middle-income countries to high-income countries [1, 7, 8, 9, 10, 11]. The underexplored intersection of these policy spheres, with different concerns and goals, has significant implications for future scenarios in both health and sustainability.

For millennia, wild capture fisheries were often viewed as inexhaustible resources. Today, scientists believe that the global fishing fleet reached its "peak catch" in 1996 at 130million metric tons (MMT), with subsequent harvests declining by about 1.2 MMT each year [12]. Others have challenged the idea that global fisheries collapse is as imminent as predicted, and authors of both competing views found a patchwork of recovery and decline in different fisheries and regions around the world [13].A combination of factors is driving many (but not all) fish stocks to decline, including: increased human demand due to population growth and rising incomes; overfishing; use of better technology to catch seafood by large fishing fleets; illegal, unregulated, unreported (IUU) fishing; poor management decisions in some fisheries; and the impacts of climate change (e.g., increasing ocean acidification and ocean temperatures) and biodiversity loss [13,14,15,16,17]. Humans have entered a new epoch, the Anthropocene, where human influence permeates the planet and decreasing wild harvests affect global food security [12,18].

Aquaculture, or aquatic animal and plant farming, has helped fill the gap between demand for seafood and declining wild fisheries [19].Aquaculture has buoyed the global edible seafood supply and now contributes equal amounts to wild capture, in terms of seafood used for human consumption [20]. However, despite the different trajectories for aquaculture and wild capture fisheries, fed aquaculture (i.e., species requiring feed inputs) cannot be considered independently from wild capture fisheries or terrestrial crop production. Carnivorous farmed species like Atlantic salmon, trout, and marine shrimp still rely on feeds containing ingredients from wild-caught forage fish, such as anchovies, herring, and sardines, which tightly couples fisheries and aquaculture and introduces natural resource trade-offs [21, 22] that may impact food system resilience. The use of fishmeal and oil has greatly decreased in carnivorous farmed species, and the use of alternative feeds from terrestrial sources (e.g., soybean meal and oil, corn, other vegetable oils, animal byproducts) has increased in herbivorous, omnivorous, and carnivorous species. This links aquaculture with terrestrial agricultural systems and creates trade-offs that potentially impact the resilience of global food systems [23, 24]. Given these multiple connections, national policies related to fisheries and agriculture will have ripple effects on aquaculture production costs, and the human food supply and nutrition [23].

A recognition of the complex and interrelated set of challenges in the commercial seafood sector provides the contextual foundation for this paper. We follow these challenges as a means of exploring how current fisheries and health policies are additive, synergistic, or antagonistic with regard to seafood. Admittedly, we are necessarily juxtaposing different institutions' ideas and practices about best uses for seafood; as we will discuss later, ideas and practices in one US-based institution can be inconsistent with another, or even internally inconsistent, due in part to differing priorities reflecting divergent values and social beliefs. Throughout this paper, we take the perspective that fish (including finfish and shellfish) is food and, more specifically, that seafood is a key ingredient in a food system [25]. Our aim is to highlight the policy intersections of fisheries and health and raise important questions while exploring this juncture.

In this study, we ask two questions: (1) how would US federal fisheries policy be different if our fisheries were also managed with health outcomes for Americans as clear management objectives; and (2) how would US health policy be different if one of its goals was to support sustainable domestic fisheries and aquaculture? As we begin to explore these questions, looking at past and current activities within US public health and fisheries management, we provide illustrative examples of collaboration and disconnects between health and fisheries management communities at different levels of the food system (federal and state policies, corporate partnerships, and civil society). Finally, we discuss opportunities to align fisheries and health policy and suggest priority areas for further research.

2. METHODS

The study began with an exploratory phase consisting of six months of regular dialogue among an interdisciplinary research team, composed of two public health experts (Love and Fry) and three fisheries experts (Pinto da Silva, Olson, and Clay). This dialogue gave study team members an opportunity to better understand each others' disciplines and identify gaps in knowledge across the disciplines. Following the exploratory phase, the study team formulated two overarching questions related to the interactions between fisheries and public health. To answer the questions, the research team identified relevant background information about health policy and fisheries policy through federal agency Web sites, communicating with experts/stakeholders, reading literature available on PubMed and Google Scholar, and sharing our expert knowledge. This information was compiled in the "Background," "Health policy," and "Fisheries policy" sections.

In the health policy section, data in Table 1 on US nutrition assistance programs were compiled using the US Department of Agriculture Web site and in consultation with Janice Fitzgerald, Household Programs Operations Branch Chief at the US Department of Agriculture, Food, and Nutrition Service. To identify American seafood consumption trends for Table 2, a search was performed in PubMed for the literature on seafood consumption using the NHANES dataset from the 1980s to present, and data from five sources were extracted and compiled. In "Fisheries policy" section, the annual commercial catch of forage fish in the USA was found in the National Marine Fisheries Service (NMFS) "Commercial Fisheries Statistics" database (in metric tons and value) and we report averages of yearly data from 2003 to 2013. Data on US edible and industrial seafood supply were procured from the 2013 edition of the NMFS annual report "Fisheries of the United States, 2013."

We then developed five case studies to provide readers with tangible examples of synergy between fisheries and public health. These case studies were carefully selected to represent different levels of the food system and different entry points into the supply chain. To write the case studies, the team conducted background research and informal interviews with an environmental advocate, a fishmonger, a seafood wholesaler, a university food service provider, a canned fish business owner, and a fishing industry representative. Notes from these interviews were compiled and drafted into case reports. We followed up with some interviewees to check facts and ask additional questions.

Finally, the study team jointly developed a set of recommendations and conclusions based on the research conducted. The team shared these recommendations with external reviewers, a fisheries economist and an agriculture/health policy expert, for comments and revisions.

3. RESULTS AND DISCUSSION

3.1 Health policy

3.1.1 Nutritional benefits and seafood safety

To understand health policy related to seafood, it is useful to understand its scientific underpinnings in two of the most important disciplines: nutrition and food safety. Among the most significant health benefits of seafood are those derived from the omega-3 polyunsaturated fatty acids (PUFAs) present in many types of seafood. Fish high in omega-3 fatty acids include herring, mackerel, sardines, and salmon. In infants, there is strong evidence of gains in childhood development when pregnant and nursing mothers consume seafood, and limited evidence for a protective effect for asthma and eczema [3]. For adults, some of the best evidence that seafood consumption provides health benefits is the prevention of cardiac deaths in individuals with preexisting coronary heart disease [4]. There is limited evidence that seafood consumption later in life improves bone health and is protective against dementia; however, more studies are needed [2]. Seafood is a lean protein and can serve as a replacement for meat high in saturated fat. Importantly, fish oil supplements are not always associated with the same health benefits as eating a meal of a whole seafood product [3,26]. In addition to nutrition, there are psychological, psychosocial, cultural, health, and food security benefits of consuming locally produced seafood [27, 28,29].

Table 1 US nutrition assistance programs that support seafood purchasing, 2013

Federal program	Approved seafood	People participating annually (millions)	Total cost (million US dollars)	Seafood cost (million)	Seafood quantity purchased (million pounds)
Supplemental Nutrition Assistance Program	Any seafood[a]	47.4	79,936	n/a	n/a
Child Nutrition Programs					
National School Lunch Program	Numerous products[b]	30.6	12,220	7.5	1.6
School Breakfast Program	n/a	13.2	3514		
Summer Food Service Program	n/a	2.4 (daily)	428		
Child/Adult Care Program	n/a	3.6 (daily)	2994		
Women, Infants, and Children Program	Canned chunk light tuna, pink salmon, sardines	8.7	6488	n/a	n/a

The Emergency Food Assistance Program		n/a	693	13.7	3.8
	Canned pink salmon			3.8	1.7
	Unbreaded catfish fillets			9.9	2.1
Commodity Supplemental Food Program	Canned pink salmon	0.58	203	3.0	1.3
Food Distribution on Indian Reservations Program	Canned pink salmon	0.09	119	0.43	0.2
National Services Incentive Program (elderly feeding)	Catfish, frozen fillet strips	0.076	100	0.19	0.04
Total		100.6	106,695	24.8	6.9

Data source: [102]

[a]Except hot food or food that can be eaten in the store

[b]Fresh/frozen clams, crabs, crawfish, fish fillets and steaks, fish portions, fish sticks, mackerel, oysters, scallops, squid, canned mackerel salmon, chunk light tuna, sardines, shrimp

Table 2 Consumption of seafood by Americans from 1988 to 2010, NHANES

Study	Per capita mean intake (g/week)		Among seafood consumers (past 30 days) mean intake (g/week)		Year
	Men (+19 years)	Women (+19 years)	Men (+19 years)	Women (+19 years)	
NCI [104]	138.9	99.2	–	–	2007–2010
Jahns et al. [62]	–	–	179.2	138.6	2005–2010
Papanikolaou	140.1	101.2	–	–	2003–

et al. [103]					2008
Tran et al. [105]	128.7	86.3	186	127.8	1999–2006
Wang et al. [106]	115.5	84.7	–	–	1999–2004
Wang et al. [106]	141.4	100.8	–	–	1988–1994

Some of the health benefits of seafood are offset by food safety risks, such as naturally occurring biotoxins, metals, microbiological agents, other pollutants, and allergens present in some seafood. The most well-known contaminant associated with seafood is methylmercury (MeHg), a neurotoxin that enters the ocean primarily by deposition from burning fossil fuels and accumulates in the marine food web. Individuals eating large amounts of long-lived marine fish (e.g., shark, swordfish, king mackerel) and bottom-feeding fish (e.g., tilefish, catfish) are most at risk of negative health effects, particularly fetuses and infants exposed via their mother's intake of contaminated seafood [3]. Persistent organic pollutants (POPs), metals, and pesticides can be present in both saltwater and freshwater fish near urban centers, and people who practice subsistence fishing can be exposed to unsafe levels of pollutants [30,31]. Biotoxins can include toxins produced by algae and incorporated into shellfish (i.e., paralytic shellfish poisoning) or fish (i.e., ciguatera), and naturally occurring marine bacteria (i.e., *Vibrio* spp.) that produce toxins in shellfish. Foodborne diseases from seafood are not uncommon, occurring most frequently through consumption of filter-feeding shellfish, raw fish products, or improperly handled or stored processed or imported seafood [32,33].

On balance, the benefits of seafood outweigh the risks [34], which undergirds many health policies related to seafood. In fact, many policies are intended to increase seafood consumption to meet national dietary guidelines.

3.1.2 Policies related to seafood and health

There are many US health policies regarding seafood, primarily related to access and use of seafood but also food safety issues related to production and processing. To start this section, we familiarize readers with the federal agencies involved in seafood as it relates to human health, and describe each type of policy—be it dietary advice, labeling, or nutrition assistance. Most of these policies are inclusive of fisheries and aquaculture, and we note where differences exist. Additional detail, specifically on fisheries policy that affects seafood availability, is provided in the "Fisheries policy" section.

3.1.3 Overview of federal agencies involved in seafood as a food source

Federal oversight of fisheries and aquaculture production is split among three agencies. The National Marine Fisheries Service (NMFS), which is a line office within the National Oceanic and Atmospheric Administration (NOAA), has regulatory purview over marine fisheries and aquaculture in federal waters and the Great Lakes [35, 36]. Freshwater and inland aquaculture is overseen by the US Department of Agriculture (USDA) [37]. Freshwater fishing outside of the Great Lakes is overseen by the Fish and Wildlife Service; however, their focus is not on food but rather on conservation.

Specific to aquaculture, the Food and Drug Administration (FDA) regulates veterinary drug use in farmed aquatic animals [38]. Aquaculture operations receive assistance from the USDA Animal and Plant Health Inspection Service (APHIS) for disease control, and live organisms originating from fish hatcheries are inspected by APHIS for diseases before sale across state or national borders [39]. The USDA is also in the process of developing organic standards and a USDA Organic label for aquaculture producers, which would limit the use of synthetic chemicals, certain feed ingredients, and some production methods; however, the draft regulation has not been released as of time of publication for this article [40]. Organic standards for aquaculture have come much later than organic standards for meat, dairy, fruits, and vegetables, in part because of the complexity and variety within aquaculture production methods and the relatively small size of the aquaculture industry in the US.

More generally, for all fisheries and aquaculture products, the FDA—along with support from NOAA, and USDA (for catfish)—inspects a small proportion of domestic and imported seafood and seafood processing plants [41]. NOAA will implement a new Seafood Import Monitoring Program focusing on traceability for 13 commercial species beginning in 2018 [42]. The USDA specifies grades and standards for various processed seafood products (such as canned or breaded processing standards) [43]. In some cases, physical or chemical inspection occurs, and in other cases, inspection of paperwork such as trade documents, permits, and hazard analysis and critical control points (HACCP) plans are reviewed by one of these agencies [44, 45]. Because shellfish carry a higher risk for causing disease, the FDA, in collaboration with states and the National Shellfish Sanitation Program, has special regulations concerning where shellfish can be grown or harvested in the US, and specific permits and regulations for shellfish processing and interstate trade [46]. Finally, at the end of the supply chain, state and local health departments inspect food preparation, restaurant, and retail locations to make sure they are meeting regulations for safe food handling. The Centers for Disease Control (CDC) also plays a role in food safety by assisting with outbreak investigations for foodborne disease [47].

3.1.4 Consumption-focused policies

The Departments of Health and Human Services (HHS) and Agriculture (USDA) produce the US Dietary Guidelines every five years with dietary advice related to foods including seafood [5, 6]. The Dietary Guidelines are developed based on input from an advisory committee that reviews the current health and nutrition literature and makes recommendations for dietary patterns [48]. The Dietary Guidelines indicate that seafood intake is below recommended levels for all age groups and should be increased. In 2015, the Dietary Guidelines Advisory Committee made recommendations about sustainability, which mainly focused on reducing meat consumption but also included some discussion of fisheries and aquaculture. Sustainability language and recommendations were not included in the final report, despite broad support from the US public [49]. The Dietary Guidelines have an impact on Americans' diets via nutrition education, food purchasing done using federal funds, and nutrition assistance programs (described below).

Product labeling is an example of a consumer-facing policy. The only government-required form of product labeling for seafood (outside of nutrition and ingredient labeling) is country of origin labeling (COOL). COOL labels report whether the product is wild-caught or farm- raised and its country of origin. COOL is required for all minimally processed seafood sold in retail locations, and some consumers infer quality attributes from these labels [50]. Third-party labels, such as eco-labels, complement or fill gaps left by federal labeling laws. Food labels are important because they inform consumers about products in ways that are more direct and tangible than other approaches, such as methylmercury warnings, which are not present on seafood products.

Beginning in the early 2000s, the FDA, Environmental Protection Agency (EPA), and states began working together to advise consumers about methylmercury (MeHg) in both commercial and self-caught seafood. To help recreational anglers, the EPA worked with all 50 states to develop recreational seafood consumption advisories for the general public and women of childbearing age to reduce human exposure to MeHg and certain pesticides [51, 52]. The FDA and EPA also produced seafood MeHg advisories for retail seafood that target women of childbearing age and young children to reduce MeHg exposures [53]. Unfortunately, the original MeHg advisories had the unintended consequence of decreasing maternal seafood consumption, thereby reducing the potential benefits for these populations of eating seafood [54]. A more recent study indicates that the MeHg advisories' message may now be better received, as women of childbearing age are eating seafood but switching to products with lower levels of mercury [55]. The FDA and EPA are in the process of revising seafood advisories for MeHg in light of the expert consensus that the benefits of seafood consumption outweigh the risks [34,56].

3.1.5 Nutrition assistance programs

The USDA Food and Nutrition Service administers 15 nutrition assistance and nutrition education programs. Table 1 presents the amount and cost of seafood purchased under nutrition assistance programs by individuals (i.e., cash purchases) and by the USDA (i.e., entitlement commodity purchases). The largest program is the Supplemental Nutrition Assistance Program (SNAP, commonly known as food stamps), which in 2014 supported 46 million Americans with low incomes [57]. Individuals can spend SNAP funds to purchase any type of fresh, frozen, or processed seafood, as long as it is not a hot or prepared food that can be eaten in the store. The Nutrition Program for Women, Infants, and Children (WIC) distributes funds to eight million pregnant, low-income women and their children. WIC participants can use electronic benefit transfer (EBT) cards (similar to debit cards) to purchase canned light tuna, pink salmon, and sardines, among other food items [58].

There are also several entitlement programs in which state agencies select what types of commodity foods they want from a list of allowable products that the USDA then purchases and distributes to the states. Examples of these entitlement programs are the Emergency Food Assistance Program, which distributed over 600 million pounds of products to food banks in 2015 [59], the Food Distribution Program on Indian Reserves, which supports 276 tribes [60], and the Commodity Supplemental Food Program, which supports approximately 500,000 low-income elderly Americans annually. Rules require that the USDA purchase only domestically produced and processed foods under these entitlement programs. The only seafood product purchased by these entitlement programs currently is canned pink salmon from Alaska, with $7.2 million (3.2 million pounds) purchased in 2013 (Table 1). Canned tuna will be approved for purchase beginning in 2016. While catfish is an allowable item, it is not typically selected by states due to its expense; instead, items like catfish tend to be purchased using the USDA bonus program, which is used to correct oversupply in the market and does not count against state entitlement spending. In 2013, the USDA purchased $17.6 million (3.7 million pounds) of catfish (Table 1).

The Child Nutrition Program (CNP) is a hybrid program that provides both cash reimbursements for students to purchase school food and entitlement funds for schools to purchase that food. The National School Lunch Program (NSLP) is one example within the CNP; it provides five billion free and reduced-price lunches to 30.5 million school-age children through the age of 18 [61]. The list of allowable seafood products purchased through the NSLP is larger than any other nutrition assistance program (Table 1), and while schools are asked to purchase domestic food with their funds, they may choose imported products that are cheaper or more widely available. Cost is a major barrier for seafood purchasing by nutrition assistance programs, which may cause commodity purchasing programs and participants of SNAP or WIC to purchase more economical animal and plant proteins. It would be useful to identify the barriers, in addition to cost, preventing nutrition assistance programs from incorporating more domestic seafood into food banks and feeding programs. In addition, given the low consumption of seafood among women and low income Americans, consideration should be given to what role the federal government can play in encouraging seafood consumption by SNAP and WIC participants [62].

3.1.6 Consumption trends in the USA

Americans make food choices based on a variety of reasons. Seafood purchasing is primarily driven by consumer taste preference, cost, income, gender, age, and for some an interest in meeting national dietary recommendations [63, 64]. American seafood consumption is as diverse as our population; however, in aggregate, the most commonly eaten seafood products are whitefish (pangasius, domestic catfish, cod, pollock, tilapia), shrimp, farmed salmon, and canned tuna. The US imports 88% of the seafood it consumes, which is equally split among wild-caught and farmed sources [65].Most popular seafood products (with the exception of tuna, tilapia, and pollock) have farmed and wild-caught versions available in stores. Yet among the top seafood products, only salmon is high in omega-3 PUFAs, the nutritional element that drives seafood consumption guidelines. Forage fish such as anchovies, herring, and sardines are high in omega-3 PUFAs, low in contaminants like mercury, and low on the food chain; however, these products rarely appear on nutrition assistance program lists and have decreased in popularity among American consumers over the last 50years.

Indeed, the sheer number of nutrition and health policies described above appears to encourage access to seafood that is safe, affordable, low in mercury, and shelf-stable for key groups such as pregnant and breastfeeding women, children, the elderly, and individuals with low incomes. Nonetheless, from 1988 to 2010, Americans have consistently eaten much less seafood than government-recommended levels of 2–3 meals per week (227 g per week, or 8–12 oz per week) (Table 2) [5]. One study found that while four out of five Americans eat seafood on a monthly basis, most people (80–90%) were not meeting dietary guidelines [62]. A very low proportion of Americans meet dietary recommendations in other areas, too, such as fruit and vegetable intake [66]. Women, young people (19–30 years old), and individuals with lower levels of income and education all consume less seafood than the average American [62]. These findings highlight potential inequalities in access to seafood (and/or related factors such as preferences and familiarity) among certain groups, which could contribute to health disparities.

Moreover, federal health policy does not mention the goal to support sustainable domestic fisheries and aquaculture, or include language that indicates how food systems are linked to healthy fisheries. In this sense, fisheries and health policies are intrinsically linked while being effectively "siloed." Below, we explore US marine fisheries policies to understand how US wild seafood resources are currently managed, and consider what changes to federal fisheries policies/management might lead to optimal food production, human nutrition, and health outcomes for Americans.

3.2 Fisheries policy

3.2.1 Fisheries policy and management

Early efforts to develop US fisheries policy began over a century ago. In 1871, Congress passed the first legislation recognizing a Federal role in conservation of natural resources: the Joint Resolution for the Protection and Preservation of the Food Fishes of the Coast of the United States [Act of February 9, 1871, Sess. III, Res. 22, 16 Stat. 593–94 (1871)]. The resolution recognized that "the most valuable food fishes of the coast and the lakes of the United States are rapidly diminishing in number, to the public injury, and so as materially to affect the interests of trade and commerce." The US Fish Commission was created to protect, preserve, and study these "food fishes." After many decades of organizational changes and realignments, duties were divided between the U.S. Fish and Wildlife Service (FWS), which governs freshwater species, and the National Marine Fisheries Service (NMFS), which governs saltwater species. In 1970, NOAA was established and "directed to improve our understanding of the nation's living marine resources, the environment in which they are found, and the interaction between the two," and NMFS (thereafter also known as NOAA Fisheries) was placed within NOAA [67].

The primary law governing marine fisheries management in federal waters is the Magnuson–Stevens Fishery Conservation and Management Act (MSA;16 U.S.C. ß 1801 et seq.). While the MSA refers to the importance of fish as food in its definition of optimum yield (OY),[2] the first of ten National Standards included in the document, fisheries regulations generally focus on rebuilding overfished stocks and ensuring that overfishing is not occurring through controls on inputs (gear restrictions/area closures) and/or outputs (limits on catch/landings) [25], but do not consider what happens to fish after it is landed or how regulatory actions might impact nutrition or food systems [68]. The US Fish and Wildlife Service regulates freshwater fishing outside of the Great Lakes and operates fish hatcheries, fishery resource offices, and ecological services field stations [69], where the primary concern is conservation of resources and not fish as food. All federal agencies must comply with the National Environmental Policy Act (NEPA), which requires agencies to assess the environmental impacts of and alternatives to proposed actions. These assessments are conducted and shared with stakeholders and the public as Environmental Assessments (EA) or, when it is determined that a Federal action will have a significant effect on the human environment, Environmental Impact Statements (EIS). For example, EPA reviewed a large gold mine planned upstream of Bristol Bay, Alaska, and found reason for concern due to potential impacts to fisheries and the livelihood or cultural identity of fishermen. Human health could be impacted by stress from loss of income or identity; however, health issues related to how the fish are marketed after landing was not considered. In this case, the EPA conducted an EA and determined that a more detailed EIS was not required.

Fisheries policy extends to aquaculture as well, especially regarding use of wild-caught forage fish as aquaculture feed (discussed below); however, our discussion of federal aquaculture-specific policy (i.e., developed by the NOAA Office of Aquaculture or in the NMFS Aquaculture Strategic Plan) is limited in this document. NMFS directs attention to aquaculture in federal waters, where it is primarily concerned with "developing and implementing policies that enable marine aquaculture and works to ensure that aquaculture complies with existing federal laws and regulations that NOAA enforces under its marine stewardship mission" [70]. The USDA gathers data on economics and animal health of US aquaculture products, conducts a census of aquaculture operations every five years [71], and tracks US imports/exports of aquaculture products [72]. It is important to note that US government agencies are working to expand the domestic aquaculture industry, which currently contributes less than one percent of global aquaculture production [20, 73, 74]. Therefore, an opportunity exists to incorporate sustainability and health outcomes, as well as alternative supply chains, as the industry grows. In addition, integration with food systems, public health, and alternative supply chains may also serve to expand the number of stakeholder groups supporting aquaculture development in the US.

3.2.2 Outcomes and markets

Seafood is the most valuable traded food commodity in the global marketplace, more valuable than coffee, tea, or spices [1]. The US is a major player in global seafood trade, routinely buying and selling to/from over 100 countries in multiple regions of the world [75]. The major trading partners for the US are Asia and Canada [76]. This trade includes two types of products that are important to differentiate: edible products and industrial products. Edible products are what we commonly think of as the seafood that is sold in grocery stores and restaurants. Industrial fisheries products, made primarily from forage fish and fish scraps, are used as animal feed, fish feed, or fertilizer. Forage fish are small oily fish that live in schools within the water column along coasts, oceans, and lakes and are caught with large nets. In the US, over 800,000 metric tons of forage fish are caught each year, which is a fifth of US landings by weight, but just 4% by value (Table 3). Forage fish caught in the US are often turned into fishmeal and fish oil. It is difficult to convert quantities of forage fish into amounts of fishmeal and oil produced because rendering removes water weight, but NMFS reports that the US produces over 200,000 metric tons of fish meal and 70,000 metric tons of fish oil each year [1].

Table 3 Annual catch of forage fish in the US, mean of 2003–2013

Forage fish species (region)	Metric tons	Value (USD)
Anchovy (Northern)	6796	883,554
Mackerel (Scomber)	4	4750
Menhaden (Atlantic) (Brevoortia)	659,579	93,930,294
Sardine (Pacific)	80,682	12,286,746
Sardine (Spanish)	591	228,673
Herrings (Atlantic)	72,694	32,398,361
Herring (Atlantic Thread)	645	236,890
Herring (Blueback)	9	10,550
Herring (Lake or Cisco)	397	554,303
Herring (Pacific)	38,761	17,262,876
Herrings	487	217,018
Total forage fish	860,645	158,014,015
Total US landings	4,174,308	4,399,172,575

Data source: [107]

Once aquatic animals are removed from a body of water, they enter seafood supply chains consisting of hundreds of processors, wholesalers, distributors, and retailers that employ 1.3 million full- and part-time workers in the US [77]. Edible and industrial fisheries products are both traded, but in different ways. For example, the US imports more edible products than it exports (i.e., a negative trade balance), while the opposite occurs with industrial products—the US exports three times more industrial products than it imports (Table 4). Overall, these trading patterns have led to a food system where nine-tenths of the US edible seafood supply is imported, sometimes as US fish that has been processed abroad, and a feed system where large amounts of forage fish are extracted from US territorial waters and exported. As we will explore later in this paper, these complex interactions between supply and demand in the use of natural resources have important outcomes for human health, as well as implications for society, the economy, and ocean ecosystems.

Table 4 US edible and industrial seafood supply in 2013 (million metric tons)

Fisheries and aquaculture product	Domestic landings	Domestic aquaculture	Imports	Exports	Total supply
Edible products	3.7	0.25	4.8	3.2	5.5
Industrial products	0.83	n/a	0.27	0.84	0.26
Total	4.5	0.25	5.0	4.0	5.7

Data source: [65]

The Atlantic herring fishery, formerly a food fishery and now a bait fishery for the New England lobster fishery, illustrates the evolution of social and economic outcomes influenced by fisheries policy. Small herring used to be canned in Maine in the nineteenth and twentieth centuries, peaking at 400 canneries that employed thousands. The last cannery in Maine closed in 2010 [78]. Changing markets and consumer preferences played some role, for example, in the dietary replacement of canned herring and sardines with canned tuna, but how Atlantic herring is caught also determines what it is used for. Fishing methods have changed since herring was used as a human food and changes in fishing methods and technology have accompanied changes in the management and use of herring. Early methods (traps called weirs and relatively small nets called seines) caught herring without damaging the fish, which was ideal for using herring as human food. Over time, regional fisheries management council regulations have allowed large trawlers and pair trawlers (trawlers working together with a net between them), to become the dominant gear types; these vessels pump large catches out of nets into onboard refrigerated seawater tanks. Fish caught this way are not marketable as human food, as they are often damaged during the process of extraction. Concurrent with these technological shifts, Atlantic herring shifted from being a food fish to a baitfish, used almost exclusively as bait for today's American lobster fishery (Pinto da Silva, Olson and Benjaman, in review). While herring was once a low-cost source of high quality protein and other nutrients, the fishery it now supports, lobster, is a high-cost (i.e., less affordable) food source. While these changes from food to bait have been exacerbated by a confluence of economic, social, technological, and cultural shifts, regulatory measures have likely contributed to this situation as well by not considering the impacts of regulatory changes on the regional and national food system.

Examining management and the use of herring and other forage fish raises the following questions: Should forage fish be managed in a way that recognizes the benefit of leaving a certain amount in the ocean to provide food for other fish higher on the food chain? Should people eat these fish instead of using it for industrial purposes, including as aquaculture feed or bait, where there is a conversion loss of protein and biomass? Herring, as well as menhaden and other forage fish, expose some of the trade-offs that are inherent in fisheries management, trade-offs that will only be increasingly highlighted as regulatory agencies begin to grapple with ecosystem-based approaches to management. The important point in this context, however, is that how we choose to manage our fisheries will also influence the food supply in the US and beyond, making fisheries policy part of US food policy and requiring more explicit recognition of the intersection of, and especially the trade-offs between, fish, food, and health policy. As the following examples help demonstrate, expanding the universe of those involved with fisheries management can shift the objectives of concern and change the questions stakeholders are asking.

3.3 Synergy between fisheries and health policy, illustrative examples

Here, we present four illustrative examples of policies, initiatives, and business models where fisheries and health policies are purposefully connected. Each example targets a different level (federal and state government, civil society, economy) as a way to acknowledge that hybrid, multi-level interventions are needed to solve complex problems. The examples we selected are: anchovy consumption in Peru; an Alaskan government program to purchase local seafood for public school lunches; a public–private partnership between a New England university, seafood distributor, and nonprofit working to support local fishers; and a California-based canned seafood business.

3.3.1 Social marketing: Peruvian anchovies, changing culture takes time

Along the West Coast of South America in the Pacific Ocean lies the Humboldt Current, where a rich upwelling of nutrients supports an abundance of marine life [79]. As a result, Peru has some of the best fishing grounds for small pelagic fish, including anchovies and sardines. At the peak of anchovy fishing in 1970, 12 million metric tons of anchoveta, a type of anchovy, were harvested off the coast of Peru, constituting an astounding 20% of all global fish harvests [80]. Peruvian anchoveta stocks have fluctuated due to El Niño conditions and overharvesting, but most recent statistics indicate that they provide a third to half of the world supply of both fishmeal and oil [81]. Peruvian anchoveta are almost exclusively harvested for export as a feed ingredient (as fishmeal and oil). Only 2% of Peruvian anchoveta are used for human consumption [79] despite food insecurity and micronutrient deficiencies in Peru and surrounding countries [11,79].

In 2006, the Oceana Center for Environmental Sustainability (CES) at the Cayetano Heredia University in Lima set out to reshape the way anchoveta were used in Peru by promoting them as human food via the campaign "Semana de la Anchoveta" (Anchovy Week) (personal communication, Patricia Majluf, Oceana). Working with Peru's top celebrity chef on a nationally recognized TV cooking show, CES was able to broadcast their message about the sustainability, health benefits, and flavor of Peru's national fish, the anchoveta. The story caught the attention of the president of Peru, and after a meal with him at the Presidential Palace, the campaign had the backing of the government. A subsequent law allocated 8% of Peru's government spending on food toward anchoveta. Anchovy Week was repeated in 2007, 2009, and 2012, though the founder of the campaign admits it never reached her expectations in terms of public participation (personal communication, Patricia Majluf, Oceana). A lesson we can draw from Anchovy Week is that enacting food system changes requires adjustments not only in policy, but also throughout the supply chain, and among consumers. For example, the anchoveta fishing quota is solely based on reduction fishmeal uses, not producing food on food-grade anchovy. If vessels in the industrial fleet could use their catch for human consumption, the ensuing supply could be at a scale that would make production for human consumption economically viable. Fresh and frozen product forms, and new more convenient presentation formats (i.e., dried, minced) are needed to attract more consumers beyond those currently eating canned anchovies. These issues may include environmental, social, and economic trade-offs [81] and require unraveling and replacing current incentives in the fisheries policy that maintain the status quo [82].

3.3.2 State-level policy: Farm to School programs

In the section on health policies, we describe several nutrition assistance programs that include seafood. One innovative method for introducing seafood into nutrition assistance programs is through the National School Lunch Program (NSLP) using a concept called "Fish to School" (related to "Farm to School" programs). The State of Alaska has been proactive in navigating the bureaucratic requirements that must be fulfilled in order to source, purchase, and serve local fish like pollock, salmon, cod, halibut, and rockfish in school cafeterias [83]. Like other states, the Alaska state government requests that USDA purchases bulk quantities of commodity seafood (mainly canned pink salmon and catfish) for entitlement programs and through schools with CNPs. Alaska also provides grants to reimburse schools that purchase Alaskan-caught seafood through the Nutritional Alaskan Foods in Schools (NAFS) program. The NAFS helps schools better align meals with the national dietary guidelines and has its underpinnings in a 2010 state law to promote local agriculture in public schools [84]. The Alaska Fish to School effort aims to improve student health via seafood consumption, help students understand where their food comes from, provide traditional foods for schools, and acknowledge the cultural and economic significance of purchasing local seafood [83]. A study of the Alaska Fish to School program found that school fish meals were well aligned with the local foodways of the community, and purchasing local seafood engendered goodwill with parents and the community [84]. Challenges for schools implementing Fish to School in Alaska were finding seafood suppliers, ordering appropriate quantities of fish, and finding suitable recipes [84]. Implementation of the program also requires that there is demand for new foods, that they are available domestically from a pool of vendors, and that there is capacity within industry to provide necessary quantities. Oregon, Massachusetts, and New Hampshire are also working on similar Fish to School programs [85,86,87], which highlights an opportunity for repeating these innovative programs in other coastal states.

3.3.3 Food procurement: connections among universities, hospitals, and regional fisheries

Universities, hospitals, and other groups with institutional food procurement can play a key role in connecting local fisheries to improved diets [88]. Universities and hospitals are increasingly interested in the sustainability of their physical campuses as well as of the services they offer. Sustainability has become a "core value" of these institutions that requires mainstreaming throughout their operations. As part of this process, most universities have developed sustainability plans that cover a range of issues from procurement and building practices to improving energy and water use efficiency and reducing their overall carbon footprint. Food service procurement is often seen as a component of these plans.

Some universities are rethinking their relationships with local economies as well as their ability to provide fresh local health food options for students. While the focus of these efforts has been on the local and regional procurement of agricultural products, some institutions are connecting with local fisheries.

In Boston, Massachusetts, improved technology around fish marketing is enabling regional universities (such as Boston University, Harvard University, and the University of Connecticut) to buy thousands of pounds of locally caught fish at competitive prices. Red's Best, an innovative seafood dealer, provides fish caught on day boats from Rhode Island to Maine directly to universities through negotiated partnerships [89]. By using QR codes that track where, when, and who caught the fish, universities are able to confirm the chain of custody and ensure that they are purchasing seafood caught locally or regionally. While universities continue to purchase much of their seafood from national distributors, purchases from Red's Best are growing as a share of total university seafood purchases in participating institutions.

Hospitals are also reflecting on the social and environmental impacts of their purchases. The quality as well as the source of hospital food can influence regional agricultural systems and markets for regionally caught seafood, not to mention health outcomes for patients. Like universities, hospitals are also developing sustainability plans, which often include increasing the amount of locally sourced food as a component of sustainability. Healthcare Without Harm is an international coalition of hospitals, community groups, environmental health organizations, and others that are working together to improve the overall contribution to health that a hospital can make beyond the medical services offered to patients [90]. The food procurement components of these plans have most often focused on agricultural products, leaving seafood sourcing to traditional mechanisms. However, through partnerships with local NGOs, some hospitals in New England are engaging with their suppliers on where their seafood comes from, as well as purchasing seafood directly through local community-supported fisheries (CSFs) [91], which are currently being encouraged in the Northeast USA through support from the Northwest Atlantic Marine Alliance (NAMA). These partnerships are helping to achieve "optimum yield" in fisheries by encouraging the purchase of US caught seafood in regional institutions.

3.3.4 Private sector: business opportunity

Bill Carvalho is the founder of Wild Planet, a canned seafood company. In 2001, a tour of the Monterey Bay Aquarium led to a realization of the economic opportunities available in exclusively selling sustainably harvested seafood, including forage fish, that are low in mercury and high in omega-3 PUFAs (personal communication, Bill Carvalho). In 2004, Wild Planet began selling canned tuna that met these standards, and later added canned sardines, anchovies, and mackerel. In 2008, Wild Planet received backing from a private equity firm involved with the Packard Foundation, enabling it to grow into one of the only American companies selling domestic forage fish, among other products, to the domestic mass market. Wild Planet products are now featured in many major grocery store chains in the US alongside traditional canned seafood.

The company's emphasis on sustainability has meant that product sourcing and supply chain management require constant attention. Tuna is purchased directly from US tuna fishermen or from the Japanese pole and troll fleet. Mackerel is also purchased from Japan. Originally, sardines were purchased from West Coast bait vendors; however, unreliable sizes of fish and a recent moratorium on sardine harvests meant the company had to source sardines primarily from Japan, and the USA as available. Most of Wild Planet's fish are canned in Vietnam, in part because the labor prices are more competitive, but also because Carvalho has had difficulty trying to revive the US West Coast seafood canning industry. Wild Planet also sources and cans anchovies in Peru, in an effort to support the Peruvian food fish canning industry. The story of Wild Planet highlights the international nature of the seafood supply, and the challenges for one company that attempts to source and process environmentally sustainable domestic seafood that is of high nutritional value.

4. RECOMMENDATIONS

US national health policy can have a significant impact on seafood access, utilization, and consumer seafood purchasing and consumption patterns. However, many of these policies, regulations, and guidance documents have been developed without any cross-integration to fisheries policies and regulation. As a result, government efforts to improve population health miss many opportunities for sourcing domestic, sustainably caught or farmed seafood. Others have raised similar concerns about misalignment of nutrition policy and agriculture policy [92].

Fisheries policy is also detached from, but integral to, public health. Fisheries and aquaculture policies refer frequently to the importance of fish as food for Americans; however, the emphasis of fisheries regulations and the science produced to support it have been on sustaining the resource and reducing impacts on fishing businesses without considering the broader outcomes, such as the production of food for the nation, that are defined in the MSA as necessary to achieve not just maximum sustainable yield but the overarching goal of optimum yield, or OY [MSA Sect. (3)(33)].[3]

There is also a general disconnect between fisheries and consumers, such that the average citizen is often unaware of where their seafood comes from [93].

Aligning fisheries and health policies is imperative to achieving food security. Change will need to occur at multiple levels of policy and across various domains and dimensions of the food system (e.g., social, political, biophysical, economic) [94]. Institutions including seafood-related businesses, hospitals, state government agencies, NGOs, and academia can also play a role in better weaving together fisheries and health priorities (as we describe in "Synergy between fisheries and health policy" section). Drawing from these examples and our own professional experiences, we have developed a set of recommendations for integrating fisheries, aquaculture, and health policies.

The following bullets describe a number of ways to connect US health policy to fisheries and aquaculture policy, within the context of food production.

Strengthen federal agency linkages Government agencies often work on similar topics but approach them from different disciplines or have divergent outcome measures. We see the following opportunities for stronger agency linkages: (1) integrate seafood consumption and nutrition guidance from other federal agencies with NOAA Fish Watch, a national database for sustainable seafood [95]; (2) include sustainable sourcing in the US Dietary Guidelines, with input from NMFS and others on seafood sustainability; (3) coordinate activities between FDA's seafood import inspection program and NOAA's seafood traceability initiative; (4) integrate state advisories for recreational fishing with state guidelines for self-caught fish consumption with support from EPA, FDA, NMFS, and FWS [52]; (5) connect seafood to the Obama White House backed initiative "Local Foods, Local Places" that connects multiple agencies around agricultural products [96]; and (6) move toward a national food policy that recognizes how fisheries and agriculture policy can support human health and the environment [97].

Use federal agency purchasing power The federal government has substantial purchasing power, which could be leveraged to purchase sustainable seafood through: (1) increasing the diversity of bulk domestic seafood products included in USDA Nutrition Assistance Programs; (2) supporting states in developing "Fish to School" programs to purchase domestic sustainably harvested and farmed seafood with National School Lunch Program funds; and (3) changing institutional food procurement polices [88] (e.g., the Federal Food Service Purchasing Guidelines) for food sold at federal buildings and in the military so they support sustainably harvested and farmed seafood.

Foster federal agency investments in people Investments are commonly made by NMFS on the harvesting side of fisheries—vessel buybacks, and payments for involvement in cooperative research. But sustainable fisheries and sustainable seafood require new skills and relationships for fishermen and others involved in the supply chain in order to build new connections that will enable selling locally, regionally, to institutions, or using other supply chains that serve as an alternative to the current high import/export model. Innovations in fisheries, such as traceability technology, community-supported fisheries, and direct marketing, could be expanded rapidly with targeted government funds.

Shift assumptions about fisheries management (1) Expand the scope of fisheries management's general consideration of fish as being a food to more specifically include food systems and supply chains that form part of the broader socio-ecological connections integral to the viability and resilience of fishing businesses [98] and to think of fishing communities as part of this system rather than simply the home of fishermen or a landing site for vessels. This is a radical shift from the knowledge and training that most individuals who work in fisheries possess, and would require new collaborations. (2) Recognize that markets are not always the most effective way to distribute seafood to achieve MSA objectives beyond revenue and profit. In some fisheries, markets may serve objectives of food production and regional distribution better than others. Fisheries assessments will need to recognize that achieving optimum yield requires consideration of factors beyond fish stock sustainability and social and economic impacts to the harvest sector and fishing communities. Given that food is one of the elements of optimum yield, NMFS must consider the broader food system and seek partners who can operate in realms beyond its limited regulatory purview.

Change how success is measured How we measure success in fisheries management matters. Focusing on ecological sustainability (the Fisheries Stock Sustainability Index or FSSI, for example) is an incomplete picture of whether the greatest benefits to the nation are being achieved. A stock might be in good health but might be under performing in terms of its contribution to the food system. Fishery performance measures could overlap with other government agency performance measures to reinforce and clarify connections between food policy and fisheries policy, expanding beyond the harvest sector to include seafood distribution, food access, supply chain, and consumption patterns.

Foster new social science research Fisheries social science research has focused primarily on the harvesting sector and fishing communities. New areas of research that combine traditional social science fisheries research with food systems, health policy, and social justice research are necessary. Additional studies are needed on, for example: consumer preferences and consumption patterns with respect to lesser-known seafood products harvested from domestic sustainable fisheries and domestic aquaculture operations; interventions for reducing consumer seafood waste, such as clear package labeling or education [99]; and policy studies of how the Affordable Care Act can be leveraged to purchase local seafood for private hospitals and healthcare facilities [100]. We need to better integrate data on seafood access, consumption, and local markets with fisheries EAs and EISs. These research projects will necessarily involve collaborations across disciplines and research specialties.

Create new relationships The perspectives used to inform the fisheries management process will need to be broadened to reflect the range of stakeholders associated with the larger fisheries system and food system. Currently, Fishery Management Council meetings are attended mainly by harvester organizations, environmental NGOs and, to differing degrees by region, processing representatives. Consumer groups, public health professionals, civil society organizations, and local seafood marketers are largely absent from the collaborative management process, in part at least because they are not targeted for announcements of these meetings and may therefore be unaware of them. Increased outreach by Regional Fishery Management Councils might help to alleviate this. One promising evolution in the Northeast Region is the creation of a Seafood Marketing Group within the NMFS Greater Atlantic Regional Fisheries Office [101] that brings together processors, harvesters, chefs, scientists, educators, state food system representatives, and others to consider how to add value to the fish harvested in the region. Another new relationship worth exploring and supporting in a broader context is a joint policy statement from health professional associations (American Public Health Associations, American Medical Association, and American Society of Nutrition) and fisheries and aquaculture associations (American Fisheries Society, US Aquaculture Association) about the importance of synergy between the fisheries and health fields.

5. CONCLUSIONS

Managing fisheries to promote optimal nutrition and efficient food production likely requires a different approach to fisheries management—new outcomes will need to be monitored, new approaches found and fisheries, aquaculture, and health policies better integrated. The Magnuson–Stevens Act as well as NMFS strategic guiding documents all highlight the importance of fish as food, as do the USDA/HHS nutritional guidelines. However, the implementation of the MSA has not focused on food production outcomes. What happens to fish beyond the harvesting and processing level is currently beyond the scope of factors considered by the councils. Council decisions are informed by impact assessments that focus on harvesters, fishing businesses, and to some degree on impacts to fishing communities.

Simultaneously, while health policies such as the US Dietary Guidelines indicate the need to consume more fish, little consideration is given to the sustainability or source of fish, their connections with US fisheries systems, and global distribution of seafood. Nonetheless, some states have implemented "Fish to School" programs in school districts, and more opportunities for bulk purchasing of sustainably caught or raised seafood exist in nutrition assistance programs and institutional food-buying programs. Seafood businesses, universities, hospitals, NGOs, and other actors are also beginning to make connections between public health, fisheries, and aquaculture. These efforts are a positive sign because change will need to occur at multiple levels and across various domains (federal, state, and local) and dimensions of the food system (e.g., social, political, biophysical, and economic). Aligning fisheries, aquaculture, and health policies should begin first where the most promising opportunities exist. There are numerous small steps to begin this process with shared responsibility among government, industry, and civil society. We also call on researchers to continue addressing questions and issues raised here, and develop and test theories of change that move us in a direction toward a more just, equitable, and sustainable food system.

ACKNOWLEDGEMENTS

The authors would like to thank the following individuals for their helpful discussion: Dr. Patricia Majluf, Vice President of Oceana Peru; Paul Johnson, owner of Monterey Fish Market; Bill Carvalho, owner of Wild Planet; and Janice Fitzgeral, Household Programs Operations Branch Chief at the US Department of Agriculture, Food, and Nutrition Service. The authors also thank Eric Thunberg and Michael Simpson of NOAA Fisheries, Shawn McKenzie and Robert Martin at the Johns Hopkins Center for a Livable Future, and Anthony So at the Johns Hopkins Bloomberg School of Public Health for reviewing a draft of the manuscript.

REFERENCES

1. Food and Agriculture Organization of the United Nations. The state ofworld fisheries and aquaculture 2014. 2014: Rome.
2. Weichselbaum E, et al. Fish in the diet: a review. Nutr Bull. 2013;38(2):128–77.

3. Mahaffey KR, et al. Balancing the benefits of n-3 polyunsaturated fatty acids and the risks of methylmercury exposure from fish consumption. Nutr Rev. 2011;69(9):493–508.
4. Sanders TA. Protective effects of dietary PUFA against chronic disease evidence from epidemiological studies and intervention trials. Proc Nutr Soc. 2014;73(01):73–9.
5. U.S. Department of Health and Human Services and U.S. Department of Agriculture. 2015–2020 Dietary guidelines for Americans. 2015.
6. U.S. Department of Health and Human Services and U.S. Department of Agriculture. Dietary guidelines for Americans, 2010. 2010.
7. Hilborn R, et al. State of the world's fisheries. Annu Rev Environ Resour.2003;28(1):359.
8. enkins DJ, et al. Are dietary recommendations for the use of fish oils sustainable? Can Med Assoc J. 2009;180(6):633–7.
9. Brunner EJ, et al. Fish, human health and marine ecosystem health: policies in collision. Int J Epidemiol. 2009;38(1):93–100.
10. Greene J, et al. Fish oils, coronary heart disease, and the environment.Am J Public Health. 2013;103(9):1568–76.
11. Golden C, et al. Nutrition: fall in fish catch threatens human health. Nature. 2016; 534(7607):317.
12. Pauly D, Zeller D. Catch reconstructions reveal that global marine fisheries catches are higher than reported and declining. Nat Commun. 2016;7:10244. doi:10.1038/ncomms10244.
13. Worm B, et al. Rebuilding global fisheries. Science. 2009;325(5940):578–85.
14. Worm B, et al. Impacts of biodiversity loss on ocean ecosystem services. Science. 2006;314(5800):787–90.
15. Rockström J, et al. A safe operating space for humanity. Nature.2009;461(7263):472–5.
16. Pauly D, et al. Fishing down marine food webs. Science1998;279(5352):860–3. Costello C et al. Global fishery prospects under contrasting management regimes. In: Proceedings of the national academy of sciences, 2016. p. 201520420.
17. Waters CN, et al. The Anthropocene is functionally and stratigraphically distinct from the Holocene. Science. 2016;351(6269):aad2622.
18. Tidwell JH, Allan GL. Fish as food: aquaculture's contribution. EMBO Rep. 2001; 2(11):958–63.
19. FAO. State of the world fisheries and aquaculture in 2014. Rome: United Nations FAO Fisheries and Aquaculture Department; 2014.
20. Naylor R, Burke M. Aquaculture and ocean resources: raising tigers of the sea. 2005;30:185–218.
21. Metian AGTM. Fishing for feed or fishing for food: increasing global competition for small pelagic forage fish. AMBIO J Hum Environ. 2009;38(6):294–302.
22. Fry JP, et al. Environmental health impacts of feeding crops to farmed fish. Environ Int. 2016;91:201–14.
23. Troell M, et al. Does aquaculture add resilience to the global food system? Proc Natl Acad Sci. 2014;111(37):13257–63.
24. Olson J, Clay PM, Pinto da Silva P. Putting the seafood in sustainable food systems. Mar Policy. 2014;43:104–11.
25. He K, et al. Accumulated evidence on fish consumption and coronary heart disease mortality a meta-analysis of cohort studies. Circulation. 2004;109(22):2705–11.
26. Loring PA, Gerlach SC. Searching for progress on food security in the North American North: a research synthesis and meta-analysis of the peer-reviewed literature. Arctic. 2015;68(3):380–92.
27. Kittinger JN, et al. From reef to table: social and ecological factors affecting coral reef fisheries, artisanal seafood supply chains, and seafood security. PLoS ONE. 2015;10(8):e0123856.
28. Islam D, Berkes F. Indigenous peoples' fisheries and food security: a case from northern Canada. Food Secur. 2016;8(4):815–26.
29. Tilden J, et al. Health advisories for consumers of Great Lakes sport fish: is the message being received? Environ Health Perspect. 1997;105(12):1360.

30. Steinback S, Wallmo K, Clay P. Saltwater sport fishing for food or income in the Northeastern US: statistical estimates and policy implications. Mar Policy. 2009;33(1):49–57.
31. Gould LH, et al. Surveillance for foodborne disease outbreaks—United States, 1998–2008. MMWR Surveill Summ. 2013;62(2):1–34.
32. Lipp E, Rose J. The role of seafood in foodborne diseases in the United States of America. Revue Scientifique et Technique—Office International des Epizooties. 1997;16(2):620–40.
33. Nesheim MC, Yaktine AL. Seafood choices: balancing benefits and risks, ed. I.o. Medicine. 2007: National Academies Press, Washington, DC.
34. NOAA Fisheries. 2016 April 12, 2016. http://www.nmfs.noaa.gov/.
35. NOAA Office of Aquaculture. 2016. http://www.nmfs.noaa.gov/ aquaculture/.
36. USDA. Aquaculture. 2016 April 12. 2016. http://www.usda.gov/wps/ portal/usda/usdahome?navid=AQUACULTURE. FDA. Animal and veterinary. 2016. http://www.fda.gov/ AnimalVeterinary/.
37. USDA-APHIS. Focus on fish health. 2016 April 12, 2016. https://www. aphis.usda.gov/focusonfish/aphis-aquaculture-program.php.
38. USDA. Organic aquaculture. 2016 April 12, 2016. https://afsic. nal.usda.gov/aquaculture-and-soilless-farming/aquaculture/ organic-aquaculture.
39. Love DC, et al. Veterinary drug residues in seafood inspected by the European Union, United States, Canada, and Japan from 2000 to 2009. Environ Sci Technol. 2011;45(17):7232–40.
40. NOAA, Magnuson–Stevens Fishery Conservation and Management Act Seafood Import Monitoring Program. 2016. pp. 88975–88998.
41. USDA. Meat, poultry, fish, and shellfish CIDs. n.d. April 12, 2016. https:// www.ams.usda.gov/grades-standards/cid/meat.
42. FDA. The imported seafood safety program. 2016 April 12, 2016. http:// www.fda.gov/Food/GuidanceRegulation/ImportsExports/Importing/ ucm248706.htm.
43. FDA. Guidance for industry: HACCP regulation for fish and fishery products; questions and answers for guidance to facilitate the implementation of a HACCP system in seafood processing. 2016 April 12, 2016. http://www.fda.gov/Food/GuidanceRegulation/GuidanceDocumentsRegulatoryInformation/ Seafood/ucm176892.htm.
44. National Shellfish Sanitation Program. Guide for the control of molluscan shellfish 2013 revision. 2013.
45. FoodSafety.gov. Selected federal agencies with a role in food safety. n.d. April 12, 2016. http://www.foodsafety.gov/about/federal/index.html.
46. Millen B, Lichtenstein AH, Abrams S, Adams-Cambell L, Anderson C, Brenna JT, Campbell W, Clinton S, Foster G, Hu F, Nelson M, Neuhouser M, Perez-Escamilla R, Siega-Riz AM, Story M. Scientific report of the 2015 dietary guidelines advisory committee. 2015.
47. Wood-Wright N. New U.S. dietary guidelines ignore broad support for food sustainability. 2016, Johns Hopkins HUB.
48. Lagasse LP, Love DC, Smith KC. Country-of-origin labeling prior to and at the point of purchase: an exploration of the information environment in Baltimore City grocery stores. Ecol Food Nutr. 2014;53(1):58–80.
49. EPA. National listing of fish advisories. 2016 April 12, 2016. http://fishadvisoryonline.epa.gov/General.aspx.
50. Love DC, Hawes M, Harding J. State-level recreational fishing regulationsand fish consumption advisories in the United States: identifying opportunities for improved interagency collaboration. J Public Health Manag Pract. 2013;19(4):E11–9.
51. FDA. What you need to know about mercury in fish and shellfish. 2004 April 12, 2016. http://www.fda.gov/food/resourcesforyou/consumers/ ucm110591.htm.
52. Oken E, et al. Decline in fish consumption among pregnant womenafter a national mercury advisory. Obstet Gynecol. 2003;102(2):346.
53. Birch RJ, et al. Trends in blood mercury concentrations and fish consumption among US women of reproductive age, NHANES, 1999–2010. Environ Res. 2014;133:431–8.
54. FDA. Fish: What pregnant women and parents should know. 2014.

55. USDA. Supplemental nutrition assistance program (SNAP). 2016 April 12, 2016. http://www.fns.usda.gov/pd/ supplemental-nutrition-assistance-program-snap.
56. USDA. Women, infants and children (WIC): links to state agency WIC approved food lists. 2016 April 12, 2016. http://www.fns.usda.gov/wic/ links-state-agency-wic-approved-food-lists.
57. USDA. The emergency food assistance program (TEFAP). 2016 April 12, 2016. http://www.fns.usda.gov/tefap/about-tefap.
58. USDA. Food distribution program on Indian reservations (FDPIR). 2016 April 12, 2016. http://www.fns.usda.gov/fdpir/about-fdpir.
59. USDA. National school lunch program factsheet. 2016. http://www.fns. usda.gov/sites/default/files/NSLPFactSheet.pdf.
60. Jahns L, et al. Intake of seafood in the US varies by age, income, and education level but not by race-ethnicity. Nutrients. 2014;6(12):6060–75.
61. Oken E, et al. Which fish should I eat? Perspectives influencing fish consumption choices. Environ Health Perspect. 2012;120(6):790.
62. Verbeke W, Vackier I. Individual determinants of fish consumption: application of the theory of planned behaviour. Appetite. 2005;44(1):67–82.
63. NOAA. Fisheries of the United States 2013. Silver Spring: National Marine Fisheries Service Office of Science and Technology; 2014.
64. Moore LV, Thompson FE. Adults meeting fruit and vegetable intake recommendations—United States, 2013. Morb Mortal Wkly Rep. 2015;64(26):709–13.
65. Abbott-Jamieson S, Clay PM. The long voyage to including sociocultural analysis in NOAA's national marine fisheries service. Mar Fish Rev. 2010;72(2):14–33.
66. Loring PA, Gerlach SC, Harrison H. Seafood as local food: food security and locally caught seafood on Alaska's Kenai Peninsula. J Agric Food Syst Commun Dev. 2013;3(3):13–30.
67. FWS. About the U.S. fish and wildlife service. 2016. http://www.fws.gov/ help/about_us.html.
68. NOAA. Aquaculture policy and regulation: NOAA's role. n.d. April 12, 2016. http://www.nmfs.noaa.gov/aquaculture/policy/13_policy_and_ reg_homepage.html.
69. USDA. Aquaculture. n.d. April 12, 2016. http://www.usda.gov/ wps/portal/usda/usdahome?navid=AQUACULTURE.
70. USDA-ERS. Aquaculture Data. n.d. April 12, 2016. http://www.ers.usda.gov/data-products/aquaculture-data.aspx.
71. NOAA. Marine aquaculture strategic plan FY 2016–2020. 2015.
72. Interagency Working Group on Aquaculture. National strategic plan for federal aquaculture research (2014–2019). 2014.
73. Kirkley JE et al. International trade in seafood and related products: An assessment of U.S. trade patterns. 2006, National Oceanographic and Atmospheric Administration, National Marine Fisheries Service. pp. 1–147.
74. McClenachan L, et al. Do community supported fisheries (CSFs) improve sustainability? Fish Res. 2014;157:62–9.
75. National Marine Fisheries Service. Fisheries economics of the UnitedStates, 2012, E.a.S.A. Division, Editor. 2014.
76. Seelye KQ. In Maine, Last Sardine Cannery in the U.S. is Clattering Out. In: New York Times. 2010.
77. Pikitch E et al. Little fish, big impact: managing a crucial link in ocean food webs., Lenfest Ocean Program, Editor. 2012: Washington, DC. p. 108.
78. Chavez FP, et al. The northern Humboldt Current System: brief history, present status and a view towards the future. Prog Oceanogr. 2008;79(2):95–105.
79. Avadí A, Fréon P, Tam J. Coupled ecosystem/supply chain modelling of fish products from sea to shelf: the Peruvian anchoveta case. PLoS ONE. 2014;9(7):e102057.
80. Christensen V, et al. Valuing seafood: the Peruvian fisheries sector. Mar Policy. 2014;44:302–11.
81. Gagnon T. A guide to serving local fish in School Cafeterias, S.C. Society, Editor. n.d.
82. Izumi BT, et al. Serving fish in school meals: perceptions of school nutrition professionals in Alaska. J Child Nutr Manag. 2015;39(1):n1.

83. Profita C. Oregon delivers local seafood from boat to school. In: Oregon Public Broadcasting. 2013.
84. Horwitz S. Sea to school: models of local, sustainable seafood for schools. 2015. http://www.farmtoschool.org/news-and-articles/ sea-to-school-models-of-local-sustainable-seafood-for-schools.
85. Zeiber R. Program seeks to bring local seafood back to the school lunch menu. In: New Hampshire Sea Grant. n.d.
86. Fitch C, Santo R. Instituting change: an overview of institutional food procurement and recommendations for improvement. The Johns Hopkins Center for a Livable Future, Editor. 2016.
87. nd. Red's Best. 2016 June 23, 2016. https://www.redsbest.com.
88. Goad M. Maine hospitals look to local fish and vegetables to improve health and patient satisfaction, in Portland Press Herald. ME: Portland; 2015.
89. Brinson A, Lee M-Y, Roundtree B. Direct marketing strategies: the rise of community supported fishery programs. Mar Policy. 2011;35(4):542–8.
90. Siegel KR et al. Association of higher consumption of foods derived from subsidized commodities with adverse cardiometabolic risk among us adults. JAMA Intern Med. 2016.
91. Da Silva PP, Fulcher C. Human dimensions of marine fisheries: using GIS to illustrate land-sea connections in the northeast US herring, Clupea harengus, fishery. Mar Fish Rev. 2005;67(4):19–25.
92. IOM (Institute of Medicine) and NRC (National Research Council). A framework for assessing effects of the food system, N.A. Press, Editor. 2015.
93. NOAA. Seafood watch. 2016 April 21. http://www.fishwatch.gov/.
94. na. Local Food, Local Places: Bringing Expertise and Creative Thinking to Community Economic Development. 2014 June 23, 2016. https://www. whitehouse.gov/ blog/2014/12/03/local-food-local-places-bringingexpertise- and-creative-thinking-community-economic-.
95. Bittman M et al. How a national food policy could save millions of American lives. In: Washington Post. 2015.
96. Stoll JS, et al. Expanding the 'geography' of resilience in fisheries by bringing focus to seafood distribution systems. Ocean Coast Manag. 2015;116:185–92.
97. Love DC, et al. Wasted seafood in the United States from production to consumption. Glob Environ Change. 2015;35:116–24.
98. Smith T, Palmer A, Martin R, Fitch C. Achieving the triple aim in health care reform: the importance of the food system. The Johns Hopkins Center for a Livable Future, Editor. 2016.
99. GARFO. Member links. 2016 May 17, 2016. http://www.greateratlantic. fisheries.noaa. gov/educational_resources/seafood/links/index.html.
100. USDA-FNS. 2015 explanatory notes. 2015.
101. Papanikolaou Y, et al. US adults are not meeting recommended levels for fish and omega-3 fatty acid intake: results of an analysis using observational data from NHANES 2003–2008. Nutr J. 2014;13(1):1.
102. National Cancer Institute (NCI). Usual intake of total seafood 2007–2010. Epidemiology and Genomics Research Program. Online. https://epi. grants.cancer.gov/ diet/ usualintakes/pop/2007-10/table_a27.html.
103. Tran NL, et al. Estimated long-term fish and shellfish intake—national health and nutrition examination survey. J Eposure Sci Environ Epidemiol. 2013;23(2):128–36.
104. Wang Y, et al. Trends and correlates in meat consumption patterns in the US adult population. Public Health Nutr. 2010;13(09):1333–45.
105. NMFS. Commercial fisheries statistics. nd June 13, 2016. http:// www.st. nmfs.noaa.gov/commercial-fisheries/commercial-landings/ annual-landings/index.
106. Dueri S, Guillotreau P, Jiménez-Toribio R, Oliveros-Ramos R, Bopp L, Maury O. Food security or economic profitability? Projecting the effects of climate and socioeconomic changes on global skipjack tuna fisheries under three management strategies. Glob Environ Change.016;41:1–12.

ENDNOTES

1 We have limited our focus to federal fisheries because we are connecting federal fisheries policies to federal food policies. Furthermore, we concentrate on commercial fisheries because much recreational and subsistence fishing is regulated at the state level. To the extent federal recreational fishing regulations exist, comparable recreational catch data are not yet available for all states. Meanwhile, subsistence fishing that is governed by federal regulations is often governed under special community-oriented programs that are not easily comparable with standard fisheries governance structures

2 The amount of fish which will provide the greatest overall benefit to the nation, particularly with respect to food production and recreational opportunities, and taking into account the protection of marine ecosystems; is prescribed as such on the basis of the maximum sustainable yield from the fishery, as reduced by any relevant economic, social, or ecological factor; and in the case of an overfished fishery, provides for rebuilding to a level consistent with producing the maximum sustainable yield in such fishery [16 United States Code (USC) §1802(33)].

3 Some economists have also argued for the inclusion of maximum economic yield (MEY) in definitions of OY. However, given that MEY leads to lower fishery yields than MSY, this risks contradicting the requirement to consider food production (see [108]).

CHAPTER 11

Small-scale fisheries in the context of traditional post-harvest practice and the quest for food and nutritional security in Nigeria

Shehu Latunji Akintola* and Kafayat Adetoun Fakoya

Fisheries Department, Lagos State University, Badagry Expressway, P.O. Box 10123, Lagos, Nigeria

ABSTRACT

Background

Existing scholarly works have paid little or no attention to the problems small-scale fisheries encounter within the context of food and nutritional security. Therefore, this work has examined the problem. Small-scale fisheries (SSF) are the dominant source of local fish production in Nigeria and, like many fisheries worldwide, receive inadequate attention for their significance within the context of food and nutritional security. A central theme to guaranteeing small-scale fisheries' contribution to sustainable food and nutritional security in Nigeria is addressing post-harvest losses (PHL). Considerable fish loss occurs between production and consumption.

A narrative review was conducted to expose PHL practiced by SSF and discuss factors that prevent the adoption of locally available interventions.

Finding

Our study revealed that sun-drying and smoking are the major interventions practiced to mitigate PHL. Unfortunately, these methods of reducing PHL are constrained by gross under-capacity and improper handling during peak fishing periods and issues related to product safety. There are a handful of potential intervention mechanisms to address issues of PHL. Many of these PHL-reducing interventions have been designed bearing in mind local contexts and situations. However, general inertia and poor policy implementation currently hamper progress for reducing PHL.

Conclusion

A plethora of strategies are proposed to reduce PHL from SSF as contained in national and regional documents such as the National Policy on Food and Nutrition (NPFN), National Policy on Food and Nutrition in Nigeria (NPFNN), National Energy Policy (NEP), and Comprehensive Africa Agriculture Development Programme (CAADP). Recommendations are offered on identified gaps in areas of technological adoption, safety, and product quality issues, which are detrimental to traditional fish processing systems.

Keywords: Small-scale fisheries, Food security, Nutrient, Smoking, Sun-drying Policy, Nigeria Gender

1. BACKGROUND

The quest to achieve or sustain fish food security and nutrition will continue to be difficult to realize when efforts are directed only at increasing production without corresponding efforts at reducing post-harvest losses (PHL) in fish. In Nigeria, the burden of supplying local fish needs to ensure food security and nutrition is borne mostly by small-scale fishers (Table 1). While there are considerable efforts aimed at increasing production from SSF, an important aspect of reducing fish loss, especially PHL, has not been given desired attention. There are huge fish losses at PHL levels in the artisanal fisheries sector as the SSF are not able to keep pace with harvest and preservation.

Table 1: Nigeria fish production by sectors (2000–2012) tones

S/no.	Sectors/year	2000	2001	2002	2003	2004	2005	2006	2007	2008	2009	2010	2011	2012
1	Artisanal: sub-total	418,069	433,537	450,965	446,203	434,830	490,594	518,537	504,226	511,382	598,211	616,981	638,486	668,754
	Coastal and brackish water	236,801	239,311	253,063	241,823	227,523	259,831	269,878	260,098	264,988	309,981	328,332	346,381	370,918
	Inland: rivers and lakes	181,268	194,226	197,902	204,380	207,307	230,763	248,659	244,128	246,394	288,230	288,649	292,105	297,836
2	Aquaculture (fish farm)	25,720	24,398	30,664	30,677	43,950	56,355	84,533	85,087	143,207	152,796	200,535	221,128	253,898
3	Industrial (commercial trawlers)	23,308	28,378	30,091	33,882	30,421	32,595	33,778	26,193	29,986	29,698	31,510	33,485	45,631
	Fish (inshore)	13,877	15,792	16,065	17,542	16,063	19,724	19,129	18,040	18,585	18,820	19,261	19,736	27,977
	Shrimp (inshore)	8056	12,380	12,797	11,416	12,469	10,946	13,767	5995	9881	10,878	12,249	13,749	17,654
	EEZ	1375	206	1229	4924	1889	1925	882	2158	1520	–	–	–	–
	Grand total	467,098	486,313	511,720	510,762	509,201	579,544	636,848	615,507	684,575	780,705	849,026	893,099	968,283

Unpublished FDF, Federal Department of Fisheries

Traditional methods of sun-drying and smoking (using mud kiln) remain popular especially where the facilities supporting ice production are inaccessible due to poor electricity supply and the prohibitive cost of ice-making machines. In tropical environments such as Nigeria, fish spoils within 10–20 h. Fish by nature is subject to rapid deterioration once outside water because of its chemical composition, which may be made up of 70–84% water, 15–24% protein, 0.1–22% fat, and 1–2% minerals [1, 2]. Depending on the fish type, factors including high moisture and fat and protein contents promote rapid changes in quality in fish. Losses in fish without interventions could be as high as 27.3 ± 14.3% or more, while loss can be minimized as much as 14.7 ± 11.9% with interventions [3].

2. PURPOSE OF STUDY

Traditional fish processing is regarded with contempt and excluded largely in the food and nutrition schemes in Nigeria. This paper provides information on the importance of traditional fish processing interventions by the small-scale fishers to achieving food and nutrition security in the country. It seeks to redirect attentions of stakeholders towards activating the many provisions in the food and energy policy, as well as other documents that impact on food and nutrition. And which adopted will ginger a general renewal, standardization and safer product and at the same time remedy misgivings against traditional fish processing interventions.

3. RESEARCH CONTEXT

In the sections following introduction, we present the definitions of key concepts in this paper including fish, food, and nutrition security. Explanations were offered on the principles of the pillars of food security with necessary connections that elucidate the concept of nutrition within the food and nutrition precepts. Next, fisheries as a food security mechanism were presented in relation to provision of important nutrients such as protein and omega-3 fatty acids crucial to sustaining good health in human. Other sections were presented that help concretise the narrative of traditional fish processing by the SSF in the context of food and nutrition security in the country.

4. DEFINITIONS OF FISH, FOOD, AND NUTRITION SECURITY

The terms "fish" and "seafood" are defined as finfish (vertebrates) and shellfish (invertebrates), whether of marine or freshwater origin, farmed or wild [4]. At the global level, fish and seafood consumption has gained increasing attention compared to other food sources. This development is easily adduced to greater understanding of the unique qualities of the nutrients found in fish.

Food security refers to the condition in which all people, at all times, have physical, social, and economic access to sufficient, safe, and nutritious food that meets their dietary needs and food preferences for an active and healthy life [5]. Food availability, stability of supplies, and food access are related determinants of food security. In addition, the term "food and nutrition security" is used to emphasize the access and appropriate utilization of micronutrient-rich foods, including the process through which they are cooked and absorbed in the body, and then used in physiological functions at the individual level [6]. This outlook was adopted in this review. Nutrition security is achieved for a household when secure access to food is coupled with a sanitary environment, adequate health services, and adequate care to ensure a healthy life for all household members.

5. FISHERIES AS A FOOD SECURITY MECHANISM

The importance of fish in seeking food and nutrition security for all is a frequently overlooked but extremely important part of global food and nutrition security [7]. Fish is nutrient rich and provides high-quality protein that is low in saturated fat and rich in polyunsaturated fats, especially the omega-3 fatty acids eicosapentaenoic acid (EPA) and docosahexaenoic acid (DHA) [8].

The contribution of both fish protein in general and small-scale fisheries in particular to food and nutrition security has been recognized and advanced at the national and international level [6]. According to the Food and Agriculture Organization (FAO), all West African fishing grounds are either fully exploited or over-exploited [9]. As a result, many development projects since the 1950s aimed at expanding fishing effort in West Africa have often ended in failure [10]. However, neither domestic fish stocks nor the sources of imported fish can be expanded to the drastic extent needed to meet at least the global average of 20 kg per capital consumption in a short timeframe. Moreover, the focus on fish within the food security context has been mainly limited thus far to questions of accessibility and stability, leaving huge gaps in policy implementation with respect to utilization and availability. Therefore, there is a need to consolidate management efforts towards directly reducing fish loss specifically within SSF in order to guarantee availability and utilization, social equity, and sustainability of the fisheries resource.

Using the United Nations International Children's Emergency Fund (UNICEF) conceptual framework [11] as a basis for analysis, this review discussed determinants of nutritional security of fish as a food resource at the level of SSF. The relationship linking traditionally processed fish and nutritional well-being is not causal, but is determined by many basic and underlying factors. The study provides insights into a number of intervention technologies available locally which currently are not being adopted but could ameliorate fish loss if adopted. Furthermore, this study recommends the adoption of creative and sustainable strategies for bridging identified gaps and enacting enduring strategies with strong potential for increasing the competitiveness of traditionally processed fish.

6. CONTEXT OF FOOD SECURITY AND FISH CONSUMPTION IN NIGERIA

Individual nutritional status is determined by various factors. In this review, the UNICEF [11] conceptual framework for nutritional status is employed (Fig. 1). The technology available at the national level defines the ability to translate or change the status of a resource from potential to available. The availability of a resource, particularly food-related resources, determines national nutritional outcomes in terms of survival, growth, and long-term development. Specifically, the level of technological development to harvest the fisheries resources is just localized and crude (Table 2). On a broader conceptual level, the political, ideological superstructure, and economic structure of Nigerian fisheries act to prevent traditional fish processing technology from realizing its own potential. For these reasons, there is a high level of post-harvest fish losses within the SSF sector. Within all fisheries, but especially SSF, resource constraints, institutional controls, and technological limitations hinder the full development of capture fisheries.

Food and nutrition security can only be achieved if adequate food (quantity, quality, safety, and sociocultural acceptability) is available, accessible, and satisfactorily used by all individuals at all times to live a healthy and active life. Household food security, maternal and child care, and health services and healthy environment are three pillars of underlying determinants of nutritional status. Diversifying livelihoods to increase household income sources is a way of improving household food security. Although fisheries-related literature often does not focus these other pillars, a consideration of children's nutritional status requires the examination of determinants such as child care, especially breastfeeding, and their importance in ensuring child nutrition. Access to health services and the living environment are also important factors in determining the risk of disease.

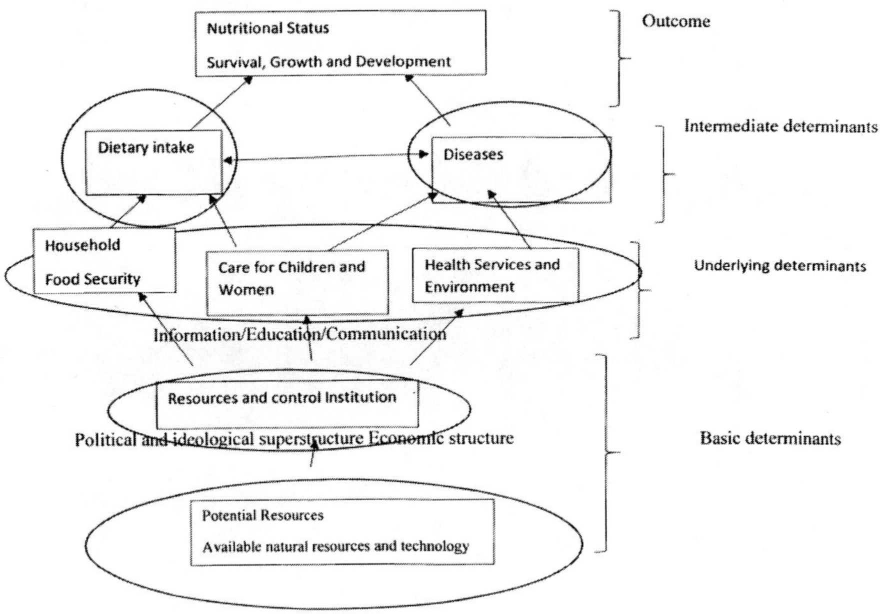

Figure. 1 Determinant of nutritional status [34]

Therefore, adequate dietary intake and the absence of disease are immediate determinants of nutritional status. The three factors of household food security, care for children and women, and access to quality health services and environment are necessary but not sufficient conditions for adequate nutrition. If all three are fulfilled, however, it is likely that dietary intake will be satisfactory, disease will be controlled, and adequate nutrition will be secured. Individuals with adequate dietary intake are able to develop appropriate immune systems and are thus better equipped to fight infectious diseases. As a result, breastfeeding mothers with adequate dietary intake will be able to pass on adequate and quality nutrients to their children during the first six months of exclusive breastfeeding, which required as foundation preparatory to subsequent life stages.

Table 2 Intervention areas in traditional fish processing and national development

Activities	Basic determinant	Underlying determinant	Immediate determinant
Issues to address	Rustic technology and safety Per capital fish consumption	Utilization	Hidden hunger (micronutrient deficiency)
Focus	Inculcate tastes for fish/seafood especially processed forms	Improving on the chain management of traditionally processed fish to enhance household access Increasing household income	Increasing micronutrient intake by promotion of traditionally processed seafood and artisanal fisheries
Interventions	Modernization, adoption, and promotion of traditional fish processing technology by government agencies, viz. Federal Department of Fisheries, Energy Commission of Nigeria, Federal Department of Forestry, Ministry of Science of Tech Mass communication: National Committee on Food and Nutrition, National Agency for Food and Drug Administration and Control, Academic and Research Institution	Engaging traditional fish processor towards adoption more efficient and effective technology Granting repayable credit facilities guaranteed by the processors' cooperative and thrift societies for purpose of securing modern and improved local facilities Improving the quantity of smoked and sun-dried fish sourced from the cultured sources (marine and inland water)	Utilizing smoked and sun-dried fish as complementary food to improve nutritional status of children Encouraging consumption of smoked and sun-dried prawns, shrimps especially by children and women at household levels Improving diet of children to providing fish as a school lunch
Expected effects	Decrease in fish loss Increased production by fish processors	Safer and increased consumption of sun-dried and smoked fish products	Reduction in incidence of hidden hunger
Indicators	Better regulated traditional fish processing sector	Increased prosperous individuals	More virile and healthier citizens Reduced cases on nutrition related diseases such as obesity

In several of the diverse cultures of Nigeria, a meal is not considered complete without some form of fish or fish product as the main animal protein component, or as a condiment. Fisheries' contribution to animal protein in the average Nigerian diet has been rising steadily and currently is at 28% [12]. Fish' contribution to animal protein consumption is 35% for urban and rural dwellers in Nigeria (national figure).[1] The value is very low compared to the Ghanaian average value of 63.2% [13]. Investment aimed at reducing food losses, therefore, offers an important pathway towards improving food security, alleviating poverty, and improving nutrition. In addition, reducing PHL in food generally has positive impacts on the environment as it enhances farm-level productivity and reduces the utilization of production resources or expansion into fragile ecosystems to produce food that will be lost and not consumed [14].

Nigeria's per capita fish consumption of 9.38 kg and population of 170 million classify it as a modest fish-consuming nation, yet in the NPFN 2001[2] document, fisheries were mentioned only in passing as a means of improving food intake compared to the greater emphases on grains or crops. Similarly, the National Plan of Action on Food and Nutrition (NPAN)[3] in Nigeria barely made mention of fisheries despite highlighting the need to improve food harvesting, processing, and preservation as a strategy towards enhancing food and nutrition security. The document also noted the need to improve the nutritional status of all Nigerians and particularly the most vulnerable groups such as children, women, and the elderly.

Through indigenous knowledge (IK) and traditional knowledge (TK), humans globally have employed salting, frying, sun-drying, and smoking for centuries as means by which the process of spoilage in fish is controlled [15]. However, the literature on SSF has largely overlooked the reliance of this category of fisheries on these kinds of traditional preservation methods. These simple methods remain the prevailing means for distributing fish products for either household consumption or sale in local markets within the fishing community [15], as well as distant ethnic markets in Europe, Asia, and North America, and regional fish trade within Africa. Also, these methods are the precursors for advanced food preservation or processing procedures, including freezing and canning, which are undertaken to prevent fish spoilage and reduce PHL.

7. CASE STUDY OF SMALL-SCALE FISH PROCESSING IN NIGERIA

Traditional fish processing methods enhance fish availability, affordability, and choice, thereby promoting nutrition, health, and well-being. Fish processing and sales in fishing communities in West Africa, including Nigeria, have been reported as a "gendered" activity comprising mainly women [16]. Governance structure for the traditional fish processing sector is a descriptive and adaptable framework which underscores the importance of these gender dimensions. Access to fish after harvest is determined by the relative position and scale of operation of women in fish processing. Scales at which women play these roles often attract different description by fishermen and male folks. Richly endowed women are able to influence happening within the value chain downstream dictating fish supply direction over the less fortunate women processor and attract better patronage from the fishers. Women are also able upset local arrangements with regards to credit scheme facilitated between the fisher and the female processors.

8. OVERVIEW OF SMALL-SCALE FISHERIES IN NIGERIA

Nigeria's fisheries sector is almost exclusively artisanal or small-scale, grossly underdeveloped, and highly heterogeneous. It contributes over 70% to domestic fish production, providing livelihoods to approximately 6.4 million fisher folk [12]. The main fishing and fishing-related activities are carried out in fishing settlements located in the Atlantic coastal area of the southern part of the country. Species-rich brackish water and estuarine canoe fisheries occupying about 12,904 km^2 exist in the creeks, estuaries, lagoons, and mangrove wetlands. Women fishers play prominent roles in this fishery [12]. Marine fishing is dominated by the coastal mechanized and canoe fisheries operating within five nautical miles of the sea shore, which is statutorily designated as non-trawling zone and 39,644 km^2 continental shelf area adjacent to the country's 853 km coastline.

Migrant fishers from neighbouring West African nations exploit fisheries resources 120 nautical miles within the extensive frontiers of the country's 216,325 km^2 Exclusive Economic Zone in the Gulf of Guinea [17]. Further inland, freshwater fisheries thrive consisting of nationwide network of rivers such as the Niger and Benue with several tributaries, natural lakes (e.g. Lake Chad), man-made lakes (e.g. Lake Kainji), reservoirs, and flood plains totalling an estimated 12.5 million hectares [17].

The SSF sub-sector is also multi-gear and multi-species in character. The craft and type of net depend on the type of fishery operated. Gears such as purse-seine net, set gill net, cast nets, encircling nets, traps, fyke nets, bag nets, pots, wounding gears, baskets and traps, drift and set gill nets, long lines, and trawl nets are used to target different fish species. Crafts (3–13 m long) used range from paddled, small-sized planked, or dugout canoes to larger, motorized canoes (half dugout/half plank) with outboard engines ranging from 15 to 45 hp, which are more like the vessel type commonly employed in Ghana.

Major freshwater fishes are fished year-round. Target species commonly caught include Nile perch (Lates spp.); tilapias (Oreochromis and Hemichromis spp.); catfishes (Clarias, Heterobranchus and Synodontis spp.): silver catfishes (Chrysichthys and Bagrus spp.); elephant snouts (Gnathonemus and Momyrus spp.); trunk fish (Gymanarchus spp.); tongue fish (Heterotis spp.); and moon fish (Citharinus) spp. [18]. The fish community inhabiting the estuaries, creeks, and coastal waters consists of small pelagics of the family Clupeidae. Bonga (Ethmalosa sp.) dominates the coastal pelagic fishery. Sardines (Sardinella sp.), shad (Ilishaafricana) various jacks (Caranx spp.), and Atlantic bumpers (Chloroscombrus chrysurus) are also caught in modest quantities. Croakers (Pseudotolithus spp.) constitute the main commercial demersal stock [18] in coastal and brackish fisheries. Other exploited species include grunters (Brachydeuterus sp.), soles (Cynoglossus spp.), marine catfish (Arius sp.), brackish water catfish (Chrisichthys), Sphyraena spp. (barracuda), snappers (Lutjanus sp.), threadfins (Galeoides decadactylus, Pentanemusquinquarius), and groupers (Epinephelus). Common exploited shellfish resources are estuarine/white shrimps ("crayfish") (Palaemon sp.) which supports a major creek fishery in the Niger Delta, Macrobrachium vollenhovenii (African river prawn); M. macrobrachion (Brackishwater Prawn); marine and estuarine Penaeus notialis (Southern pink shrimp); Parapenaeopsisatlantica and (Guinea shrimp). Other shellfishes include crabs of the genus Callinectes in the families Portunidae and Geryonidae, molluscs such as the mangrove oyster (Crassotreagasar), periwinkles (Tympanostomus spp.), and certain bivalves [19].

Actual production figures from SSF could be grossly undervalued. There is limited information on the state of fisheries resources nationwide. Many water bodies have not been assessed for their fisheries potentials, and stock assessments available are very few in number and out-dated. There is potential underestimation in catches ranging between 100,000 and 180,000 mt annually for artisanal shrimp fisheries, part-time fishermen, and brackish water areas [20]. When combined with poor availability of data from freshwater fisheries, this implies an enormous gap in overall fish landings for the country. Fishing communities are numerous and generally highly scattered in remote, inaccessible settlements due to poor or non-existent access roads and susceptibility to flooding during the rainy seasons. They usually do not have access to basic social infrastructures. Facilities such as cold storage and processing plants are very poorly developed. There is, therefore, heavy reliance on traditional fish processing and preservation methods which, unfortunately, are bedevilled by gross under-capacity and improper handling particularly during the peak fishing period.

PHLs are characteristics of SSF in developing nations, and an accurate estimation of total losses is problematic. In Nigeria, these have been reported to exceed 30% of the total catch [21]. A seasonal glut occurs between November to May for sardines and Bonga (Clupeids). Under these circumstances, fisherfolk are unable to cope with heavy fish catches, resulting in substantial quantities of fish being wasted, leading to putrefying gluts of spoilt fish which litter the sandy beaches and buried spoilt fish in the sands [22, 23]. The difficulty experienced in the evacuation, distribution, and marketing of fish products further contributes to a high percentage of PHL along the entire value chain from harvest to consumption. Ultimately, regardless of the quantity of fish wasted, this waste also translates into huge financial losses, reduction in the quantity of available fish supplied, and animal protein for human consumption, respectively, thus threatening food security.

9. FISH TRADE AND FOOD AND NUTRITIONAL SECURITY: NIGERIA SCENARIO

Nigeria is ranked among 55 nations classified as being low-income food-deficit countries (LIFDCs). This list, created by the FAO in 2014, is based on three criteria. First, a country should have a per capita gross national income (GNI) below the "historical" ceiling used by the World Bank to determine eligibility for International Development Association (IDA) assistance and for 20-year International Bank for Reconstruction and Development (IBRD) terms, applied to countries included in World Bank's categories I and II. The 2014 LIFDC list is based on the GNI for 2011 (estimated by the World Bank using the Atlas method) and the historical ceiling of USD 1 945 in 2011.

The second criterion is based on the net (i.e. gross imports less gross exports) food trade position of a country averaged over the preceding three years for which statistics are available. Trade volumes for a broad basket of basic foodstuffs (cereals, roots and tubers, pulses, oilseeds and oils other than tree crop oils, meat and dairy products) are converted and aggregated by the calorie content of individual commodities. Thirdly, the self-exclusion criterion is applied when countries that meet the above two criteria specifically request to be excluded from the LIFDC category.

This designation means that the country must pursue a policy of balancing the quest for fish trade and food security in relation to fish as a food or commodity. Fish is an important component of the National Economy and

Trade, since Nigeria is both a fish exporting and importing nation. Fish has an inherent tendency to be commoditized, and seafood have been commodities that were preserved and traded since the Bronze Age [24]. Therefore, fish have played an important role in the first phase of the emergence of a global economy and a singular role in the international expansion of trade before becoming an internationally traded commodity in its own right.

Specific to fish, Nigeria has a positive net food trade averaged over the previous three years [12]. Products emanating from the industrial fisheries, and shrimp in particular, are destined for the European and American markets where they attract premium values. The protein needs for the domestic markets are, therefore, met largely by supply from the artisanal fisheries, importation, and, by far less, aquaculture. The poor state of infrastructure, especially electricity supply and irregularity of fish supply from SSF occasioned by unpredictability of catch, fuel the need to resort to traditional means of processing by fishmongers. Generally speaking, infrastructure required driving both the quest for fish food security and trade has been reported [25] and will include fish landing centres, processing facilities, link roads, stable electricity, portable water supply, housing, and sanitary and environmental engineering works. These infrastructures can be broadly grouped as livelihood and trade related.

The scenario we have today is that investment directions in infrastructure development have been tilted largely in favour of trade-related infrastructure development favouring exportation and importation. Although the traditional food processing business invariably benefits from increased link roads and improved electricity, it presently suffers from neglect over the years to ensure broad-based investment in provision of processing facilities, storage, good hygienic practices (GHP)/good manufacturing practice (GMP), or sanitation standard operating procedures (SSOP), or prerequisite programmes for food safety. In contrast, food safety and quality management have become standards in the formal fish trade markets. The fish processing business requires that effective hygienic, quality, and safe food be promoted. A good healthy environment is crucial if food supplies are to ensure food security and nutritional well-being.

Fish in the Nigerian Food Policy have been largely directed towards increasing production, with lessening emphasis on getting the processing and distribution right. Efficient and effective domestic fish marketing is the ground rules for achieving sustainable fish food, nutritional well-being, and, ultimately, sustainable development. Presently, traditional fish processing techniques of salting, smoking, and sun-drying are at the centre of guaranteeing fish protein, food security, and nutritional well-being. The more lucrative export fish trade is the focus of medium fish processors whose major supply is from the aquaculture supply. The traditional fish processing methods are faced with serious challenges in view of the increased consumers' awareness on matter of food safety and qualities.

There is a general sense of apathy from government for monitoring, regulating, and intervening in the business of traditional processing by means of interventions such as the modernization of PHL reduction technologies, product standardization, and regulation. The need for government to give direction is of the utmost importance, given that the small-scale operators are not organized to create the necessary structures that must be put in place to bring about the required changes. They do not enjoy the economic stimulus offered by foreign international markets.

10. NUTRITIONAL STATUS OF CHILDREN IN NIGERIA

The 2013 Nigeria Demographic and Health Survey provided new insight on the nutritional status of children and adults in the country [26]. About 37% of children under 5 years old in Nigeria are stunted, thus reflecting chronic malnutrition. National figures show that approximately 10–20% encountered acute malnutrition and 29% are described as underweight, reflecting chronic or acute malnutrition or a combination of both [26]. Under-nutrition is described as an outcome of insufficient food intake and repeated infectious diseases, while malnutrition refers to both under-nutrition and over-nutrition [27].

The problem of micronutrient deficiencies is different in nature as it results from an inadequate quality of diet. Micronutrient deficiencies increase the general risk of infectious illness (e.g. measles) and non-infectious diseases such as malaria and pneumonia, or even diarrhoea. Adequate intake of essential micronutrients, especially by pregnant women and young children, is important to attaining nutritional security. Appropriate nutrition includes feeding children with a variety of foods to ensure that nutrient requirements are met. It is well known that plant-based complementary foods (food given to a child after exclusive breastfeeding) by themselves are insufficient to meet the needs for certain micronutrients [1, 2]. It has been advised that meat, poultry, fish, or eggs should be eaten daily, or as often as possible. Fish has been variously suggested to be an important part of infant feed or complementary food [1, 2].

11. CONTRIBUTION OF FISH AND SSF TO FOOD SECURITY AND NUTRITION IN NIGERIA

To understand the importance of SSF in contributing to food security, it is necessary to grasp the significance of underlying factors which affect fish consumption patterns. In 2002, the FAO estimated that fish provide about 19% of the protein intake in developing countries with the following qualification: "the figure, however, represents an average at a global level and does not reflect the very large heterogeneity at the national or, even more importantly, at the local level. Similarly, fish contribution to animal protein consumption can be as low as 23% in Liberia and as high as 63.2% in Ghana" [13].

In Nigeria, fish is reported to be the most frequently consumed animal protein consumed by households [28]. Per capita fish consumption in Nigeria increased from an estimated 13.71 kg in 1993 to 14.49 kg in 2000 [20]. However, comparing Nigeria's per capita fish consumption of 13.9 kg to the global average of 18.9 kg in 2011 [29] shows that fish consumption is still inadequate. Disaggregating national data [25], (see footnote 1) the percentage of fish as animal protein at the national level is 35, or between 41.15 and 43.1 of animal source protein and 39.7 of animal-source food [29]. While cultural standards appear to have no bearing on motivations to consume fish [30], decisions around the consumption and choice of fish product form is governed by factors such as income and market values. Other important factors are the urban–rural divide, non-fishing households versus fishing households, and relative distance or proximity to common-pool resources. However, much of the sector's contribution to rural household consumption is often undervalued, and their importance ignored. In the Nigerian context, small-scale fisheries supply as much as 75% of animal protein intake in coastal or inland fishing communities [31]. Generally, fishing households have greater access to fish for direct consumption. Fish from common-pool resources or open-access regimes are literally free for harvesting and contribute in large part to nutritional security at the local level, particularly in the diets of poor people. Subsistence fishing of small fish, crustaceans, and molluscs, though almost marginal in terms of quantity, plays an important role in the food and nutritional security of household members. Consequently, per capita annual consumption of fish in fishing households (an average 230 g per day) is usually higher compared to non-fishing households [31].

12. PATHWAYS LINKING TRADITIONAL FISH PROCESSING AND NUTRITIONAL SECURITY TO NATIONAL DEVELOPMENT

The national aggregate wealth generated from inland fish production and consequently traditionally processed fish are estimated between US$ 230–330 million and US$ 280 million per year.[4,5] Processing is an important activity within the fishing industry, although it is often ignored. However, it is very evident that traditional processing like small-scale fisheries, in general, is an activity that is too big to ignore. As a sub-sector, its economic output provides an important dimension and interlinked pathways to national development in many ways.

Directly, traditional fish processing contributes to food and nutritional security in terms of the utilization, availability, and stability of price at the individual or household, regional, and national levels. Increased levels of protein, carbohydrates, fat, and energy were recorded in boiled, grilled, and fried fish in comparison with the raw forms [32]. Smoked products offered the healthiest advantage with the lowest values of saturated fatty acids, index of atherogenicity, index of thrombogenicity, and the highest omega-3/-6 ratios, which are important for reducing coronary heart disease (CHD) in food consumption [1]. Furthermore, smoking increases the concentration of alanine, threonine, tyrosine, and cysteine and accounts for a sweeter taste of the shrimp over other product forms, while sun-dried specimens were found to have increased levels of histidine and arginine [1, 2].

In a country with high levels of malnutrition among children in many regions, the eradication of malnutrition is a critical national development goal. It is well established that fish products as a form of animal protein can effectively be used as complementary foods to fight malnutrition. In addition, since nutrients from seafood are cheaper than other sources of animal nutrients, consuming processed fish products (which are reported to compare even better than raw fish) provides affordable high-quality protein. It is cheaper to eat processed fish forms such as sun-dried, smoked, grilled, and boiled (authors' experience).

The present disdain and apathy with which traditional fish processing methods are viewed require a change in policy direction to ensure increased fish availability, particularly in a country where every gram of protein and energy is crucial for reducing considerably the nation's high protein–energy malnutrition (PEM). The national average protein and energy intake of 53 g and 2071 kcal is below the World Health Organization (WHO) standard. Thus, it is crucial to provide support to the traditional fish processing sector and to improve the access of citizens, especially children, to processed fish protein.

13. TRADITIONAL FISH PROCESSING ACTIVITIES AND GENDER

Fish value chains are highly gendered and fraught with inequalities, creating food security and nutrition challenges at local, regional, and national levels. It is well known that of workers in SSF, women working in processing outnumber men. Unfortunately, much of the work performed by women in post-harvest processing is not compensated or is poorly recorded. Hence, to understand the gendered dynamics between actors in the value chain in the production of traditionally processed fish, it is essential to consider the governance structure for this important sub-sector.

In Nigeria, the traditional culture of the people is the basis of community-level governance structures, including in fishing communities [15, 33]. Artisanal fish processing and preservation activities are primarily carried out by women. However, in rare instances cultural norms restrict the participation of women in post-harvest activities at landing sites, thus leaving post-harvest activities to men. Women often engage in fish processing as their sole economic activity, or combine this with tasks upstream or downstream the value chain, or outside of fishing altogether [15]. Fish processing is carried out at different scales, depending on the quantity of fish being handled. The producer–processor group, which occurs the most, is prevalent in rural fishing communities and may involve male fishers [34]. Women processors obtain their supplies from within their family (usually the husband), while a small number of women engage in fishing themselves with the use of both trap setting and gear netting. Sometimes, women fishers obtain fish under different purchase arrangements from their male counterparts. Processor–sellers/traders form another category which mostly consists of women who combine functions of processing and marketing [35]. Lastly, women engaged solely in the trading of processed fish buy their products at either wholesale or retail markets. They also sell via commission agents or brokers to different market intermediaries. Sometimes commission agents or brokers serve as links between the many intermediaries along the value chain.

14. EVALUATION OF TRADITIONAL FISH PROCESSING TECHNOLOGIES

Technology drives many human endeavours at all levels of human activities and history, from rudimentary tools to advanced modern-day technology. Community-based knowledge, which has evolved in every culture over a long period of time in many generations, is referred to as indigenous knowledge (IK) or traditional knowledge (TK). However, TK/IK may have borrowed extensively from the process of technological adaption as people began to borrow from other knowledge outside of their geographical location and traditional practices. In the case of fisheries and fish processing in particular, women employ the use of TK/IK in preserving fish in order to increase its shelf life and ensure value addition. Common traditional methods include gutting, washing, splitting, filleting, and sticking the fish, cooking, salting (brining), smoking, curing, fermentation, and sun-drying (drying).

15. COOKING

It is the most common and simple method requiring no more than basic household equipment which provides short-term preservation of fish usually for a few days before any deterioration becomes noticeable. Although warming after cooking does extend the period, a loss in value quickly becomes discernible. Different methods are used for cooking fish, but the principle of these processes is similar. The flesh of the fish softens, enzymes become inactivated, and the process kills many of the bacteria present on the surface of the fish.

16. SALTING

Salting is by far the most influential form of fish preservation and trade, serving as the precursor for international commodity trading. The principle is based on the knowledge that food-poisoning bacteria cannot live in salty conditions and a concentration of 6–10% salt in fish tissue will prevent bacterial activity, thereby impacting a longer shelf life. However, a group of micro-organisms known as "halophytic bacteria" are salt-loving and will spoil the salted fish even at a concentration of 6–10%. Traditional methods of using salt usually involve removing the guts and gills and cutting the flesh into pieces before rubbing salt into the flesh or making alternate layers. The recommended levels of salt usage are 30–40% of the prepared weight of the fish.

However, the concentration of salt in the flesh is not sufficient to preserve the fish if it is not uniformly applied. Brining offers a better preservation over the direct application of salt to the fish muscle and tissues. This process involves immersing the fish into a pre-prepared solution (36% salt). The advantage is that the salt concentration can be more easily controlled and salt penetration is more uniform.

In most fish preservation, it is a usual practice to use salt in combination with other means of preservation. Nevertheless, the use of salt curing for food preparation must be approached with a high degree of caution, since high salt intake is deleterious to good health. The growing concern is founded in reports of increased heart palpitations in some individuals. This occurrence could predispose pre-hypertensive and hypertensive individuals to additional health risks. Thus, there is a need to investigate and ensure safe concentrations of domestic salt in fish preservation.

17. SUN-DRYING

This process consists of the use of the sun and movement of air to remove moisture and preserve the fish. In order to prevent spoilage, the moisture content needs to be reduced to 25% or less. The time it takes to dry fish products depends on the nature of the product, the intensity of the sun, and the surfaces used for drying. The simplest form of drying involves exposing whole small fish or split large fish to heat from the sun by placing products either directly on the ground, roofs, nets, and mats placed on the ground or on racks. However, this technique makes the fish susceptible to predations by animals. This method is commonly used in the Northern parts of the country, due to the sun's intensity and other favourable conditions such as dry weather, low humidity, and clear skies [2]. Fish is typically sun-dried for three to ten days, but drying periods of one to three days are more common [36].

Sun-drying has found favour with fishmongers for centuries, producing fish meat that is condensed, saturated with oil, translucent and amber in colour, dense in consistency, and pleasant in taste. Impediments to the use of this method are related to product quality as well as safety. Sun-drying does not allow very much control over drying times, and it also exposes the fish to attack by insects or vermin and there are all possibilities of contamination by sand and dirt. Other constraints include considerable product losses, lower fish quality because of contamination by foreign materials, reduced protein quality, insects and micro-organisms as well as discolouring by ultraviolet radiation [36].

During the drying process, the fish surface dries faster and hardens, thereby locking moisture inside, which slows the drying process and encourages degradation of protein and fatty acid oxidation. Degradation of protein is accelerated when fish products are subjected to high temperatures for an extended period. Since traditional sun-drying is weather dependent, some losses in quality also result from inadequate drying.

18. SMOKING

Smoking is used mainly to preserve fish, partly by drying and partly by adding naturally produced anti-microbiological constituents such as phenols from the smoke to the fish. Smoking is a preferred method of preservation because it dries the fish, melts some fat out of the fish, and reduces microbial growth. When this is achieved, smoking is expected to extend the shelf life of most fish products to several weeks. However, in recent times it is being used in addition to preservation to achieve the characteristic taste and appearance of the smoked fish. Heat from the fire causes drying, and if the temperature is high enough, the flesh becomes cooked. Both of these factors prevent bacterial growth and enzyme activity, thus preventing spoilage.

Fish smoking differs according to the type of kiln technology used and can be categorized as the following types:

- Cold smoking: In this method, the temperature is not high enough to cook the fish. It is not usually higher than 35 °C.
- Hot smoking: In this method, the temperature is high enough to cook fish. Hot smoking is often the preferred method. Traditionally, in West Africa in particular, fish smoking is the most extensively practiced fish preservation method in Ghana and Nigeria, which uses the traditional kiln with wood burning temperatures between 300 and 700 °C (for wood combustion) usually 70–80 °C of the oven temperature. This is because the process requires less control than cold processing and the shelf life of the hot-smoked product is longer because the fish is smoked until dry. Hot smoking consumes more fuel than the cold smoking method.

The wholesomeness of smoked fish products using the traditional kiln depends on the following factors: the type of wood used for the smoking process; the temperature used; the duration of smoking; the type of kiln used; the proximity of the fish from fire; the type of fish being smoked; and the fat content of the fish [37]. The primary challenge with the traditional kiln technology common across Nigeria, and West Africa in general, has to do with the temperature regulatory system of the design, which poses difficulties for the commercial fishmongers operating this type of technology.

Smoking temperatures result in the release of a wide range of antioxidant and antimicrobial chemicals like polycyclic aromatic hydrocarbons (PAHs), phenols, aldehydes, and acetic acids, some of which are known to be carcinogenic, mutagenic, and endocrine disrupting [38, 39]. PAHs produced in wood smokes are known to originate from the thermal pyrolysis (depolymerization) of lignin and subsequent condensation of the lignin components in lignocelluloses at temperatures above 350 °C. Charcoal used as filters in modified traditional smoking kilns is effective in removing PAHs from the smoke generated, thus potentially posing little or no health effects on consumers. Therefore, the adoption of this method may help reduce the levels of PAH-associated health effects like cancer and other cancer-related ailments which is reported to be on the increase in Ghana [40]. A typical system for traditional fish smoking in Nigeria is shown in Fig. 2.

Certain traditional fish processing practices constitute occupational and health hazards to consumers, processors, and the environment. A study in Nigeria indicated that smoked catfish/sole fish from commercial fishmongers using a traditional smoking kiln showed elevated levels of PAHs compared to the modern kiln [41]. The study postulated that this trend may be associated with increasing cases of cancer and cancer-related ailments in the nation. Generally, fish that is smoke-cured with hardwoods (e.g. acacia and mangrove) for a longer duration (≥4 h) using the traditional kiln may be unsafe for consumption, but sugarcane bagasse-cured fish using the traditional kiln was both the safest and the best fish smoking practice for short-duration smoking [37]. There may be elevated risks of cancer and non-cancerous diseases associated with lifelong (70 years) consumption of mackerel and sardine that is smoke-cured with hardwoods, especially at longer smoking durations due to their high lipid content.

Long hours and many years spent smoking fish have negatively impacted the health of women processors. Exposure to smoke from the burning of biomass fuels, including firewood, is a major risk factor for several respiratory ailments, such as asthma, chronic obstructive pulmonary disease (COPD), respiratory tract infections, and lung cancer [42]. A study in the rural fishing community of Obaka has shown that women processors are exposed to increased risk of developing COPD compared to women who do not work in the fish smoking business [43]. The smoke also causes redness/swelling of the eye, poor vision, rashes, eczema, pile, rheumatism, diarrhoea, dysentery, and heat [44]. Fish processors could also be exposed to smoke particles that contain potential or confirmed carcinogens such as

PAHs. Milder forms of occupational hazards linked to fish processing include sting from fish spines, minor cuts, and scrapes [45]. Overall, poor socio-economic status, poor nutrition, exposure to other sources of indoor air pollutants such as mosquito coils and kerosene lamps, and risk of injuries, combined with poor access to sanitation and healthcare facilities, increase risks of developing health hazards linked to fish smoking [45].

Figure. 2 : Typical traditional smoking systems are shown in **a** wood fuel, **b** smoked fish, **c** drum kiln

In addition, the use of firewood as the major energy source in traditional smoking kilns has drawn some other criticism on the bases of economic and environmental consequences. The supply of fuel wood is characterized by scarcity, rendering it costly for fish smoking [46]. The practice is also time-consuming and considered wasteful due to the small quantities of fish that are smoked at a time [45]. Exploitation and consumption of firewood for fish drying and other uses is a contributory factor to continued deforestation in the country, particularly in the arid Northern region where this practice is contributing to desertification [47]. Although heat treatments may impact on the nutritive values of traditionally preserved fish, the overall quality has been demonstrated not to have significant differences from the fresh or raw fish [1, 2]. In many instances, the reduction in nutritive value may be due to the method used for the analyses rather than due to impact of intervention [1].

19. CURING

Curing involves the techniques of drying, dry salting/brining (soaking in salt solution), or smoking. These may be used alone or in various combinations to produce a range of products with a long shelf life. For example:

- Drying–smoking–drying.
- Brining–smoking–drying.
- Salting–drying.
- Salting–drying–smoking.

These techniques reduce the water content in the flesh of the fish, thereby preventing the growth spoilage caused by micro-organisms. Curing is not widely used in Nigeria, along with fish fermentation and deep frying.

20. STRATEGIES FOR BRIDGING IDENTIFIED GAPS

Traditional fish processors lack the financial and economic capacity to adopt new technological interventions which have been developed and tested at pilot levels but have since scaled up, therefore leaving these processors behind in the space of technological adoption dynamics and processes. The need for cost-effective loss reduction had been highlighted in the Voluntary Guidelines for Securing Sustainable Small-Scale Fisheries in the Context of Food Security and Poverty Eradication[6] and the Code of

Conduct for Responsible Fisheries.[7] Specifically, these documents mention value addition, building on existing traditional and local cost-efficient technologies, the adoption of local innovations and culturally appropriate technology transfers, and the promotion of environmentally sustainable practices within an ecosystem approach.

In addition, the SSF Guidelines advocate for the inclusion of post-harvest actors in relevant decision-making processes and special support for vulnerable and marginalized group in the recognition of unequal power relationships between actors in the value chain. The document also highlights the need for improvements in the sub-sector by provision of amenities, services, appropriate infrastructures, promotion of cooperatives, and capacity development, which will support the post-harvest sub-sector in producing good quality and safe fish and fishery products for both export and domestic markets, and also enhance income and livelihood security.

21. WALKING THE TALK: IMPLEMENTING THE STRATEGIES ON FOOD PROCESSING IN THE NATIONAL POLICY ON FOOD AND NUTRITION IN NIGERIA

The NPFNN document is over a decade old. Detailed study of the six-chapter document shows an excellent understanding of the situation, but it is unclear why not much has been achieved since its release. Clearly, the main challenge is not putting these commitments to practice; in other words, there has been no desire to walk the talk. Chapter 4 specifically deals with Institutional Arrangements:

The National Food and Nutrition Policy requires an effective institutional arrangement to ensure a results-oriented programme implementation. Past implementation efforts have been principally sectoral (health, agriculture, science and technology, and education, etc.], uncoordinated, inadequately funded and limited in scope and coverage. Such efforts, therefore, fell short of the desired outcome of solving the problems they were designed to tackle.

The agency in charge of implementation is the National Planning Commission (NPC), with the National Committee on Food and Nutrition (NCFN) being the Siamese (being both connected to deliver food and nutrition security). There is a lot to be done in order to align these bodies to ensure the delivery of their mandates, especially regarding the utilization dimension of food security. These two agencies must devise ways by which fish: (a) is fully integrated into inter-sectoral national food security and nutrition policies and programmes and (b) is more systematically included in national nutritional programmes and interventions aimed at tackling micronutrient deficiencies, especially among children and women. The NCFN should pursue the multi sectoral approach in dealing with the issue of food and nutrition programming and bring to bear the utilization dimensions of fish in the context of food security. This approach will require regular monthly meetings of identified stakeholders within the food and nutrition spectrum to review their success at meeting departmental mandates and ensure the larger involvement of developmental agencies internationally and locally.

22. VOICE AND VISIBILITY TO THE ROLE OF FISH AND FISHERY PRODUCTS

Aside from the promotion of the accessibility, availability, and, perhaps, stability of food and nutrition security (as the cardinal principle which respective governmental agencies and ministry are following to make progress), there is an urgent need to promote the consumption of fish products as part of the overall objective of promoting food and nutrition security at the household and national level.

The 35% fish consumption rate out of all animal protein at the household, regional, and national level is grossly inadequate and far below some countries of Africa and Asia, where fish accounts for more than 50% of the total animal protein intake. There is the urgent need for the NCFN to work assiduously in promoting required change. The present dietary habit which favours consumption of staple plant-based food over that of animal protein has been proven to be grossly inadequate in meeting the present nutritional needs of the present generation in Nigeria, especially in the case of children.

In the same vein, the high preference for meat over fish consumption suggests that we are not doing enough to prevent, rather than treat, the malnutrition that affects a large population of children, as well as non-communicable diseases of aged adult such as diabetes, cancer, coronary heart diseases, and obesity. The activities of NPC and NCFN related to food and nutrition security need to be made transparent to the public. There must be concise efforts to promote the culture of fish consumption as well as the establishment of a national policy on dietary food intake in favour of fish products by means of establishing an Expert Consultative Forum on fish and fish products as food with an emphasis on: the neurological development benefits of fish consumption to the children of women of childbearing age; the reduction in mortality from coronary heart disease (and the risks of mortality from coronary heart disease associated with not eating fish) among the adult population; and the potential to use dried fish products, especially shell fish, as an ingredient in complementary or weaning food items for infants following the period of exclusive breastfeeding.

The Fisheries Act of 2014 pragmatically gave a holistic approach to understanding fish and fisheries governance. The framing of fish as food is an important step that is necessary for achieving food and nutritional security on a national level and strengthening the various pillars of food security. However, it is imperative that the Fisheries Act 2014 be signed into law. The 8th National Assembly should consider initiating this proposed legislation. NPC and NCFN should consider incorporating the voice of relevant fisheries and nutrition-based Civil Society Organizations (CSOs) and coordinate public engagement activities to include these groups in lobbying to pass the Act into law.

Proper nutrition, especially in terms of ensuring the standard protein–energy ratio (PER) balance, has been known to be an appropriate prevention method for malnutrition and diseases associated with old age. The current situation in Nigeria points to the need to revise the stale definition of food security in terms of caloric intake but widen food security indices to include such as food availability, affordability, percentage of fish caught with maximum sustainable yield, land areas affected by desertification as percentage of total land mass, per capita protein and energy intake, income indices, level of unemployment, average real income, and income distribution index. There is a need for a paradigm shift in the operational terminology of what constitutes food security and its unit of measurement.

23. ADOPTION OF HOMEGROWN TECHNOLOGICAL PACKAGES IN SMALL-SCALE FISH PROCESSING

Small-scale fish processors are critical in fulfilling any mandate of making fish and fishery products, playing a vital role within the concept of fish as food and promoting household and national food security. The NPFNN document while proselytizing strategies for achieving food and nutrition policy objectives towards food security stated to:

1) Promote and support research on food processing and preservation technologies at the village and household levels; and

2) Introduce and expand such technologies and training through their inclusion in the curricular of schools and training institutions.

There has been remarkable progress in implementing certain aspects of these strategies. Researchers in academic institutes and tertiary institutions have helped in responding to this call. There are many homegrown technological packages that can modernize the traditional fish processing technologies of the sun-drying and smoking of fish. There have been many innovative and spectacular ingenuity creations of solar dryers (Kainji Solar Tent Dryer (KSTD), Plastic Driers, Mosquito Net Driers, Glass Driers, Aluminium Driers, and Glass Driers (which contain black stones). Localized fish drying technologies include the Kanji Solar Tent Dryer (Fig. 3) [48]. Solar and cabinet dryers, plastic drier (constructed using a thermopile plastic material), mosquito net drier [constructed by using plywood for the frame (edges) and subsequently covering the dryer with mosquito net all around the wooden frame), glass drier (made of transparent glass), aluminium drier (constructed from aluminium sheets and coated both inside and outside with black paint), and glass drier containing black stones. This is similar to the glass drier in every respect but with a black (igneous rock) stone placed in it. There have also been local variants of kiln which seek to replace the traditional mud and drum kiln which dominate the landscape.

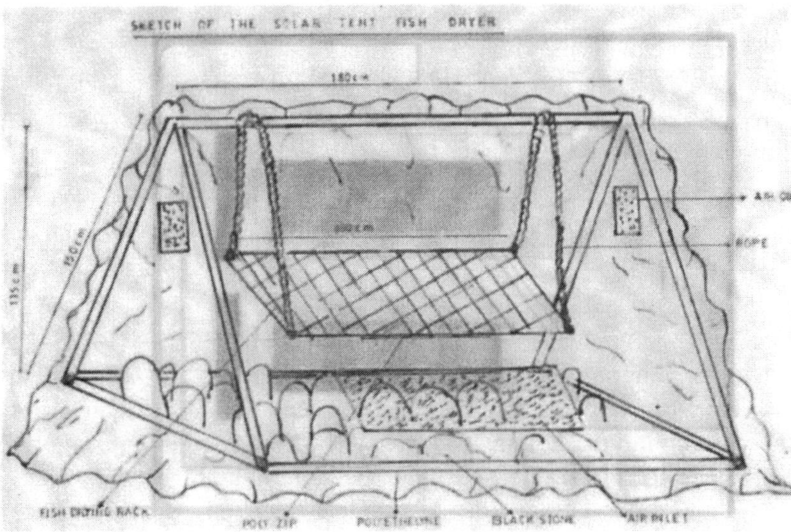

Figure. 3 The sketch of Kainji Solar Tent Dryer [48]

The challenge has been that many of these technologies have not been adopted in the field. Thus, there is a need to demonstrate and situate the applicability of many of these technologies and their variants at the level of small-scale fishmongers in relation to their socio-economic status. Present outcomes of these research outputs offer a suite of opportunities which can be adopted based on the circumstances of state and local government areas. Once the fishmongers buy into any technology package, there is a need to have a credit scheme in place that is co-managed by the operators' thrift and cooperative societies as well as interested financial bodies.

24. DEFINITE ACTION IMPLEMENTING THE ENERGY ISSUE WITH REGARDS TO FISH SMOKING

Climate and energy smart food processing ultimately is the direction to move. Most of the modern kiln is not able to meet the disposition of the small-scale fish processors. The practice of traditional fish processors throws up many issues on environment, deforestation, health, and energy use. The adoption of the modern smoking kiln (Fig. 4) [49] is non-existent at the level of the small-scale fish processors, and therefore, emphasis should be laced on wood fuel technology. The NEP document provides enough strategic direction to take in ameliorating the conflict between conservation and utilization of wood fuel. However, no direct mention of the small-scale fish processors was made. In our views, they are too big to be ignored. Among the strategies on fuel wood proposed by NEP include:

1) Encouraging the establishment of private and community woodlots for supply of fuel wood in the short term.

2) Establishing micro-credit facilities for entrepreneurs, especially for women's groups, for the establishment and operation of commercial fuel wood lots and the production of renewable energy devices and systems.

While not being unaware of any efforts at implementing these freely imposed guidelines even when these options are to be followed, it is important to exercise discretion on the choice of wood fuel in order to produce the minimum amount of toxic PAHs. This should be promoted in line with the use of such fuel woods with the minimum risk of cancer and cancer-related diseases. The Code of Conduct for Responsible Fisheries an important international instrument in fisheries stated that activities are related to fish as food or commodity:

...do not result in environmental degradation or adversely impact the nutritional rights and needs of people for whom fish is critical to their health and well-being.

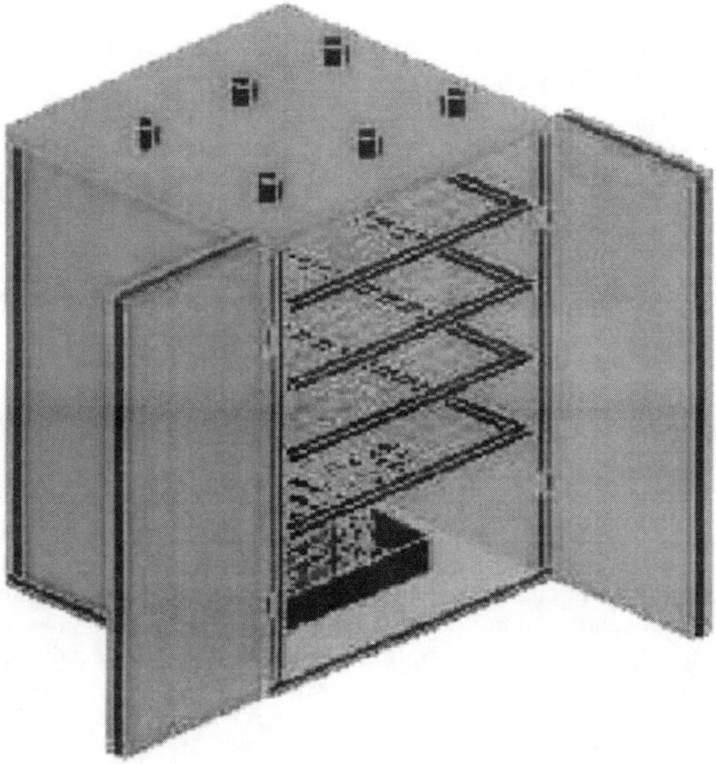

Figure. 4 Isometric view of smoking kiln [49]

The National Food and Drug Administration (NAFDAC) must step up its control mandate on food at the level of small-scale fish processors. This measure would ensure that the operators have safe fish while guaranteeing good health and well-being.

At the African level, the Federal Government adopted the CAADP. A highlight of the declaration is the resolve to halve the current levels of PHL, by the year 2025 [50]. Two years after the Malabo Declaration, it is encouraging that significant progress is being made towards including the development of strategies and key guidelines to support the CAADP country process and extensive engagements made at country, regional and continental levels.

Arising from findings from this study, the following recommendations are made:

1) The NPC and the NCFN are encouraged to generate sufficient political will to give deserve attention to fish and PHL mitigation strategies

contained in NPFN, NEP, CAADP, and other national and international documents as concerned fish utilization within context of food and nutrition security.

2) The Fisheries Act 2014, yet to be assented, makes provision for the establishment of the Fisheries Commission, charged with sustainable utilization of fisheries resources, in order to meet the needs of present and future generations. There is the need to review urgently and passed to law the Fisheries Act.
3) Consolidate the traditional fish processing methods (with an emphasis on sun-drying and fish smoking) in terms of quality assurance and food safety within the food system at all levels of government.
4) With respect to solar drying technologies for fish drying, there is a need to determine and adopt the most efficient local technology.
5) Identified local technology should be promoted with the overarching objective of replacing traditional sun-drying with solar drying technologies.
6) Government at all levels should urgently implement the establishment of private and community woodlots for supply of fuel wood in the short term, as contained in the National Energy Policy, and establish micro-credit facilities for entrepreneurs, especially women's groups, for the establishment and operation of commercial fuel wood lots and the production of renewable energy devices and systems.
7) Promote the culture of fish consumption and set a national policy on dietary food intake in favour of fish products by means of establishment of Expert Consultative Forum on fish and fish products as food.
8) Emphasize the benefits of fish consumption on reducing mortality from coronary heart disease (and the risks of mortality from coronary heart disease associated with not eating fish) for the general adult population.
9) Promote the use of fish products especially shell fish as ingredient of complementary or weaning food items for infants post-period of exclusive breastfeeding.

25. CONCLUSION

Nigeria is endowed with a wide array of aquatic systems for producing fish. However, production capacity is presently limited and the gap in the local fish demand is filled by importation. In the light of the fact that fish supply locally cannot be drastically expanded to meet the continually increasing demand, it is more appropriate to preserve and production by small-scale producers. Fish is highly perishable. There is need to ensure efficient and effective processing and preservation of the valuable fish protein towards achieving the objectives of the national food policy on food security. Small-scale fish processors using traditional methods of fish smoking and sun-drying dominate the fish processing business and are key to guaranteeing fish consumption. Medium- to large-scale processors target the more lucrative foreign markets.

Traditional fish processing is linked to food security in all dimensions of availability, accessibility, stability, and utilization, as well as national development in terms of healthy citizens, job creation, and income generation. In relation to utilization, traditional fish processing method provides seafood products with adequate nutrient to combat the challenge of malnutrition among children and other vulnerable groups. It can be modernized to overcome the challenges being faced with issues of safety and product qualities, and there are arrays of technologies to be adopted.

Fish should be managed as food ingredients whose consumption and preference over other sources of animal protein should be pushed to the national consciousness. National policy documents examined which have strong direct bearing or tangential to subject of food and nutrition indicated that our policy substantially is adequate, dynamic and meets the requirement of modern challenges, but we have substantially been too slow or failed at implementations. The thrust of this paper is pivoted on the notion that building a virile, prosperous, and healthy society requires that citizens must have food which makes them healthy. There is the need to inculcate the taste for fish in our diets and culture.

FOOTNOTES

1 Values were calculated based on figures from Maziya-Dixon, B., I.O. Akinyele, E.B. Oguntona, S. Nokoe, R.A. Sanusi, and E. Harris, 2004 Nigeria Food Consumption and Nutrition Survey 2001–2003 Summary. IITA, Ibadan.

2 The National Food Policy documents provide information on the government understanding of the challenges inherent in the nation's quest for food security and nutrition. We treat this document as the guideline for the operation of government 2001–2015. See http://www.foodbasketfoundation.org/downloads/download/?.

3 This document is an improved document on the earlier version. It specified the actions government intend taken to solve the problems and

challenges related to food security and nutrition. See https://extranet.who.int/nutrition/gina/sites/default/files/NGA.

4 Assumption: 60% of the imported and local production is consumed in process form.

5 Assumption: average volume of fishes sold as smoked/sun-dried. Exchange rate of US$ 1.00 to N 100.

6 Voluntary Guidelines for Securing Sustainable Small-Scale Fisheries in the Context of Food Security and Poverty Eradication 2015 accessed 16 January, 2016, at 20 p.m. http://www.fao.org/fishery/ssf/guidelines/en.

7 FAO Code of Conduct for Responsible Fisheries Special Edition 2011 accessed 3 February, 2016, at 16.30 p.m. http://www.fao.org/docrep/013/i1900e/i1900e00.htm.

ACKNOWLEDGEMENTS

We thank immensely the Chairman and members of the Governing Board Centre for Human Security Olusegun Obasanjo Presidential Library (OOPL) especially Professor. Peter A. Okebukola for the invitation to attend the colloquium. It was the presentation made at the colloquium that provided inspiration to deal in more elaborate way the issues raised in this manuscript. The contributions of the anonymous reviewers are acknowledged. Also, we appreciate the immense contributions of Brennan Lowery for the editing and the Too Big To Ignore (TBTI) for providing required supports in this direction. This is a contribution to TBTI Fish as Food Research Cluster.

REFERENCES

1. Akintola SL, Brown A, Abdullahi B, Osowo OD, Bello BO. Effects of hot smoking and sun drying processes on nutritional composition of giant tiger shrimp (Penaeusmonodon, Fabricius, 1798). Pol J Food Nutr Sci. 2013;63(4):227–37.
2. Akintola SL. Effects of smoking and sun-drying on proximate, fatty and amino acids compositions of Southern pink shrimp (Penaeusnotialis). J Food Sci Technol. 2015;52(5):2646–56.
3. Affognon H, Mutungi C, Sanginga P, Borgemeister C. Unpacking postharvest losses in sub-Saharan Africa: a meta-analysis. World Dev. 2015;66:49–68.
4. FAO/WHO: report of the joint FAO/WHO expert consultation on the risks and benefits of fish consumption. Rome, Food and Agriculture Organization of the United Nations; Geneva, World Health Organization. 2011.
5. FAO World Food Summit: Rome Declaration on World Food Security, Rome. 1996.

6. Kawarazuka N. The contribution of fish intake, aquaculture, and small-scale fisheries to improving nutrition: a literature review. The World Fish Center Working Paper No. 2106. The World Fish Center, Penang, Malaysia. 2010.
7. HLPE: sustainable fisheries and aquaculture for food security and nutrition. A report by the High Level Panel of Experts on Food Security and Nutrition of the Committee on World Food Security, P118 2014. www.fao.org/3/a-i3844e.pdf. Accessed 23 June 2015.
8. Hu FB, Bronner L, Willett WC, Stampfer MJ, Rexrode KM, Albert CM. Fish and omega-3 fatty acid intake and risk of coronary heart disease in women. JAMA. 2002;287:1815–21.
9. Food and Agriculture Organisation: Contribution of Fisheries to National Economies in West and Central Africa—policies to increase the wealth generated by small-scale fisheries, new directions in fisheries e a series of policy briefs on development issues. FAO, Rome. 2006.
10. Belhabib D, Sumaila UR, Pauly D. Feeding the poor: contribution of West African fisheries to employment and food security. Ocean Coast Manag. 2014;111:72–81.
11. United Nations International Children's Emergency Fund, UNICEF. Strategy for improved nutrition of children and women in developing countries. New York: the United Nations Children's Fund. 1990.
12. Food and Agriculture Organisation: Fact Sheet, FAO Country Profile, Nigeria; 2007. ftp://ftp.fao.org/FI/DOCUMENT/fcp/en/FI_CP_NG.pdf. Accessed 23 Feb 2015.
13. Béné C, Heck S. Fish and food security in Africa. Naga. 2005;28:1–8.
14. GIZ—Deutsche Gesellschaftfür Internationale Zusammenarbeit (GIZ) GmbH. The Ecological footprint of cassava and maize postharvest losses in Nigeria. A Life Cycle Assessment. Eschborn: GIZ. 2013.
15. Kolawole OD, Williams SD, Awujola AF. Indigenous fish processing and preservation practices amongst women in south-western Nigeria. Indian J Tradit Knowl. 2010;9(4):668–72.
16. Udong E, Niehof A, Tilburg AV. The livelihood strategies of women fish traders in adapting to cultural and institutional constraints in Ibaka, Nigeria. MAST. 2010;9(2):65–93.
17. Sea Around Us. A global database on marine fisheries and ecosystems. 2015. www.seaaroundus.org. Accessed 10 Dec 2016.
18. Akinyemi O, Balogun AM, Mabawonku AF. Investment prospects in inshore trawling in Nigeria. In: 5th Annual Conference of the Fisheries Society of Nigeria (FISON), 22–25 September, 1986, Ilorin, Nigeria; 1987. p. 58–72.
19. Ssentongo GW, Ajayi T, Ukpe ET. Report on a resource appraisal of the artisanal and inshore fisheries of Nigeria. FAO/UNDP/FNDP/NIR/77/001, Rome. 1983.
20. Falaye AE. Illegal unreported unregulated (IUU) fishing in West Africa (Nigeria & Ghana). Report of a study prepared for Marine Resources Assessment Group (MRAG) Ltd. 2008.
21. Bolorunduro PI, Adesehinwa AOK, Ayanda JO. Adoption of improved fish preservation technologies in Northwestern Nigeria. Tropicultura. 2005;23(3):117–23.
22. Akande GR, Simpa J. Cottage level production of fishmeal. NIOMR Tech. Pap. 81: 1992.
23. Jim-Saiki LO, Alhaji TA, Giwa JE, Oyerinde M, Adedeji AK. Factors constraining artisanal fish production in the fishing communities of Ibeju-Lekki Local Government Area of Lagos State. Int J Innov Res Dev. 2014;3(13):97–101.
24. Thompson SJ, Cowan JT. Durable food production and consumption in the world economy. In: McMichael P, editor. Food and agrarian orders in the world economy. Westport: Praeger; 1995. p. 35–52.
25. Kurien J. Responsible fish trade and food security toward understanding the relationship between international fish trade and food security. Report of the study on the impact of international trade in fishery products on food security Conducted jointly by Food and Agriculture Organisation of the United Nations and the Royal Norwegian Ministry of Foreign Affairs. 2004.
26. National Population Commission (NPC) [Nigeria] and ICF International: Nigeria Demographic and Health Survey 2013. Abuja, Nigeria, and Rockville, Maryland, USA: NPC and ICF International. 2014.

27. UNICEF: progress for children: a report card on nutrition. UNICEF-publications. 2006. http://www.unicef.org/progressforchildren/2006n4/index_undernutrition.html. Accessed 20 Feb 2015.
28. Adeniyi OR, Omitoyin SA, Ojo OO. Socio-economic determinants of consumption pattern of fish among households in Ibadan North Local Government Area of Oyo State, Nigeria. Afr J Food Agric Nutr Dev. 2012;12(5):6537–52.
29. Belton B, Thilsted SH. Fisheries in transition: food and nutrition security implications for the global South. Glob Food Secur. 2014;3(1):59–66.
30. Ibrahim FM, Osikabor B, Akinosho HO, Adesope AA, Ibrahim AG, Olatunji BT, Akanni OF, Ogunwale OG. Meat versus fish: health risk perception, convenience, price, ethnic value and consumption in Ibadan, Nigeria. Food Public Health. 2014;4(4):193–9.
31. Sustainable Fisheries Livelihoods Programme, SFLP Contribution of Fisheries Research to the Improvement of Livelihoods in West African Communities, Case Study: Nigeria, SFPL/DFID-FAO. 2002.
32. Bassey FI, Oguntunde FC, Iwegbue CMA, Osabor VN, Edem CA. Effects of processing on the proximate and metal contents in three fish species from Nigerian coastal waters. Food Sci Nutr. 2014;2(3):272–81.
33. Akintola SL, Fakoya KA. Governance and social-institutional arrangement of small-scale fisheries and relationship with non-fishery users in Badagry Creek, Lagos State, Nigeria. In: Song AM, Bower SD, Onyango P, Cooke SJ, Chuenpagdee R, editors. Inter-sectoral governance of inland fisheries. TBTI Publication Series, St John's, NL, Canada 2016. p. xx–xx. http://toobigtoignore.net/wp-content/uploads/2016/08/final-Akintola-and-Fakoya.pdf. Accessed 23 June 2016.
34. Madugu AJ, Edward A. Marketing and distribution channel of processed fish in Adamawa State, Nigeria. Glob J Manag Bus Res. 2011;11(4):1–7.
35. Dambatta MA, Sogbesan OA, Tafida AA, Haruna MA, Fagge AU. Profitability and constraints of three major fisheries enterprises in Kano State, Nigeria. Glob J Sci Front Res I Interdiscip. 2016;16(1):1–7.
36. Tiwari GN, Sarkar B. Greenhouse drying and other applications. Fundamentals of aquaculture greenhouse. 1st ed. New Delhi: Anamaya Publishers; 2007. p. 93.
37. Essumang DK, Dodoo DK, Adjei JK. Effect of smoke generation sources and smoke curing duration on the levels of polycyclic aromatic hydrocarbon (PAH) in different suites of fish. Food Chem Toxicol. 2013;58:86–94.
38. Okuda T, Naoi D, Tenmoku M, Tanaka S, He K, Ma Y, Yang F, Lei Y, Jia Y, Zhang D. Polycyclic aromatic hydrocarbons (PAHs) in the aerosol in Beijing, China, measured aminopropylsilane chemically-bonded stationary phase column chromatography and HPLC/fluorescence detection. Chemosphere. 2006;65:427–35.
39. Kjhallstrand J, Petersson G. Phenolic antioxidants in wood smoke. Sci Total Environ. 2001;27:69–75.
40. Essumang DK, Dodoo DK, Adjei JK. Effective reduction of PAH contamination in smoke cured fish products using charcoal filters in a modified traditional kiln. Food Control. 2014;35:85–93.
41. Yusuf KA, Ezechukwu LN, Kafaya KA, Akintola SL, Agboola JI, Omoleye TO. Influence of fish smoking methods on polycyclic aromatic hydrocarbons content and possible risks to human health. Afr J Food Sci. 2015;9(3):126–35.
42. Salvi S, Brashier B. Fish smoking and COPD: a fishy affair. Lung India. 2014;31:105–6.
43. Umoh VA, Etete P. The relationship between lung function and indoor air pollution among rural women in the Niger Delta region of Nigeria. Lung India. 2014;31:110–5.
44. Udoh JP, Okoko AC. Perception of impact of donor agencies on the socioeconomic wellbeing of fisher folks in southeast Nigeria. Croat J Fish. 2014;72:77–82.
45. Olaoye OJ, Adegbite DA, Oluwalana EO, Vaughan IO, Odebiyi CO, Adediji AP. Comparative evaluation of economic benefits of earthen fish ponds and concrete tanks in aquaculture enterprises in Oyo state, Nigeria. Croat J Fish. 2014;72:107–17.
46. George FO, Ogbolu AO, Olaoye OJ, Obasa SO, Idowu AA, Odulate DO. Fish processing technologies in Nigeria: a case study of Ibeju-Lekki Local Government Area, Lagos State. Am J Food Technol. 2014;9(6):302–10.

47. Ikurekong EE, Esin JO, Mba AC. Rural fuelwood exploitation in Mbo Local Government Area—a Nigerian Coastal Settlement. Ethiop J Environ Stud Manag. 2009;2(3):44–55.
48. Olokor JO, Omojowo FS. Adaptation and improvement of a simple Solar Tent Dryer to enhance fish drying. Nat Sci. 2009;7(10):18–24.
49. Olayemi FF, Raji AO, Oyelese OA, Oyewole SN, Omodara MA. Effective fish smoking kiln for developing country. Int J Sci Eng Res. 2013;4(1):1–7.
50. Africa Union. http://www.au.int. Accessed 12 Jan 2017.

CHAPTER 12

Feeding Ecology and Establishment of the Naturally-Colonized Freshwater Cichlid, Sarotherodon galilaeus (Pisces: Actinopterigii: Perciformes) from a Man-Made Lake, South-Benin, West Africa

Houehanou M. A. G. Gbaguidi[1], Alphonse Adite[1*], Edmond Sossoukpe[2]

[1]*Laboratoire d'Ecologie et de Management des Ecosystèmes Aquatiques (LEMEA), Département de Zoologie, Faculté des Sciences et Techniques, Université d'Abomey-Calavi, Cotonou, Bénin*

[2]*Unité de Recherches sur les Zones Humides, Département de Zoologie, Faculté des Sciences et Techniques, Université d'Abomey-Calavi, Cotonou, Bénin*

ABSTRACT

The freshwater cichlid, Sarotherodon galilaeus (Linné, 1758) is an economically and commercially important fisheries species that has naturally colonized and dominated a sand-dragged man-made freshwater lake of Ahozon (South-Benin) where the species made about 85.21% of the fish community. We investigated the feeding ecology of S. galilaeus in order to evaluate resource exploitation and the establishment of this species in

Lake Ahozon. During wet, flood and dry seasons, 1189 individuals of S. galilaeus have been monthly sampled in the open water and in the aquatic vegetation habitats. The study showed that S. galilaeus consumed about seventy (70) food items dominated by algae (52.88%) composed of 28 families and 52 genera from blue-green algae, green algae, desmids, and diatoms, sand particles (23.95%), detritus (12.27%) and protozoans (7.68%). Minor preys were rotifers, copepods, cladodera, crustacea and insect parts. Foods items with higher diet occurrence were sand particles with a percentage occurrence of 72.33%, detritus (69.47%), and some algae such as Scenedesmus (58.96%), Closterium (55.68%), and Microcystis (51.30%). Significant ($P \leq 0.05$) ontogenetic variations of empty stomachs were recorded. S. galilaeus exhibited a wide diet breadth (DB) ranging between 5.55 and 7.29 that tended to increase with fish sizes. Diet overlaps ($Ø_{jk}$) varied from 0.77 to 0.97, indicating high diet similarities between different life stage categories. The slope $b = 0.560$ obtained from Log (body weight)-Log (gut length) linear regression, and the mean ratio of $GL/SL = 6.72 \pm 2.23$ support our findings that S. galilaeus exhibited detritivorous, herbivorous and omnivorous food habits. The broad spectrum of food resources consumed/utilized and the high diet breadth recorded leading to allometric growth, the active breeding and the high propagation of the species, were some indicators of the successful establishment of S. galilaeus in the man-made lake of Ahozon. Sustainable fisheries/ aquaculture exploitation of S. galilaeus and successful valorization of the man-made lake of Ahozon require an integrated management of the lake including the prevention of domestic waste dumpings, the protection of the foraging and spawning grounds, the fertilization of the lake, the introduction of predators to control the population of S. galilaeus, the protection of lake margin to avoid erosion, a planned harvest of the fish stock and a periodic ecological monitoring of Lake Ahozon.

Keywords: Algae, Artificial Lake, Diet Breadth, Diet Similarities, Ecomorphology, Establishment, Fisheries, Foraging Grounds, Tilapine Species

1. INTRODUCTION

The freshwater cichlid, Sarotherodon galilaeus (Linné, 1758) is an economically and commercially important fish resource occurring in some natural freshwater bodies of Benin, such as Lakes Toho-Todougba, Dati, Ahouangan, the man-made lake from a sugar industry (SUCOBE), rivers such as Sô, Oueme, Niger and floodplains [1] [2]. In particular, S. galilaeus occurred abundantly in Lake Toho-Todougba (South-Benin) where the species made 73% of the total biomass [2]. In addition, in Southern Benin, S. galilaeus has naturally colonized and dominated a sand-dragged man-made freshwater lake of Ahozon Village (Ouidah City, County of Pahou) where the species reproduce actively, made about 85.21% of the fish community, and hence, appears to be well-estab- lished in this artificial lake [3] [4].

However, despite the high abundance of S. galilaeus in Lake Ahozon, and its economic and commercial importance, information on the ecology and biology of the species is scant. In particular, nothing is known about the feeding ecology of this tilapine cichlid in this abandoned and neglected artificial lake originated from a progressive accumulation of running water in an extensive hole created by sand-dragging activities. Knowledge on foods exploitation/utilization and feeding habits of fishes is crucial for a sound fisheries management and the development of aquaculture industries [5] - [9]. Indeed, foods are energy source of the various biological functions of fishes, and consistently affect fecundity, reproduction periodicity, recruitment, growth and condition factors of the species in its habitat [10] [11]. Furthermore, feeding ecology is important not only to understand the dynamic and the structure of the population, but also to evaluate the establishment of S. galilaeus in this special habitat, the sand-dragged man-made lake of Ahozon [12] [13].

Also, seasonal variations in preys availability greatly affects diet composition and the feeding patterns of the species [14] [15]. Since Lake Ahozon depends solely on running waters, the intensity of rainfall and seasonal hydrological regimes should affect the availability of marginal and open water habitats, and hence, preys availabilities [16] [17]. Diet breadth is an important feature in feeding ecology and measure the degree of trophic plasticity in fishes, that is, the capability of a fish to colonize many different habitats [18]. In trophic ecology, ontogenetic diet shifts depict the size-related patterns of feeding [19] [20]. In general the preys ingested vary according to fish development stages namely larva, juvenile, sub-adult and adult [21]. In fishes, the ecomorphological trends and reference ratios such as those from gut length could aid to predict or to confirm the food and feeding habits of a species [22]. Evaluating the prominence and establishment of S. galilaeus in the man-made lake of Ahozon requires extensive qualitative and quantitative knowledge on the exploitation and utilization of the food resources considered as energy sources for species survival, growth and active spawning.

The present research was carried out to investigate the feeding ecology of the freshwater cichlid, S. galilaeus, in order to evaluate its establishment in the sand-dragged artificial lake of Ahozon, and to contribute to the fisheries and aquacultural management of this special habitat. Specifically, the study aims to investigate on 1) diet composition and seasonal dietary variations of S. galilaeus in Lake Ahozon, 2) diet breadth (DB) variations, 3) diet overlap ($Ø_{jk}$) between different size classes, and ontogenetic diet shifts, 4) ecomorphological correlates and food habits and 5) implications for fisheries, aquaculture development, species conservation and the overall establishment of S. galilaeus in the man-made lake of Ahozon.

2. MATERIAL AND METHODS

2.1. Study Region

The man-made lake of Ahozon is located in Ouidah city (South-Benin), and lies between two water bodies, Toho-Todougba (15 km^2), a freshwater lake, and the coastal lagoon (60 km^2), a brackish water, both about 5 km apart from Lake Ahozon (06°22'52"N; 002°10'34"E). The artificial lake of Ahozon is a relatively small water bodies extending on about 0.165845 km^2 (Figure 1) and originated from the accumulation of running waters in a hole created by sand-dragging activities. In general, after the exploitation of the sand, these artificial and unmanaged lakes are neglected and abandoned by the land owners.

The study region comprises some sandy soils, red ferric soils and a vast swampy soil located at the coastal zone and extended from Cotonou City (South-Benin) to Ouidah. The Southern Benin experiences a major wet season from April to July with a peak usually recorded in June, a minor wet season from mid-September to October, and two (2) dry seasons, a major one from December to March and the minor one from mid-August to mid-September [9]. Ambient temperatures varied between 25°C and 33.6°C and monthly evaporation ranged between 59.2 - 145 mm [23]. Annual mean rainfall was about 1307.3 mm.

According to Gbaduidi et al. [4], Lake Ahozon showed depths varying between 16.2 and 240 cm (mean: 80.19 cm) and transparencies between 16.2 and 60.5 cm (mean: 36.63 cm). Water temperatures ranged between 28.2°C and 38.7°C (mean: 33.25°C), pH between 6.7 and 9.7 (mean: 7.51). The dissolved oxygen concentrations ranged between 0.73 and 11.8 mg/l (mean: 5.43 mg/l), the percentage of dissolved oxygen saturation between 10.5 and 208.8% (mean: 82.9%) and conductivity between 50 and 560 μ/cm (mean: 240 μ/cm).

Dominant phytoplankton genera in Lake Ahozon were Navicula, Peridinium, Scennedesmus, Pinnularia, Spirogyra, Cosmarium, Melosira, Synechocystis, Microcystis, Oscillatoria, Euglena, Phacus, Surirella and Lychmophora and zooplanktons were dominated by Trichocerca, Keratella, Brachionus and copepods. Benthic macro invertebrates recorded were chironomid larvae and a Gasteropod mollusk, Melanoïde tubercularis.

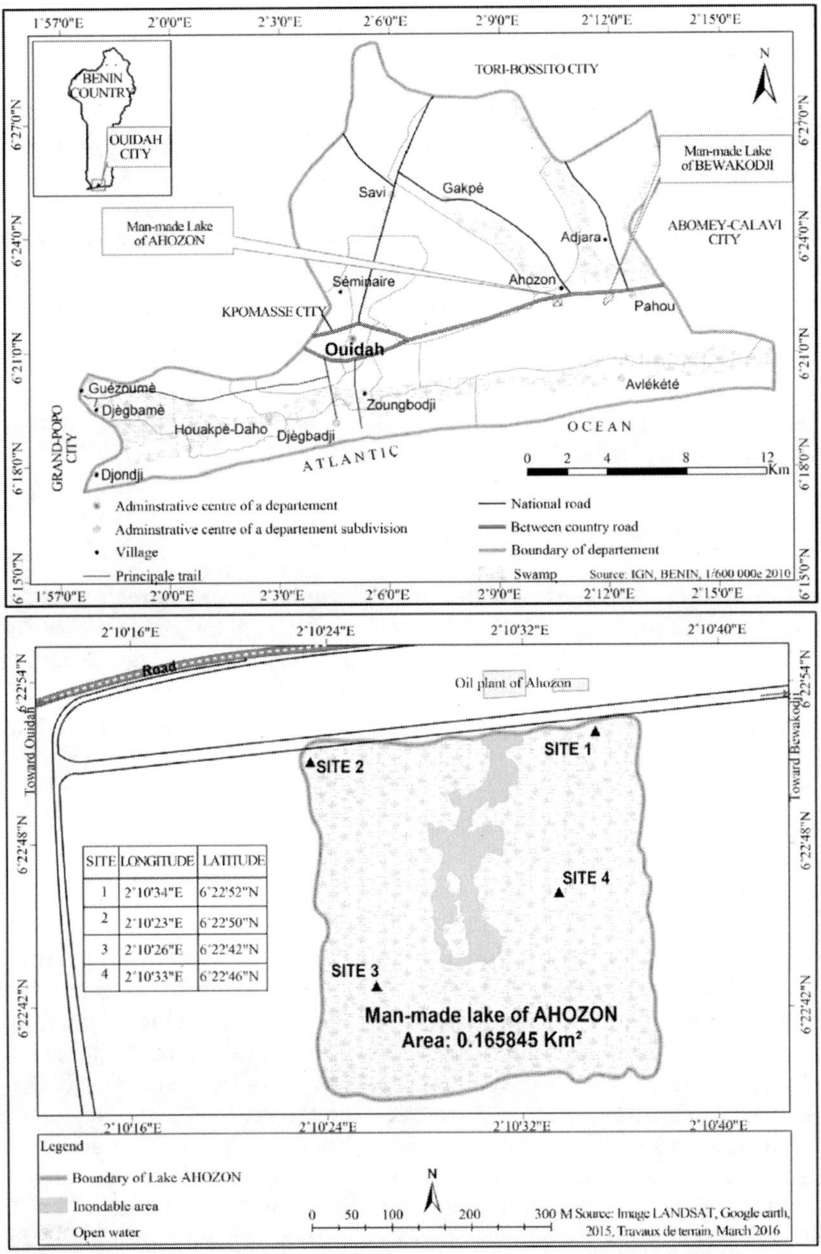

Figure 1. Map showing: (a) Ouidah City in South-Benin and the study location, the man-made lake of Ahozon, and (b) the sampling sites: Site 1 and Site 2 = Aquatic vegetation habitats; Site 3 and Site 4 = Open water habitats.

Aquatic vegetation mainly comprised Cyperus crassipes, Cyperus rotundus, Fuirena umbellata, Andropogon gayanus, Ludwigia perennis, Emilia praetermissa, Eleocharis complanata, Enydra fluctuans et Mariscus ligularis. The ichtyofauna of Lake Ahozon comprised three tilapine cichlids, Sarotherodon galilaeus, the dominant species, Oreochromis niloticus and Tilapia guineensis, a claroteid, Chrysichthys nigrodigitatus, the African bonytongue, Heterotis niloticus (Osteoglossidae), and the African catfish, Clarias gariepinus (Clariidae) (Gbaguidi et al., in press). Subsistence multi species fisheries sporadically occurred in the man-made lake of Ahozon and were practiced by a couple of migrant fishermen. In term of agricultural exploitation, some plantations such as those of Elaeis guineensis (palm tree) and Coco nucifera (coconut tree) occurred around Lake Ahozon [2]. In addition, the water of Lake Ahozon is utilized for sporadic subsistence and commercial agriculture involving beans, carrots lettuce, tomatoes, cucumbers and salads.

2.2. Sampling Sites

Four sampling sites were selected for the study, two (2) sites in the "open water" habitat and two (2) sites in the "aquatic vegetation" habitat (Figure 1). The "open water" habitat exhibited a relatively high depth and high water velocity, but exempt of vegetation whereas the "aquatic vegetation", the edge of the lake, is shallow and characterized by a low water velocity and relatively dense vegetation.

2.3. Aspects of Systematic and Biology

Taxonomically, Sarotherodon galilaeus is a tilapine cichlid belonging to the genus Sarotherodon, family Cichlidae, Percoidei Sub-order, Perciformes Order and Acanthopterygii Super Order. Synonimies are Sparus galilaells, Chromis galilaeus, Tilapia galilaea. The species exhibited a grew silvery color with juveniles showing black vertical lines on the sides. According to Leveque et al. [24], head length is 32.5% - 39.0% the standard length (SL), body height is 43.0% - 56.5% SL and pectoral length is 36.0% - 50.0% SL. Meristic counts showed 29 - 32 scales on the lateral line, 4 - 8 lines of teeths on the jaws, and the lower pharyngeal bone possesses numerous small teeth. S. galilaeus exhibited diverse reproductive behaviors such as mouth brooding, nest construction, nest guarding and parental care. In Lake Ahozon, the species is characterized by an early reproduction and sizes at sexual maturation (L50) were 131 mm-TL and 106 mm-TL, respectively for males and females. In the International Institute of Tropical Agriculture (IITA) Lake of Ibadan, Fagade [9] [10] reported the species as foraging mainly on algae.

2.4. Fish Sample Collection

Specimens of S. galilaeus were sampled twice a month from August 2014 to July 2015 in the "aquatic vegetation habitat" and in the "open water" using various fishing gears such as cast net (9.80 m-diameter, 4.90 m- height, 40 mm-mesh), seine (4.15 m-length × 1.77 m-width, 3 mm-mesh), hooks (90 m-length) and experimental gill net (40 m × 1.05 m, 40 mm mesh). Seinings were done in the marginal aquatic vegetation by setting the seine stationary, and kicking the vegetation to drive the fish in to the net [2]. Cast nets, gill nets and hooks were used in the "open water". After collection, fish samples were preserved in 10% formalin and then transported to the "Laboratory of Ecology and Management of Aquatic Ecosystems". In the lab, the fish samples were removed from the formalin and preserved in 70% ethanol [4].

2.5. Laboratory Procedure and Stomach Content Analysis

In the laborabory, the ethanol preserved-fishes were identified using references such as Leveque et al., [24], Van Thielen et al. [1], and Lowe McConnell [25] [26]. S. galilaeus were then measured with a graduated measuring board for total length (TL) and standard length (SL) to the nearest 0.1 mm with an ichtyometer, and weighed to the nearest 0.1 g with an electronic balance (Philips).

Each specimen of S. galilaeus was then dissected, and the digestive canal was removed and its length was measured as the distance from the distal end to the anus [27]. The gut was then opened and all the stomach contents were removed and spread on a transparent container for examination first under a binocular (model: Pierron) to identify large preys. In case of high level of digestion, fragments or parts of items were identified and inferences were made on the food resources (algae, microcrocrustacea, insects etc.) consumed. Sometimes, water was added to facilitate separation of small items. To identify the phytoplankton, a sub-sample of known volume of the stomach contents was taken and the total gross volume was estimated by water displacement using an appropriately sized graduated cylinder. Before measuring the total gross volume, the food was blotted on a paper towel to remove excess moisture. The volumes of the sub-samples were estimated by spreading the food materials on a glass slide then visually estimating the collective volume by comparison with a sample of known volume (<0.002 ml) [13]. The sub-sample is then examined under a photonic microscope. Preys were identified to the lowest possible taxonomic level using the identification key of Needham [28] for phytoplankton, zooplankton and aquatic invertebrates. After identification, the volumetric percentage of each prey was estimated for the sub-sample and then for the total gross stomach content [9].

2.6. Data Analysis

Overall, 1189 stomachs of S. galilaeus were dissected. For each individual, the estimated volume of each prey item was recorded on Excel software spreadsheet and the volumetric proportion of each prey was computed following the formula:

$$p_i = \sum_{i=1}^{n} vi/Vt$$

where p_i is the volumetric proportion of food item *i* in the diet, n is the number of stomachs, *vi* is the volume of the food item *i* in a single stomach, *Vt* is the total volume of food ingested by n stomachs. Seasonal (wet, flood, dry) variations in volumetric proportions were depicted with analysis of variances (ANOVA) using SPSS [29]. Also, volumetric proportions of each food item ingested were computed for different sizes classes of S. galilaeus to explore ontogenetic diet shifts. Niche breadth was calculated following Simpson's diet breadth formula [30]:

$$\text{Diet breadth}\left(\text{DB}\right) = 1 \Big/ \sum_{i=1}^{n} p_i^2$$

where p_i is the proportion of food item *i* in the diet, and *n* is the total number of food items in the diet. DB ranges from 1, when only one resource is used, to *n*, when all resources are consumed in equal proportions. The values of computed indices were submitted to one-way analysis of variances (ANOVA) using SPSS software computer program [29]. Diet overlap was calculated using Pianka's overlap index ($Ø_{jk}$) [31]:

$$Ø_{jk} = \frac{\sum_{i=1}^{n} P_{ij} P_{ik}}{\left(\sum_{i=1}^{n} P_{ij}^2 \times \sum_{i=1}^{n} P_{ik}^2 \right)^{1/2}}$$

where $Ø_{jk}$ is dietary overlap between species *j* and species *k*, p_{ij} is the proportion of resource *i* used by species j, p_{ik} is the proportion of resource i used by species k, and n is the number of resource categories utilized.

To explore the ecomorphological correlates, the linear regression analysis between body weight (W) and gut length (GL) were examined as well as the ratio of gut length and standard length (GL/SL).

3. RESULTS

3.1. Diet Composition

Table 1 shows the matrix of food resources ingested by S. galilaeus in the man-made lake of Ahozon and their respective volumetric proportions. In Lake Ahozon, S. galilaeus consumed about seventy (70) food items from benthic and pelagic habitats. Dominant diet items were algae accounting for 52.88%, sand particles (23.95%), detritus (12.27%) and protozoans (7.68%) (Figure 2). Aggregated algae were composed of 28 families and 52 genera including blue-green algae (24.48%), green algae (11.77%), desmids (8.29%) and diatoms (8.35%).

Blue-green algae comprised eighteen (18) genera, with three (3) Chroococcaceae Mycrocystis (12.25%), Synechocystis (3.87%), Coelosphaerium (3.15%) and Nostoc (1.92%), a Nostococacea, the most abundant items. The remaining 14 blue-green algae individually, made less than 0.84%. Green algae comprised nine (9) genera with Scenedesmus (6.97%), a Scenedesmaceae and Botryoccocus (2.99%) a Dictyosphaeriacae the most abundant green algae items. The remaining seven (7) green algae, Ankistrodesmus, Crucigenia, Coelastrum, Tetraspora, Ulothrix, Binuclearia, Ophiocytium individually made less than 0.70%. Thirteen (13) genera of desmids, Colacium, Closterium, Euastrum, Staurastrum, Gonatozygon, Tetmemorus, Cosmarium, Euglena, Phacus, Micractinium, Selenastrum, Peridinium and Cymatopleura were recorded in the diet of S. galilaeus with Closterium (5.97%) a Desmidiaceae and Phacus (1.16%), an Euglenaceae, the most abundant. Twelve (12) genera of diatoms, Chaetoceros, Fragilaria, Stephanodiscus, Anomoeneis, Navicula, Pinnularia, Gomphonema, Cymbella, Melosira, Asterionella, Nitzschia and Cyclotella were recorded in the diet of S. galilaeus with 2 Naviculaceae, Navicula (2.95%), Pinnularia (2.12%) and a Coscinodisceae, Melosira (2.84%), the most abundant diatoms.

In general, among algae, Chroococcaceae was the most speciose and dominant family constituated of nine (9) genera, Microcystis, Synechocystis, Coelosphaerium, Gomphosphaeria, Synechococcus, Polycystis, Chroococus, Merismopedia, Tetrapedia making about 20.34% of the diet. Other speciose algae families were Desmidiaceae with 6 genera, Closterium, Euastrum, Staurastrum, Gonatozygon, Tetmemorus, Cosmarium, Naviculaceae with 5 genera, Anomoeneis, Navicula, Pinnularia, Gomphonema, Cymbella, Scenedesmaceae with 3 genera, Scenedesmus, Crucigenia and Coelastrum. The remaining 24 algae families comprised no more than 1 or 2 genera.

Eleven (11) algae, Nostoc (Nostocaceae), Botryoccocus (Dictyosphaeriacae), Scenedesmus (Scenedesmaceae), Closterium (Desmidiaceae), Phacus (Euglenaceae), Melosira (Coscinodisceae), three Chroococcaceae, Microcystis, Synechocystis, Coelosphaerium, 2 Naviculaceae, Navicula, Pinnularia were relatively more abundant and made together 46.09%. The 41 remaining algae made 53.91% and none of them had a volumetric proportion more than 0.86%.

Protozoans were mostly Eudorina (7.58%), a Volvocaceae, the dominant genus, Uroglena (Chrysophyceae) and Chilodonella (Chilodonellideae). Minor preys (3.22%) recorded included Rotifers (Euchlanis, Colurella, Chromogaster, Rotaria, Trichocerca), microcrustacea composed of Copepods (Diaptomus, Arctodiaptomes, Tropocyclops), Cladocerans (Camptocercus), and other uncommon diet items such as Crustaceans (Eubranchipus, metazoe larvae), and insects parts.

3.2. Occurrence

In S. galilaeus, the lowest diet percentage occurrence (0.08%) was recorded for Tetrapedia (Chroococcaceae), Microchaete (Microchaetaceae), Ophiocytium (Xanthophyceae), Fragilaria (Diatomophycae), Arctodiaptomes (Naviculaceae), Gomphonema (Diaptomideae) and Crustacean (Table 1). The highest occurrence recorded was 72.33% and corresponded to 860 individuals of S. galilaeus that have ingested sand particles, which alone dominated the stomach content with a volumetric proportion of 23.95%. In addition, of the 70 foods items recorded, the ten (10) first preys with relatively higher occurrence were sand grains occurring in 860 stomachs (72.23%), detritus occurring in 826 (69.47%), Scenedesmus (Scenedesmaceae) found in 701 (58.96%) stomachs, Closterium (Desmidiaceae) found in 662 (55.68%) stomachs, Microcystis (Chroococcaceae) occurring in 610 (51.3%) stomachs, Eudorina (Volvocaceae) occurring in 496 (41.72%) stomach, Navicula (Naviculaceae) recorded in 289 (24.31%) stomachs, Melosira (Coscinodisceae) recorded in 285 (23.97%) stomachs, Nostoc (Nostocaceae) occurring in 281 (23.63%) stomachs and Synechocystis (Chroococcaceae) occurring in 262 (22.04%) stomachs.

Table 1. Volumetric and occurrence percentages (%) of food resources consumed by Sarotherodon galilaeus (N = 1189) from the artificial lake of Ahozon (South-Benin).

Prey categories	Prey families	Genera	Volumetric percentage (%)	Occurrence (number)	Percentage occurrence (%)
Bleue-green algae	Chroococcaceae	*Coelosphaerium*	3.15	196	16.48
	Chroococcaceae	*Mycrocystis*	12.25	610	51.30
	Chroococcaceae	*Gomphosphaeria*	0.01	4	0.34
	Chroococcaceae	*Synechocystis*	3.87	262	22.04
	Chroococcaceae	*Synechococcus*	0.30	24	2.02
	Chroococcaceae	*Polycystis*	0.57	24	2.02
	Chroococcaceae	*Chroococus*	0.13	15	1.26
	Chroococcaceae	*Merismopedia*	0.06	8	0.67
	Chroococcaceae	*Tetrapedia*	0.003	1	0.08
	Microchaetaceae	*Microchaete*	0.005	1	0.08
	Nostocaceae	*Anabaena*	0.84	112	9.42
	Nostocaceae	*Nostoc*	1.92	281	23.63
	Oscillatoriaceae	*Oscillatoria*	0.01	13	1.09
	Oscillatoriaceae	*Spirulina*	0.37	53	4.46
	Palmellaceae	*Sphaerocystis*	0.13	5	0.42
	Pleurococcacae	*Protococus*	0.14	5	0.42
	Rivulariaceae	*Calothrix*	0.10	54	4.54
	Rivulariaceae	*Rivularia*	0.62	68	5.72
Green algae	Dictyosphaeriacae	*Botryoccocus*	2.99	177	14.89
	Oocystaceae	*Ankistrodesmus*	0.66	95	7.99
	Scenedesmaceae	*Scenedesmus*	6.97	701	58.96
	Scenedesmaceae	*Crucigenia*	0.60	67	5.63
	Scenedesmaceae	*Coelastrum*	0.04	13	1.09
	Tetrasporaceae	*Tetraspora*	0.44	34	2.86
	Ulothricaceae	*Ulothrix*	0.01	3	0.25
	Ulothricaceae	*Binucleuria*	0.04	8	0.67
	Xanthophyceae	*Ophiocytium*	0.02	1	0.08
Desmids	Colaciaceae	*Colacium*	0.02	6	0.50
	Desmidiaceae	*Closterium*	5.87	662	55.68
	Desmidiaceae	*Euastrum*	0.02	16	1.35
	Desmidiaceae	*Staurastrum*	0.02	10	0.84
	Desmidiaceae	*Gonatozygon*	0.03	6	0.50
	Desmidiaceae	*Tetmemorus*	0.01	2	0.17
	Desmidiaceae	*Cosmarium*	0.005	5	0.42
	Euglenaceae	*Euglena*	0.03	2	0.17
	Euglenaceae	*Phacus*	1.16	98	8.24
	Micractiniaceae	*Micractinium*	0.13	18	1.51

	Oocystaceae	*Selenastrum*	0.01	3	0.25
	Peridiniaceae	*Peridinium*	0.12	10	0.84
	Surirellaceae	*Cymatopleura*	0.86	110	9.25
Diatoms	Chaetoceracae	*Chaetoceros*	0.11	20	1.68
	Diatomophycae	*Fragilaria*	0.01	1	0.08
	Coscinodisceae	*Stephanodiscus*	0.01	2	0.17
	Naviculaceae	*Anomoeneis*	0.005	7	0.59
	Naviculaceae	*Navicula*	2.95	289	24.31
	Naviculaceae	*Pinnularia*	2.12	230	19.34
	Naviculaceae	*Gomphonema*	0.002	1	0.08
	Naviculaceae	*Cymbella*	0.22	46	3.87
	Coscinodisceae	*Melosira*	2.84	285	23.97
	Diatomophyceae	*Asterionella*	0.05	8	0.67
	Nitzschiaceae	*Nitzschia*	0.01	2	0.17
	Coscinodisceae	*Cyclotella*	0.02	4	0.34
Protozoans	Chilodonellideae	*Chilodonella*	0.02	4	0.34
	Chrysophyceae	*Uroglena*	0.08	7	0.59
	Volvocaceae	*Eudorina*	7.58	496	41.72
Rotatoria	Euchlanideae	*Euchlanis*	0.03	4	0.34
	Euchlanideae	*Colurella*	1.51	115	9.67
	Gastropodideae	*Chromogaster*	0.06	7	0.59
	Philodinideae	*Rotaria*	0.02	6	0.50
	Trichocercideae	*Trichocerca*	0.21	13	1.09
Copepoda	Cyclopideae	*Tropocyclops*	0.14	9	0.76
	Diaptomideae	*Arctodiaptomes*	0.02	1	0.08
	Diaptomideae	*Diaptomus*	0.02	3	0.25
Cladocera	Cladocera	*Camptocercus*	0.02	5	0.42
Crustaceans	Chirocephalidae	*Eubranchipus*	0.14	7	0.59
	Crustacean	*Metazoe larvae*	0.01	1	0.08
Insect parts			0.63	75	6.31
Detritus			12.27	826	69.47
Sand grains			23.95	860	72.33
Unidentified preys			0.41	81	6.81

3.3. Empty Stomachs

In Lake Ahozon, empty stomachs significantly ($P \leq 0.05$) varied with the life stages categories (size classes) of S. galilaeus and ranged between 0.40% recorded for adults subpopulation and 83.33% recorded for small juveniles (Table 2). Indeed, one-way analysis on the percentage of empty stomachs gave $F_{3,11} = 3.905$ with p-value = 0.05, thus showing ontogenetic variations of empty stomachs. Overall, empty stomachs in small juveniles were higher and ranged between 28.57% - 83.33%. Those of large juveniles and subadults ranged between 11.11% - 45.59% and 2.78% - 22.73%, respectively. Adults exhibited lower empty stomachs varying between 0.40% and 11.76%. In contrast, the incidences of empty stomachs were not associated with seasons. Indeed, one-way ANOVA on the percentage of empty stomachs fail to show any significant ($F_{2,11}$ = 0.339, p = 0.721) seasonal variations.

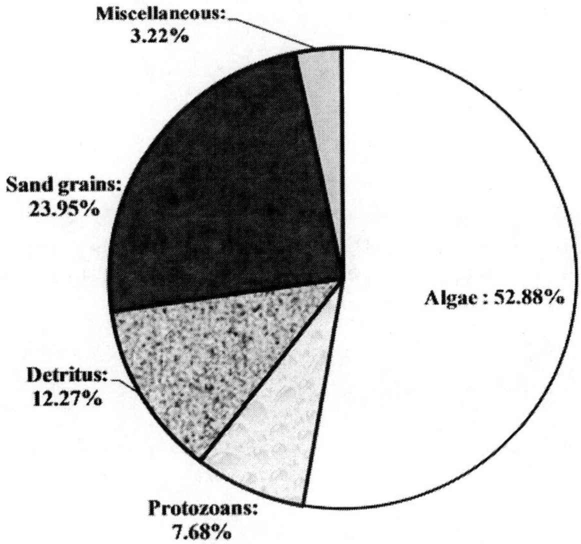

Figure 2. Volumetric percentage (%) of the foods resources ingested by Sarotherodon galilaeus (n = 1189) from the artificial lake of Ahozon (South-Benin).

3.4. Diet Breadth

In Lake Ahozon, S. galilaeus had a relatively high diet breadth (DB = 6.01) indicating that this species consumed a relatively broad spectrum of food resources. Ontogenetically, one-way ANOVA showed significant ($p \leq 0.005$) variations among computed diet breadth for different life stage categories. Indeed, the computed F-value, along with degrees of freedom and p-value was $F_{2,15}$ = 6.726, p = 0.007. In general, diet breadth tended to increase with fish sizes and ranged from DB = 5.55 for small juveniles (TL < 30 mm) to DB = 7.29 (50 mm ≤ TL < 90 mm) for subadults (Table 3). Inversely, one-way analysis of variance on diet breadth fails to show any significant (P > 0.05) seasonal variations. The computed F-value, along with degrees of freedom and p-value was $F_{2,11}$ = 0.749, p = 0.500. Nevertheless, diet breadth was higher during the dry season (DB = 6.36) compared to those of wet and flood seasons (Table 3).

Table 2. Percentages (%) of empty stomachs recorded by life stage category of Sarotherodon galilaeus from the artificial lake of Ahozon (South-Benin) during wet, flood and dry season.

S. galilaeus life stages	Wet		Flood		Dry	
	No. of individuals	Empty stomachs (%)	N° of individuals	Empty stomachs (%)	No. of individuals	Empty stomachs (%)
Small juveniles	21	28.57	6	83.33	16	43.75
Large juveniles	63	11.11	68	45.59	58	31.03
Subadults	88	22.73	26	2.78	60	11.67
Adults	306	11.76	250	0.40	227	10.13
Total	**478**	**14.44**	**350**	**10.86**	**361**	**14.96**

Total number of individuals: **1189**. Total number of empty stomachs (**161**): **Wet** = **69**; **Flood** = **38**; **Dry** = **54**. Total percentage (%) of empty stomachs: **13.54%**.

Table 3. Diet breadth (DB) by season and life stage category of Sarotherodon galilaeus (N = 1189) from the artificial lake of Ahozon (South-Benin).

Seasons	Life stage categories			
	Small juveniles	Large juveniles	Subadults	Adults
Wet	6.18	6.23	6.50	5.64
Flood	3.57	6.13	6.81	5.03
Dry	4.44	6.79	7.13	6.23
Total DB	**5.55**	**6.85**	**7.29**	**5.87**
Abundance	**43**	**189**	**174**	**783**

3.5. Seasonal Variations of Diet

Except detritus, most food items (sand particules, algae, protozoans, rotifers) consumed by S. galilaeus fail to show any significant ($p > 0.05$) seasonal variations. The calculated F and p values was $F_{2,11} = 1.611$, $p = 0.252$ for sand particles, $F_{2,11} = 0.904$, $p = 0.439$ for aggregated algae, $F_{2,11} = 0.208$, $p = 0.816$ for protozoans, and $F_{2,11} = 1.121$, $p = 0.367$ for rotifers. Nevertheless, proportional consumption of sand particles was greater during the high-water season with a volumetric proportion reaching 31.69% (Figure 3). Likewise, proportional consumption of algae and protozoans were greater during the dry season with volumetric proportions reaching 57.04% and 11.28%, respectively. Also, the highest volumetric proportion of rotifers (2.42%) was recorded during the wet season. In contract, the volumetric proportion of detritus showed significant ($F_{2,11} = 4.302$, $p = 0.049$) seasonal variations with the highest proportional consumption (16.05%) recorded during the dry season.

3.6. Pianka's Diet Overlaps and Ontogenetic Diet Shifts

In this study, diet overlaps ($Ø_{jk}$) evaluate dietary similarity between subpopulations of different size classes or life stages. Overall, diet overlaps ranged from $Ø_{jk}$ = 0.77 for the pairing "small juveniles X adults" to $Ø_{jk}$ = 0.97 for the pairing "subadult X adults", and averaged $Ø_{jk}$ = 0.87 ± 0.08, an indication of high diet similarities among different life stage categories (Table 4). In general, adjacent life stages or size classes tended to exhibit higher diet overlaps than distant pairings involving distant size classes categories. This trend indicated ontogenetic diet shifts, also shown by the increase in the proportional consumption of sand particles & detritus with S. galilaeus size classes. The associated regression equation gave a positive slope b = 0.241 along with a significant ($p \leq 0.01$) correlation coefficient r = 0.99. Also, the volumetric proportion of sand particles & detritus ranged between 23.1% for small juveniles (TL < 30 mm) to 37.04% for adults (TL ≥ 90 mm) (Table 5). Though proportional consumptions of aggregated algae by different size classes were not significantly ($p > 0.05$) different, the consumption of some algae such as Synechocystis (Chroococcaceae), Microcystis (Chroococcaceae) and Closterium (Desmidiaceae) increased with the sizes of S. galilaeus. This trend was shown by the positive slopes (b) ranging between 0.033 and 0.164 and the correlation coefficient varying from 0.79 - 0.99 recorded from the linear regression equations between the fish sizes and the volumetric proportion of these food resources.

Table 4. Matrix of diet overlaps ($Ø_{jk}$) of Sarotherodon galilaeus (N = 1189) by life stage category from the artificial lake of Ahozon (South-Benin).

S. galilaeus life stages	Small juveniles	Large juveniles	Subadults	Adults
Small juveniles	1	0.78	0.86	0.77
Large juveniles		1	0.92	0.90
Subadults			1	0.97
Adults				1

3.7. Ecomorphological Correlates

The ecomorphological trends of the trophic structure of S. galilaeus has been analyzed by plotting the volumetric proportion of each prey ingested (detritus, sand particles, algae etc.) against total length (TL) and gut length (GL). The linear regression equations indicated that TL and GL were positively (slope b_{TL} = 0.241; b_{GL} = 0.280) correlated with the volumetric proportion of sand particles with significant ($p \leq 0.05$) correlation coefficients r = 0.99 and r = 0.87, respectively, indicating that the consumption of sand particles increased with TL and GL (Table 6, Table 7).

The same trends were recorded for dominant algae of high occurrence and high volumetric proportions such as Synechocystis (Chroococcaceae), Microcystis (Chroococcaceae) (Table 6, Table 7). In addition, ecomorphological relationships were explored by plotting GL against body weight (W) and standard length (SL). To lessen skew in the dataset, logarithmic transformations were applied to GL, W and SL. Also, the ratio of SL to GL (GL/SL) was computed as a measure of relative gut length and compared to reference values. The linear regression equations obtained were as follows:

LOG (GL) = 0.560*LOG(W) + 1.011, $R^2 = 0.89$, $N = 1189$ (Figure 4)

LOG (GL) = 1.613*LOG(SL) + 0.312, $R^2 = 0.87$, $N = 1189$ (Figure 5)

The slopes, $b = 0.560$ and $b = 1.613$ of both equations were positive with significant ($p \leq 0.05$) correlation coefficients $r = 0.94$ and $r = 0.93$, respectively. The ratio of GL/SL ranged between 1 (30 mm SL) and 13.85 (68 mm SL) with a mean of 6.72 ± 2.23.

Table 5. Volumetric percentages (%) of food resources consumed by size classes (mm TL) of Sarotherodon galilaeus (N = 1189) from the artificial lake of Ahozon (South-Benin). TL = Total length.

Prey categories	Prey families	Genera	Small juvenile (TL < 30)	Juveniles (30 ≤ TL < 50)	Subadults (50 ≤ TL < 90)	Adults (TL ≥ 90)
Bleue-green algae	Chroococcaceae	*Coelosphaerium*	-	9.02	3.97	3.00
	Chroococcaceae	*Microcystis*	1.73	1.40	10.33	12.60
	Chroococcaceae	*Gomphosphaeria*	-	0.06	-	0.02
	Chroococcaceae	*Synechocystis*	0.74	2.33	3.19	3.97
	Chroococcaceae	*Synechococcus*	0.45	0.03	0.28	0.30
	Chroococcaceae	*Polycystis*	-	-	0.52	0.58
	Chroococcaceae	*Chroococus*	-	-	0.45	0.09
	Chroococcaceae	*Merismopedia*	0.19	0.04	0.04	0.07
	Chroococcaceae	*Tetrapedia*	-	-	-	0.004
	Microchaetaceae	*Microchaete*	-	-	-	0.002
	Nostocaceae	*Anabaena*	-	4.47	0.73	0.82
	Nostocaceae	*Nostoc*	3.98	3.51	1.11	2.00
	Oscillatoriaceae	*Oscillatoria*	-	0.12	0.06	0.002
	Oscillatoriaceae	*Spirulina*	0.28	6.60	0.46	0.31
	Palmellaceae	*Sphaerocystis*	-	-	0.36	0.11
	Pleurococcacae	*Protococus*	-	-	0.02	0.16
	Rivulariaceae	*Calothrix*	2.42	3.66	0.31	0.04
	Rivulariaceae	*Rivularia*	1.45	2.56	0.47	0.62

Green algae	Dictyosphaeriacae	*Botryoccocus*	-	7.16	2.56	3.01
	Oocystaceae	*Ankistrodesmus*	1.06	0.10	0.38	0.70
	Scenedesmaceae	*Scenedesmus*	13.47	3.39	6.76	7.03
	Scenedesmaceae	*Crucigenia*	-	1.52	0.24	0.63
	Scenedesmaceae	*Coelastrum*	0.41	0.13	-	0.04
	Tetrasporaceae	*Tetraspora*	-	0.07	0.29	0.46
	Ulothricaceae	*Ulothrix*	-	-	-	0.007
	Ulothricaceae	*Binuclearia*	0.85	-	0.18	0.02
	Xanthophyceae	*Ophiocytium*	-	-	-	0.02
Desmids	Colaciaceae	*Colacium*	4.13	-	-	0.02
	Desmidiaceae	*Closterium*	3.44	5.60	6.77	5.77
	Desmidiaceae	*Euastrum*	0.84	1.49	-	0.008
	Desmidiaceae	*Staurastrum*	-	0.28	-	0.02
	Desmidiaceae	*Gonatozygon*	-	-	-	0.04
	Desmidiaceae	*Tetmemorus*	-	-	0.04	0.009
	Desmidiaceae	*Cosmarium*	0.32	0.01	0.01	0
	Euglenaceae	*Euglena*	-	-	-	0.04
	Euglenaceae	*Phacus*	0.97	0.06	1.18	1.17
	Micractiniaceae	*Micractinium*	0.48	-	0.08	0.14
	Oocystaceae	*Selenastrum*	-	-	-	0.01
	Peridiniaceae	*Peridinium*	-	0.04	0.30	0.10
	Surirellaceae	*Cymatopleura*	0.80	2.83	1.92	0.71
Diatoms	Chaetoceracae	*Chaetoceros*	0.22	0.10	0.11	0.11
	Diatomophyceae	*Fragilaria*	-	-	-	0.01
	Coscinodisceae	*Stephanodiscus*	-	-	-	0.02
	Naviculaceae	*Anomoeneis*	0.45	0.10	-	0.004
	Naviculaceae	*Navicula*	4.65	3.99	3.12	2.92
	Naviculaceae	*Pinnularia*	6.84	0.44	2.23	2.12
	Naviculaceae	*Gomphonema*	-	-	-	0.002
	Naviculaceae	*Cymbella*	5.58	1.32	0.34	0.20
	Coscinodisceae	*Melosira*	1.64	4.85	4.10	2.67
	Diatomophyceae	*Asterionella*	-	-	-	0.06
	Nitzschiaceae	*Nitzschia*	-	-	0.04	0.01
	Coscinodisceae	*Cyclotella*	-	-	-	0.03
Protozoans	Chilodonellideae	*Chilodonella*	-	-	-	0.03
	Chrysophyceae	*Uroglena*	-	-	0.05	0.08
	Volvocaceae	*Eudorina*	12.97	4.09	13.02	6.93
Rotatoria	Euchlanideae	*Euchlanis*	-	-	-	0.04
	Euchlanideae	*Colurella*	0.28	0.12	1.04	1.59
	Gastropodideae	*Chromogaster*	0.50	0.10	-	0.06
	Philodinideae	*Rotaria*	-	-	0.15	0.005
	Trichocercideae	*Trichocerca*	-	-	0.11	0.22
Copepoda	Cyclopideae	*Tropocyclops*	-	-	0.08	0.15
	Diaptomideae	*Arctodiaptomes*	-	-	-	0.007
	Diaptomideae	*Diaptomus*	-	-	0.02	0.01
Cladocera	Cladocera	*Camptocercus*	4.24	-	-	0.01
Crustaceans	Chirocephalidae	*Eubranchipus*	-	-	0.22	0.13
	Crustacean	*Metazoe larvae*	-	-	-	0.02
Insect parts			0.28	2.86	1.16	0.54
Detritus			16.24	12.27	12.89	12.19
Sand grains			6.86	9.67	17.60	24.85
Unidentified preys			1.24	3.61	0.71	0.33

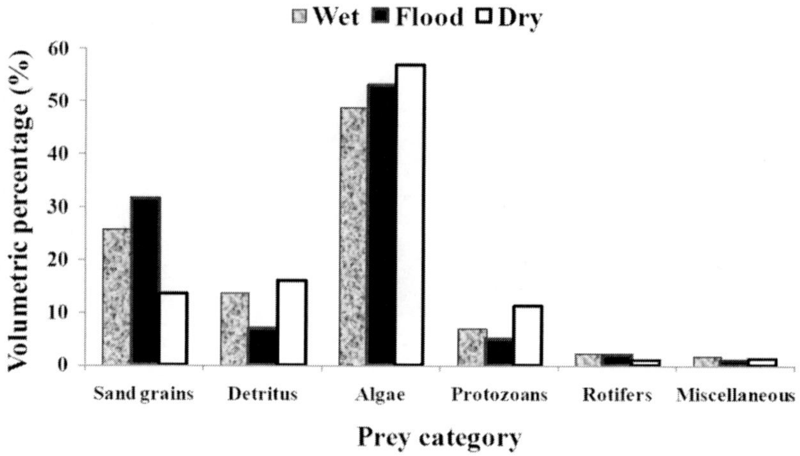

Figure 3. Seasonal trends of the volumetric percentage (%) of the foods resources ingested by Sarotherodon galilaeus (n = 1189) from the artificial lake of Ahozon (South-Benin).

4. DISCUSSIONS

4.1. Food and Feeding Patterns

The success of fisheries and aquaculture valorizations of a species in a man-made lake require knowledge on feeding ecology in order to evaluate the food resource exploitation and utilization, and the establishment of the fishes in this artificial habitat [32] [33]. The present study indicated that S. galilaeus consumed about seventy (70) food resources composed of a variety of benthic and pelagic prey items dominated by algae (52.88%) constituated of blue-green algae, green algae, desmids and diatoms, sand particles (23.95%), detritus (12.27%) and protozoans (8.69%). Though consuming about 52 genera of algae, only ten (10) of them, Microcystis (Chroococcaceae), Scenedesmus (Scenedesmaceae), Closterium (Desmidiaceae), Synechocystis (Chroococcaceae), Coelosphaerium (Chroococcaceae), Botryoccocus (Dictyosphaeriacae), Navicula (Naviculaceae), Melosira (Coscinodisceae), Pinnularia (Naviculaceae) and Nostoc (Nostocaceae) were the most abundant and gave an aggregated volumetric proportion of 44.93%. This "alguivore"-like trophic specialization of S. galilaeus probably results from anatomical structures, mainly the presence of numerous gill rakers (20 - 27), which facilitated the filtering of numerous algae species [34].

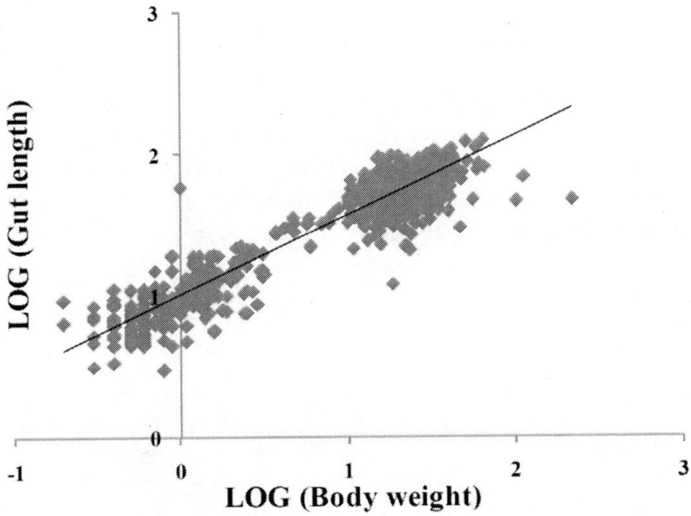

Figure 4. Relationship between Log (gut length) and Log (body weight) of Sarotherodon galilaeus (N = 1189) from the artificial lake of Ahozon (South-Benin). Regression equation and coefficient of determination were as follows: Log(GL) = 0.560*Log (W) + 1.011, $R^2 = 0.89$.

Table 6. Linear regression between the volumetric percentages (%) of food resources consumed and total length (TL) of Sarotherodon galilaeus (N = 1189) from the artificial lake of Ahozon (South-Benin).

Regressions factors	No. of fish size categories	Slope (*b*)	Determination coefficient	Correlation coefficient	Intercept
Sand particles	4	0.241	0.96	0.99	1.787
Synechocystis (Chroococcaceae)	4	0.041	0.97	0.99	0.337
Microcystis (Chroococcaceae)	4	0.164	0.88	0.94	2.337
Closterium (Desmidiaceae)	4	0.033	0.62	0.79	3.604

Table 7. Linear regression between the volumetric percentages (%) of food resources consumed and gut length (GL) of Sarotherodon galilaeus (N = 1189) from the artificial lake of Ahozon (South-Benin).

Regressions factors	No. of fish size categories	Slope (b)	Determination coefficient	Correlation coefficient	Intercept
Sand particles	4	0.280	0.75	0.87	8.407
Synechocystis (Chroococcaceae)	4	0.040	0.54	0.72	1.634
Microcystis (Chroococcaceae)	4	0.170	0.54	0.74	2.337
Closterium (Desmidiaceae)	4	0.0174	0.063	0.25	5.077

The broad spectrum of the food items found in the stomachs of S. galilaeus is comparable or similar to what have been reported from numerous studies on tilapine cichlids. Greenwood [35] recorded a mixture of phytoplanktons and zooplanktons in the gut of S. galilaeus. According to Adebisi [36], S. galilaeus fed on desmids, diatoms, blue-green algae and benthic protozoans in the Upper Ogun River. As reported by Akintunde & Imevbore [37], phytoplankton was the main food item as being present in the diet of S. galilaeus from Kaindji Lake.

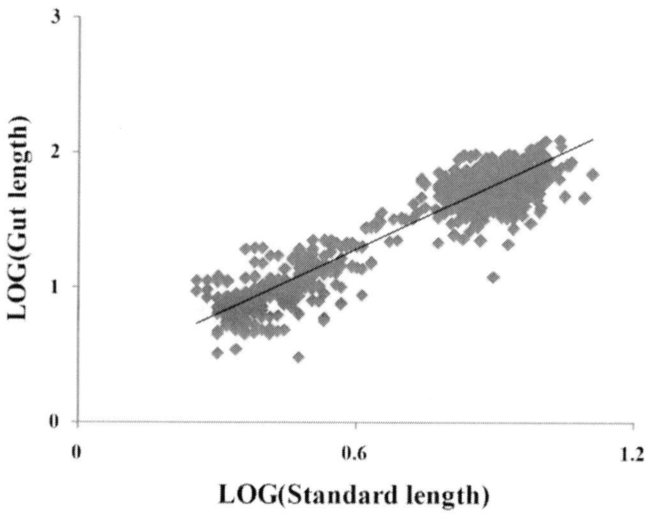

Figure 5. Relationship between Log (gut length) and Log (standard length) of Sarotherodon galilaeus (N = 1189) from the artificial lake of Ahozon (South-Benin). Regression equation and coefficient of determination were as follows: Log (GL) = 1.613*Log (SL) + 0.312, R^2 = 0.87.

Fish [38] [39] and Fagade [40] [41] reported that many tilapine species foraged mainly on diatoms. In the International Institute of Tropical Agriculture (IITA) lake of Ibadan, algae dominated the stomach content of S. galilaeus with a proportional consumption of 70% [42]. Arame [43] reported the same trends for two other cichlids, Sarotherodon melanotheron and Oreochromis niloticus which consumed respectively, about 90 and 56 species of phytoplanktons, aggregating dietary volumetric proportions of about 86.88% and 62.74%, respectively. Though not tilapine species, the bonytongue Heterotis niloticus, an omnivore, because of the presence of numerous gill rakers (42 - 94 on the first branchial arch) during its whole life, ingested various algae taxa such as diatoms (Raphoneis, Pleurosigma, Synedra, Nitzschia, Gomphonema, Melosira, Tabellaria,

Asterinella, Navicula), Polycistis, Protococcus, Phormidium, Coelosphaerium, Nostoc, Oscillatoria, Merismopodia), green algae (Rhizodonium, Botyoccocus, Ulothrix sp., Richterella, Spirogyra, Coelastrum) and desmids (Gonatozygon, Closterium) [34]. Nevertheless, in the current study, the 23.95% of sand particles and the 12.27% of detritus recorded in the diet of S. galilaeus reflected both the benthic-like and detritivore-like foraging behavior of this species. The relatively high proportion of sand particles in the diet suggests that, probably, S. galilaeus ingested sand as normal diet components. Also, sand grains may help in the maceration of food materials in the pyloric stomach as observed in feeding ecology studies of mullets [8].

Of the 70 food resources ingested by S. galilaeus, ten (10) consistently occurred in a relatively high number of stomachs ranging between 262 (22.04%) stomachs containing Synechocystis (Chroococcaceae) and 860 (72.33%) stomachs containing sand particles. When considering the aggregated algae, 1025 (86.21%) stomachs contained phytoplankton species which appeared to be the major food items occurring in most stomachs. In the IITA lake of Ibadan, Fagade [42] reported similar trends of algae occurrence ranging between 193 (18.71%) for a filamentous algae (Spirogyra) to 750 (98.81%) for diatoms with blue-green algae (Anabaena, Oscillatoria) occurring in 709 (93.41%) stomachs [34]. In Lake Toho (South-Benin), Arame [43] reported some percentage occurrences of 60.79% for Melosira (Coscinodisceae) and 58.48% for Scenedesmus (Scenedesmaceae) in the dietary of the tilapine cichlid, Sarotherodon melanotheron. Also, in the same lake, Zygnema (green algae) and Melosira (Coscinodisceae) occurred in 59.19% and 44.29% stomachs of Oreochromis niloticus, respectively [43].

The study revealed insignificant ($p > 0.05$) seasonal variations of empty stomach, that is, the number of empty stomachs recorded was not associated with seasons. In contrast, the results showed ontogenetic variations of empty stomachs which significantly ($p \leq 0.05$) varied with the life stage categories of S. galilaeus. Indeed, lower size classes (small juveniles) exhibited higher percentage of empty stomachs (28.57% - 83.33%) compared to adults exhibiting lower empty stomachs varying between 0.40% and 11.76%. The higher percentage of empty stomach recorded in small juveniles was probably due to limited access to the food resources coupled with the undeveloped digestive tractus [34].

In Lake Ahozon, S. galilaeus had a broader diet breadth resulting from the numerous food resources (70 prey items) consumed, and ranging between DB = 5.5 for small juveniles and DB = 7.29 for subadults (mean: 6.39 ± 0.87). In general, diet breadths tended to increase with the size of S. galilaeus, indicating that this species ingested a broader range of prey items as it grew [9]. However, the lack of seasonal variations in the diet breadth was probably due to the fact that Lake Ahozon was isolated and lacked connection with other water bodies such as coastal lagoons (brackish waters) or running waters (rivers, streams) which could cause intrusion of alien food resources. Nevertheless, the relatively low diet breadth (mean: 5.40 ± 1.41) recorded during the flooding season was probably due to the reduced planktonic bloom caused by high water turbidity that inhibited or limited photosynthetis. Compared to other tilapine cichlids, the diet breadth (mean: 5.40 ± 1.41) obtained for S. galilaeus in the man-made lake of Ahozon was lower than that (8.01 ± 3.30) recorded for Sarotherodon melanotheron from Lake Toho, a Southern Benin floodplain lake. Probably, this difference was the result of differential food availability in both lakes. Lake Toho is an ancient natural floodplain lake connected to Mono River and should dwell higher food resources than Lake Ahozon, a younger man-made water body isolated from the other inland waters. In addition, the difference in anatomical structures of both species could affect their foraging yields. For exemple, S. galilaeus shows 20 - 27 gill rakers on the first branchial arch compared to 27 - 30 gill rakers recorded for S. melanotheron [24].

The study consistently showed high diet similarities between size classes, indicated by high diet overlaps ranging between $Ø_{jk}$ = 0.77 and $Ø_{jk}$ = 0.97. Also, individuals with adjacent size classes tended to have higher diet overlaps than pairings of distant size classes (Table 3). These trends indicated ontogenetic diet shifts in S. galilaeus shown for exemple by the increase in sand particles consumptions as the fish grew [9].

In the artificial lake of Ahozon, proportional consumptions of most food resources, such as sand particles, combined algae, protozoans and rotifers ingested by S. galilaeus exhibited insignificant ($p > 0.05$) seasonal variations. Probably, these prey items may have been available all seasons (wet, flood, dry) so that S. galilaeus may have unlimited access and consumed them ad libitum according to their needs and food requirements. However, the proportional consumption of detritus showed significant ($p \leq 0.05$) seasonal variations, but higher consumption (17.44%) occurred in the dry season. Indeed, during the dry season, the lake's water should be less turbid and relatively calm, with bottom substrates more stable and more available for the fish, compared to the wet and flood seasons characterized by high water disturbance causing unstable bottom/substrates and high turbidities.

4.2. Trophic Plasticity

This investigation showed that S. galilaeus exhibited a wide diet breadth resulting from the broad spectrum of food consumed and composed of 18 genera of blue-green algae, 9 genera of green algae, 13 genera of desmids,

12 genera of diatoms, 3 genera of protozoans, 5 genera of rotifers, 3 genera of copepods, 1 genus of cladocera, 2 genera of crustacean, insect parts, detritus and sand particles. Although the proportional consumption of these food items were not equally represented in the diet, the wide choise of foods available to S. galilaeus suggests that when one prey was in reduced supply, the species could forage on another abundant and available prey of the man-made lake [42]. As reported by Bowen et al. and Adite et al. [9] [18] this food habit pattern is an indicator of trophic plasticity in S. galilaeus which may shift its feeding structure according to prey availability in the man-made lake of Ahozon. This trophic plasticity results from specialized anatomical structures, mainly the presence of numerous gill rakers, 20 - 27 on the first branchial arch, which facilitate the filtering of numerous algae species [9] [34].

4.3. Ecomorphological Relationships and Food Habits

In Lake Ahozon, the highly significant ($p \leq 0.05$) positive correlation coefficients, $r = 0.99$ and $r = 0.87$, recorded respectively from both regression equations of Log (TL) and Log (GL) with the proportional consumption of detritus indicated that the consumption of detritus increased with the size and the development of the digestive tract of S. galilaeus [34]. Also, some relatively abundant and dominant algae such as Synechocystis (Chroo- coccaceae), Microcystis (Chroococcaceae) and at some extent, Closterium (Desmidiaceae) followed the same ecomophological pattern. According to Adite et al. [9], small fish individuals with less developed digestive system forage mostly on live food, and as the digestive tract becomes well-developed, they take advantage to other delicate resources such as detritus [34].

In the current study, the linear regression between Log (W) and Log (GL) gave a positive slope $b = 0.560$. This value fails into Kramer & Bryant (1995) omnivore slope range (0.4 - 0.68) supporting that S. galilaeus may be an omnivore. However, the slope b = 0.560 also is close to Kramer & Bryant [44] herbivore slope range (0.58 - 0.68) which aligns S. galilaeus with herbivore. Furthermore, in S. galilaeus, the ratio of SL to GL (GL/SL) ranged between 1 (30 mm SL) and 13.85 (68 mm SL) with a mean of 6.72 ± 2.23 which also is close to Paugy's [45] African detritivorous fishes ratio (mean = 7.2). In Lake Ahozon, the dominant food resources consumed by S. galilaeus were sand particles (23.95%), detritus (12.27%) and algae (52.88%), occurred respectively in 860 (72.23%) stomachs, 826 (69.47%) stomachs and 1025 (86.21%) stomachs. Consequently, the slope $b = 0.560$ obtained from the linear regression Log (W) - Log (GL) and the mean ratio GL/SL = 6.72 ± 2.23 support our findings that S. galilaeus exhibited detritivorous, herbivorous (alguivorous) and omnivorous food habits because of the sand particles & detritus consumed (36.22%), the proportional consumption of algae (52.88%) and its balanced diet that included about 10.09% of animal materials.

4.4. Implications for Species Establishment and Valorization

The current investigation on the feeding ecology of S. galilaeus revealed that this naturally-colonized tilapine cichlid is well-established in the artificial lake of Ahozon. Indeed, the study consistently showed that S. galilaeus exploit and utilize in Lake Ahozon, a combination of plant and animal resources as source of foods. Though dominated by algae (52.88%) composed of 28 famillies and 52 genera and detritus & sand particles (36.22%), the diet also included about 10.09% of animal organisms composed of protozoans, microcrustacea (rotifers, copepods, cladocera) and insect parts. The ability of S. galilaeus to exploit and to utilize this wide variety of available food resources to satisfy its nutrional needs accounted for the prominence, the abundance and the perfect establishment of the species in the man-made lake of Ahozon. The physical structure of the lake allowed the availability of aquatic vegetation habitat at the shallower lake margin, considered not only as a potential source of food resources, but also as potential spawning ground for most fishes including S. galilaeus. Hence, the availability of food would insure rapid growth leading to early maturation, and the presence of suitable reproduction grounds would lead to a greater recruitment and high propagation of the species in the man-made lake of Ahozon. As results, the broad spectrum of food items consumed, the high diet breadth, the active reproduction of the species, the allometric/isometric growths and the greater prominence of the species were some indicators of the successful establishment of S. galilaeus in the man-made lake of Ahozon. In addition, the high ecological niche is an advantage for the aquacultural valorization of S. galilaeus, and particularly in extensive aquaculture systems where habitat conditions are close to those of natural lakes. Successful and sustainable fisheries valorization/management of the species require 1) the protection of its habitats, mainly spawning and foraging grounds, 2) the introduction of predators to control the population of the species, 3) the prevention of domestic wastes dumping 4) the planned harvest of the fish stock, 5) the fertilization of the lake by installing a park of branches in the lake, and 6) the protection of the lake margin to avoid erosion. Especially to enhance aquaculture valorization, some medium-environments such as “net enclosures” and “floating cages” could be introduced in the lake for intensive rearing of S. galilaeus and other fish species such Oreochromis niloticus, Clarias gariepinus and Chrysichthys nigrodigitatus.

5. CONCLUSION

The current study gives quantitative and qualitative information on the trophic ecology of the freshwater tilapine cichlid Sarotherodon galilaeus and its establishment in the man-made lake of Ahozon. The results consistently showed that S. galilaeus exploited and utilized a broad spectrum of food resources in the artificial lake of Ahozon. Main prey items included blue-green algae, green algae, desmids, diatoms, sand particles, detritus, protozoans and microcrustacean that conferred to the species, an herbivorous (alguivorous), detritivorous and omnivorous food's habits. The broad spectrum of food items consumed and the high diet breadth recorded leading to allometric/isometric growth, the early maturation, the active breeding and the high propagation of the species in the lake, are some indicators of the successful establishment of S. galilaeus in the man-made lake of Ahozon. An integrated management scheme of Lake Ahozon must include the periodic ecological monitoring of the lake, the protection of the reproduction and foraging grounds and the valorization of S. galilaeus in fisheries and aquaculture along with a conservation plan.

ACKNOWLEDGEMENTS

Logistic and financial assistance were provided by the "Laboratoire d'Ecologie et de Management des Ecosyst- èmes Aquatiques, Département de Zoologie, Faculté des Sciences et techniques, Université d'Abomey-Calavi". We express our gratitude to Mr Doukpo Célestin, the owner of the artificial lake to allow us to conduct this investigation on both man-made lakes. We are also grateful to Mr Houessinon Geoffroy, Houndjetin Louis and Djihouessi Bernold for their assistance in fish sampling and laboratory works. We thank the numerous reviewers for their thorough peer-review of this manuscript.

REFERENCES

1. Van Thielen, R., Hounkpe, C., Agon, G. and Dagba, L. (1987) Guide de determination des poissons et crustacés des lagunes et lacs du Bas Bénin. GTZ & Direction des Pêche, Cotonou, Benin.
2. Adite, A. and Van Thielen, R. (1995) Ecology and Fish Catches in the Natural Lakes of Benin, West Africa. Environmental Biology of Fishes, 43, 381-391. http://dx.doi.org/10.1007/BF00001173
3. Djihouessi, M.B. (2015) Ecologie et valorisation piscicole des lacs artificiels du Sud-Bénin (Bénin, Afrique de l'Ouest). Master's Thesis, FAST, Université d'Abomey-Calavi, Bénin.

4. Gbaguidi, G.A.M.H., Adite, A. and Sossoukpe, E. (2016) Ecology and Fish Biodiversity of Some Artificial Lakes of Southern Benin (West Africa): Implications for Fisheries Management and Aquaculture Valorization. Journal of Environmental Protection, In Press. http://dx.doi.org/10.4236/jep.2016.76079
5. Jimoh, A.A., Clarke, E.O., Whenu, O.O. and Adeoye, H.B. (2011) Foods and Feeding Habits of the African River Prawn (Macrobrachim vollenhovenii, Herklots, 1857) in the Epe Lagoon, Southwest Nigeria. International Journal of Fisheries and Aquaculture, 3, 10-15.
6. Nunoo, F.K.E., Sossoukpe, E., Adite, A. and Fiogbe, D.E. (2013) Foods Habits of Two Species of Pseudotolithus (Scianidae) off (West Africa) near Shore Waters and Implications for Management. International Journal of Fisheries and Aquaculture, 5, 142-151.
7. Newell, R.I.E. and Jordan, S.J. (1983) Preferential Ingestion of Organic Material by the American Oyster Crassostrea virginica. Marine Ecological Progress Series, 13, 47-53. http://dx.doi.org/10.3354/meps013047
8. Amisah, S. and Agbo, N.W. (2008) An Investigation into the Food and Feeding Ecology of a Potential Aquaculture Candidate, Sarotherodon galilaeus multifasciatus in a Meteoritic Crater Lake in Ghana. Journal of Applied Sciences and Environmental Management, 12, 15-18.
9. Adite, A., Winemiller, K.O. and Fiogbe, D.E. (2005) Ontogenetic, Seasonal, and Spatial Variation in the Diet of Heterotis niloticus (Osteoglossiformes; Osteoglossidae) in the Sô River-Lac Hlan System, Benin, West Africa. Environmental Biology of Fishes, 3, 367-378. http://dx.doi.org/10.1007/s10641-004-5563-9
10. Lauzanne, L. (1976) Régimes alimentaires et relations trophiques des poissons du lac Tchad.Cahiers ORSTOM, Sér.Hydrobiologie, 10, 267-310.
11. Hickley, P. and Bailey, R.G. (1987) Foods and Feeding Relationships of the FSH in the Sudd Swamps. Journal of Fish Biology, 30, 147-159. http://dx.doi.org/10.1111/j.1095-8649.1987.tb05741.x
12. Fagade, S.O. and Olaniyan, C.I.O. (1973) The Foods and Feeding Interrelationships of the Fishes in the Lagos Lagoon. Journal of Fish Biology, 5, 205-225. http://dx.doi.org/10.1111/j.1095-8649.1973.tb04449.x
13. Winemiller, O.K. (1990) Spatial and Temporal Variation in Tropical Fish Trophic Network. Ecological Monographs, 60, 331-367. http://dx.doi.org/10.2307/1943061
14. Winemiller, K.O. (1989) Ontogenetic Diet Shifts and Resource Partitioning among Piscivorous Fishes in the Venezuelan Llanos. Environmental Biology of Fishes, 26, 177-199. http://dx.doi.org/10.1007/BF00004815
15. Steingrimsson, S.O. and Gislason, G.M. (2002) Body Size, Diet and Growth of Landlocked Brown Trout, Salmon trutta, in the Subarctic River Laxa, North-East Iceland. Environmental Biology of Fishes, 63, 417-426. http://dx.doi.org/10. 1023/ A: 1014976612970
16. Koen Alonso, M., Crespo, E.A., Garcia, N.A., Pedraza, S.N., Mariotti, P.A. and Mora, N.J. (2002) Fishery and Ontogenetic Driven Changes in the Diets of the Spiny Dogfish, Squalus acanthias, in Patagonian Waters, Argentina. Environmental Biology of Fishes, 63, 193-202. http://dx.doi.org/10.1023/A:1014229432375
17. Barbarino Duque, A. and Winemiller, K.O. (2003) Dietary Segregation among Large Catfishes of the Apure and Arauca Rivers, Venezuela. Journal of Fish Biology, 63, 410-427. http://dx.doi.org/10.1046/j.1095-8649.2003.00163.x
18. Bowen, S.H. and Allanson, B.R. (1982) Behavioral and Trophic Plasticity of Juvenile Tilapia mossambica in Utilization of the Unstable Littoral Habitat. Environmental Biology of Fishes, 7, 354-362. http://dx.doi.org/10.1007/BF00005570
19. Garcia-Berthou, E. (1999) Food of Introduced Mosquitofish: Ontogenetic Diet Shift and Prey Selection. Journal of Fish Biology, 55, 135-147. http://dx.doi.org/10.1111/j.1095-8649.1999.tb00663.x
20. Bowen, S.H. (1983) Detritivory in the Neotropical Fish Communities. Environmental Biology of Fishes, 9, 137-144. http://dx.doi.org/10.1007/BF00690858

21. Garcia-Berthou, E. and Moreno-Amich, R. (2000) Food of Introduced Pumpkinseed Sunfish: Ontogenetic Diet Shift and Seasonal Variation. Journal of Fish Biology, 57, 29-40. http://dx.doi.org/10.1111/j.1095-8649.2000.tb00773.x
22. Adite, A. and Winemiller, K.O. (1997) Trophic Ecology and Ecomorphology of Fish Assemblages in Coastal Lakes of Benin, West Africa. Ecoscience, 4, 6-23.
23. ASCENA (2003) Données sur la pluviométrie des années 2002 et 2003. Agence Pour la Sécurité de la Navigation Aérienne, Service de la Méréorologie Nationale Cotonou.
24. Leveque, C., Paugy, D. and Teugels, G.G. (1990) Faunes des poissons d'eaux douces et saumatres de l'Afrique de l'Ouest. Tome 1, Editions ORSTOM/MDRAC, Paris, 384.
25. Lowe McConnell, R.H. (1975) Fish Communities in the Tropical Freshwaters. Longman, London, 337.
26. Lowe McConnell, R.H. (1987) Ecological Studies in the in Tropical Fish Communities. Cambridge University Press, Cambridge, 382 p. http://dx.doi.org/10. 1017/ CBO9780511721892
27. Moreau, J. (1982) Exposé synoptique des données biologiques sur Heterotis niloticus (Cuvier, 1929). Food and Agriculture Organization Synopsis de Pêches, 131, 1-45.
28. Needham, G.J. and Needham, P.R. (1962) A Guide to the Study of Fresh-Water Biology. Holden Day, San Francisco.
29. Morgan, G.A., Grieggo, O.V. and Gloeckner, G.W. (2001) SPSS for Windows: An Introduction to Use and Interpretation in Research. Lawrence Erlbaum Associates Publishers, Mahwah.
30. Simpson, E.H. (1949) Measurement of Diversity. Nature, 163, 688. http://dx.doi.org/ 10.1038/163688a0
31. Pianka, E.R. (1994) Evolutionary Ecology. 5th Edition, Harper Collins College Publishers, New York.
32. Winemiller, K.O. and Kelso-Winemiller, L.C. (2003) Food Habits of Tilapiine Cichlids of the Upper Zambezi River and Floodplains during the Descending Phase of the Hydrological Cycle. Journal of Fish Biology, 63, 120-128. http://dx.doi.org/10.1046/j.1095-8649.2003.00134.x
33. Welcomme, R.L. (1979) Fisheries Ecology of Floodplains Rivers. Longman, New York, 317 p.
34. Adite, A. (2007) Ecologie de Heterotis niloticus (Osteoglossiformes: Osteoglossidae) du Système fluvio-lacustre rivière Sô-lac Hlan (Sud-Bénin): Conservation et Intérêt pour l'aquaculture. Ph.D. Dissertation, Faculté des Scienes et Techniques, Université d'Abomey-Calavi, Abomey-Calavi.
35. Greenwood, P.H. (1953) Feeding Mechanism of the Cichlid Fish Tilapia esculenta Graham. Nature, 172, 207-208. http://dx.doi.org/10.1038/172207c0
36. Adebissi, A.A. (1978) Study on the Ecology, Growth and Reproduction of the Fishes of Upper Ogun River, Nigeria. Ph.D. Dissertation, University of Ibadan, Ibadan.
37. Kintunde, E.A. and Imevbore, A.M.A. (1979) Aspects of the Biology of Cichlids Fishes of Lake Kainji with Special Reference to Sarotherodon galilaeus. Niger Journal of Natural Sciences, 1, 35-39.
38. Fish, G.R. (1951) Digestion in Tilapia esculenta. Nature, 167, 900-901. http://dx.doi.org/ 10.1038/167900a0
39. Fish, G.R. (1955) The Food of Tilapia in the East Africa. The Uganda Journal, 19, 85-89.
40. Fagade, S.O. (1971) The Foods and Feeding Habits of Tilapia Species from the Lagos lagoons. Journal of Fish Biology, 3, 151-156. http://dx.doi.org/10.1111/j.1095-8649.1971.tb03659.x
41. Fagade, S.O. (1978) On the Biology of Tilapia guineensis (Dumel) from the Lekki Lagoon, Lagos State, Nigeria. Niger Journal of Natural Sciences, 12, 73-87.
42. Fagade, S.O. (1982) The Foods and Feeding Habits of Sarotherodon galilaeus from a Small Lake. Archiv für Hydrobiologie, 93, 256-263.
43. Arame, H. (2012) Ecologie trophique comparée de Oreochromis niloticus (Perciforme: Cichlidae) introduit dans le lac Toho (Sud-Bénin) et de Sarotherodon melanotheron (Perciforme: Cichlidae): Implications pour la conservation de l'ichtyofaune du Lac Toho.

Master's Thesis, Faculté des Scienes et Techniques, Université d'Abomey-Calavi, Abomey-Calavi.

44. Kramer, D.L. (1995) The Intestine Length in the Fish of a Tropical Stream: Ontogenetic Allometry. Environmental Biology of Fishes, 42, 115-125. http://dx.doi.org/10.1007/BF00001990
45. Paugy, D. (1994) Ecologie des poisons tropicaux d'un cours d'eau temporaire (Baoulé, haut basin du Senegal au Mali): Adaptation au milieu et plasticité du régime alimentaire. Revue d'Hydrobiologie Tropicale, 27, 157-172.

CHAPTER 13

Territorial Analysis in Environmental Studies on Fisheries: A Question of Scale and Spatial Representation

Christian Nunes da Silva, João Marcio Palheta da Silva, Clay Anderson Nunes Chagas

Federal University of Pará,(UFPA), Belém, Brazil

ABSTRACT

In recent years, several technologies have helped human society to make better use of the resources provided by nature, which can be in continental environments, such as aquatic environments. However, the use of human resources occurs in a different way: some are rudimentary and others with use of appropriate technologies, seeking for a better management of the use of resources. In activities related to fishing, this fact can be observed, with the use of new technologies which has added increasing importance to this activity. Considering that the areas where fishing is developed, extension and expression have varied—local, regional, national and even global level, however, the scales and techniques used to study fish are likely to be represented in digital media and monitored through software and hardware called geo-technologies. This paper aims to analyze the use of geo-information techniques in environmental studies—focusing on fishing activities and in what ways, the issue of territory, the scale study, and the use of these geo-technologies favoring fishing activities. Thus, we seek to demonstrate the use of these methods as a subsidy in territorial planning and management of fishery resources.

Keywords: Territoriality, Fisheries Studies, Geoinformation, Scale, Geotechnologies

1. INTRODUCTION

Fishing is one of the most productive activities for humanity, constituting a source of protein, trade, income and leisure for fishing communities, being artisanal, sports, industrial or others working with fisheries, or who develop activities directly or indirectly aimed at the extraction of resources from the aquatic environment. This activity in Brazil has undergone profound changes over the last century since the establishment of the fishermen colonies [1] , until recent technological advances in transportation and storage of fish (faster engines, refrigerator rooms on ships etc.), besides the increased demand for fish and new ways of management that are processed at the local level [2] . Nevertheless, these modifications only helped in the increase of pressure on fish stocks, causing major negative environmental impacts among the population that survive fishery.

Among these advances in fishing activity, one notices nowadays the use of new technologies for the location and assistance with fisheries management, for instance, the use of satellite images and geo-processing techniques [3] -[7] in environmental analysis and for the location of boats and shoals, as well as generating of information on the best seasons of the year to fish. The so-called geo-technologies must be understood as application of the remote sensing techniques: the use of softwares, hardwares etc., specialized in geo-processing techniques, as well as techniques for the creating of Geographic Information Systems (GIS) that focus on the interpolation and of geo-referenced spatial data. In other words, these techniques are directed towards the generation of geo-information[1], and spatial information that assists man in the detection of phenomena and objects, not only on the continent, but also in the aquatic environment.

These aspects, the geo-technologies besides the use of softwares, hardwares etc. specialized in geo-process- ing techniques allow generating different scenarios (vegetation, soils, location of paraphernalia etc.) that can be crossed with information on fishing activity. These techniques provide the user with the development of new methodologies for analyzing geo-information, besides facilitating the assessment and diagnosis of distant phenomena of the local decision making, rather than just a tool for visualization of spatial information and becoming a tool for planning and land use [8] -[10] .

It is intended with this work to discuss/talk about the following questions: How to represent spatial the territories of fishing activities, which are dynamic/changeable/unstable in computing environments, presenting them- selves as static/permanent? How to break the problem of "vertical integration" of the water column in fishing spots and its representation in a cartographic product? What is the appropriate scale—local/micro, regional/ma- cro or "multiscale" in order to analyze the activities that take place in fisheries, being the ecological relationships of the fish or the social relations of fishermen? Is there a problem in the multiscale representation of fish environmental studies? How to interpolate geographic areas—polygons, with spatially dynamic and punctual data?

Thus it is proposed that with this work, we discuss these questions. However, further analyses are necessary, since the technological advancement in the call "science of geo-information" is constant and the fishing industry cannot be deprived of such advancement. What can be verified for the future is the aggregation of such technology to all sectors of human activity, not only to fishing, which is an important tool in planning and monitoring of the actions that man performs over the geographic space.

2. TERRITORIAL ANALYSIS OF FISHING AND ITS SPATIAL REPRESENTATION

Territoriality can be proven in any study that aims to analyze how the performance of the society is in the geographic space. Thus, the ways of life and territoriality are the products and reflection of the performance of individuals in space and use of natural resources. These ways of life of fishermen and fishing territories themselves affirm, interact, otherwise do not speak in way of life and territory—humanized, but rather only the common characteristics of living animals, which interact with the space making it a mere support for life, without thinking about it and without construction, then without intentionality. This way, in the analysis of territoriality of fishermen it is verified that several mechanisms and/or processes are part of the training of their ways of life and their territoriality.

With multiple realities/territorialities, it is verified that the fishery is complex. Mechanisms and techniques such as geo-processing, tend to optimize the actions of fishermen—artisanal and industrial, and the government, despite the "open access". It is notorious that in reality the fishing areas have an "abstract" territorial delimitation that requires rules and agreements between fishermen [11] and which may be intermediated by governmental agencies. In this sense, the mapping techniques have assisted the planner for best use of space managed by him, in order not to deplete resources and optimize the uses. The cartography is an essential tool when it comes to land use for any human activity.

Therefore, with reference to fisheries management, cartography can also be very useful, because it concerns an environment that can be spatialized, so it can be represented cartographically in a computing environment. However, what one observes is that in this activity geo-information techniques are still timid, underexplored and without the visibility necessary to improve the management of fishery resources within the national territory, being necessary prior knowledge of local fishing of fishing grounds, to be able to map out the fishing activities and their territorialities [13] .

In this regard, the daily interactions of fishermen with the space where they live results in the territorialization of aquatic environments, as long as there is a relationship of possession of the same. Despite, in some instances, these territorialized spaces are not recognized by the authorities, there is the notion of respect among fishermen and/or other users. This appropriation of spaces does not occur randomly, but in accordance with the guidance of socio environmental phenomena that direct human activities for a specific purpose, either in search of natural resources, or by occupation and socio-political use of space for adding value to the "land use". When one has the appropriation of space by an individual or a group of individuals, one configures the power relations and ownership of that space, which becomes territory. In this case, the concept of territory is shown as crucial for understanding the processes of planning that is intended for a particular activity, because the notion of order is related to the question of use and power in the geographic space, as from the territorial delimitation.

The recognition of territoriality of fishing activities with all its techniques and fishing gear, as well as ownership of natural resources in general, are part of the customs, are soon among the ways of life that characterize the fishing grounds. Raffestin [14] , in his study of the territory, states that it forms from the space is the result of an action performed by a syntagmatic actor at any level and appropriating of a space, concrete or abstract, this actor "expand the territory" and that space shall make use of features of the new territory conformed. However, the case of fisheries it is important to check that whoever is expanding the territory of the fisherman because this is an action-presupposes where the fish is just the existing natural resource in the space, whereas support exists and unintentionally.

This production from the space is made through the use that society performs on the territory. The social actors accomplish their political-economic-social actions territorializing social practices for their residence on it. But not always territorial practices are revealed as desired by its social actors, because they depend on a number of trading factors and conflicts involving nearly always more of an interest on the territory.

Thus, one of the factors that enable the execution of territoriality in the fishing activity are the uses given to fishing equipment—paraphernalia, used in the capture of ichthyological species, requiring a limited work space. Therefore, according to the use of the fixture will be delimited its territory, being that if that territory is "invaded" there may be consequences for the fishermen themselves and automatically about their territoriality, as result having conflicts of the fishing activity [15] . These conflicts can occur in several ways, fisherman to fisherman, fisherman with riverside communities, artisanal fishing to commercial fishing, with tourists and others. This demonstrates that territoriality can be proven in any study that seeks to analyze how the ownership of a particular natural resource is developed in the geographic space. Figure 1 shows an example of what occurs in river fishing and in many other places of fishing on the high seas [16] .

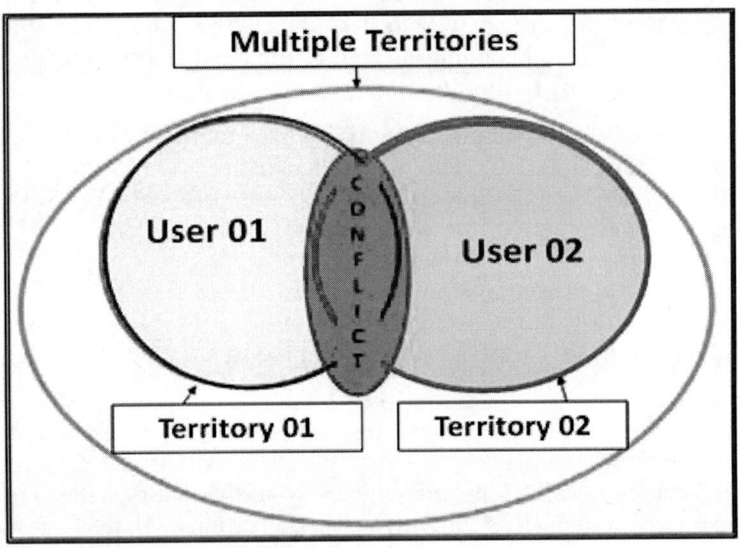

Figure 1. Simplified schematic model of using the territory and its resources (source: Elaborated by the authors).

In the simplified model above, the user is not restricted only to fisherman, but to any user of natural resources, i.e., lumberman, tourism, extractivism etc. using the rivers for transportation or resource extraction, whether they are aquatic or not and reflecting on conflicts over ownership of this resource. In this way, the territory will have a fundamental understanding of the expression in the development of activities that use or extracting natural resources make it possible to differentiate the use as a social process, which provides individualization of the characters involved in its construction, by increasing their equity-territory. At the moment the conflicting territoriality emerge among the characters that make up the space, starting from appropriations present themselves evidence that the forms of territorial management—land-use plans should take into account the diversity of actors and interests, in order to be able to understand the various existing territorialities in the appropriate space.

In the case of fishing, in aquatic environments exists a boundary recognized by fishermen, accustomed to these fish locations where "disrespect" present conflicts that may cause the loss of life. This reality is also possible when the mobility of fishermen is lower, or in coastal environments where industrial fishing is a significant competitor [16] . In river and lakes environments [13] [15] [17] [18] , no great mobility occurs, e.g. the need for fishermen to extract resources in remote locations, as it occurs in open sea because in most fisheries—regarded as the fishing grounds are located close to the houses of fishermen or their communities, making it a peculiar feature—a type of "sedentary fisheries" where the fisherman collects in a single region, close to his home, as well as the fixed gear and fishing, which are also located close to their homes.

So, how to study such complex activities where there is variability climbing and environmental? The particularities of the fishing activity—environmental ecological and economic instability, fluidity etc., become knowledgeable of the water environment from which they extract their sustenance [19] . Nevertheless, despite the complexity of the definition of fishing grounds for scholars of continental activities [20] -[22] , the exact area in fishing environments can be estimated according to the methods of analysis of spatial perception and geographical analysis of the study area.

The fishing grounds—the fishing territories/fishermen, as well as their areas of influence—adhere to a set of regulations that are created by the very fishermen, establishing the concept of individual or collective power and express themselves in own spatializations, that have natural features—the existence of resources and are regulated by habits and customs of the fishermen/users. For example, the reference of a fishery is usually identified by a natural resource, a tree, a bush, a stream, position of a heavenly body etc., where the fact that the overlap of different fishing and other areas where the demand for fish is greater and higher incidences of conflicts can occur between fishermen, considering that often are points that are not territorially demarcated by visible boundaries, but by fishing rules that can be broken with/without user intent.

In this regard, one must also bear in mind that some species of fish frequent specific regions of a river or the sea, for example, the demersal species—benthic communities and those who frequent more strongly to water column, near the surface—pelagics, which demonstrates also the spatial variability of these fish, wherein the computerized environment in most cases is not considered, once it is represented on the map merely the phenomenon as a dot or polygon, seen "from above", showing a space that can be perceived instantly as homogeneous, which is not true because as marine biology have studied for some time the catch is not located merely in one place in a body of water. In Figure 2(a) and Figure 2(b) it is hereby confirmed what was shown in Figure 1, and endeavors to show this fact alluded.

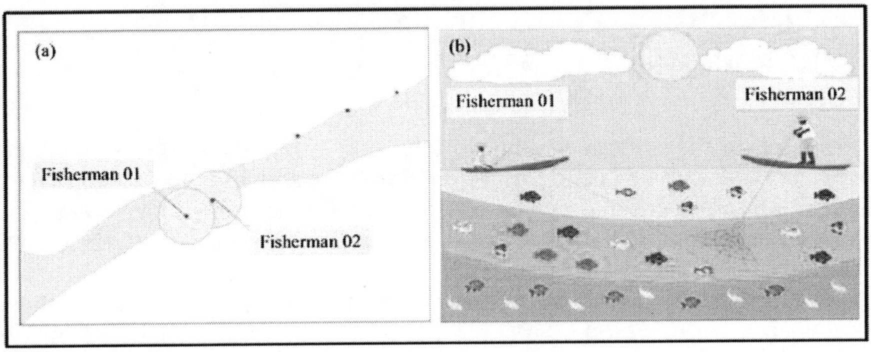

Figure 2. Two possibilities "to see the phenomenon in fishing" (vertical and horizontal). (source: Organized by the authors).

In Figure 2(a) one can see the fishing activity view "above", as commonly observed in traditional fisheries studies, where a point or polygonal pattern of activity occurs that does not consider the deepest part of the aquatic environment where the fishing phenomenon happens, but only the information of the occurrence of an activity at the time mostly their extraction. In Figure 2(b) the cartographic representation considers the different environments, according to the "vertical integration" of water column found in the aquatic environment, where it is observed—symbolically the different habitats for distinct species of fish, that can be influenced by the characteristics of the water body, such as: acidity, turbidity, luminosity, temperature etc., interacting with the depth of the water body and in most studies on fisheries, characteristics are neglected when developing cartographic products, due perhaps to the complexity of analyzing and unawareness of the ecology of most existing aquatic species [16] [20] [23] .

However, for Cruz and Menezes [24] at some moments the spatial representation of a phenomenon or object does not require precise positioning as in the case of fishing for its instability or by the occurrence of the phenomenon, as example in the case studied here, a fishing area map. Nonetheless, when it comes to fishing it becomes important that there is concern about the occurrence of a correct approximation of the distribution, because this approximate location depend on extraction productivity of fish, it is indispensable for a cartographic base precisely compatible user needs. In this sense, the scale of study and action may vary depending again of what the user expects to capture and quantity which craves, causing thereby the variability of what is perceived as a territory.

3. THE QUESTION OF THE SCALE IN THE ENVIRONMENTAL ANALYSIS ON FISHING

Research in the area of geo-information in fishing activity, commonly, do not adopt the name "geoinformation" for their activities of spatial objects and phenomena studied in fishing. However, they use techniques related to cartography and geography for handling and disclosure of their data through cartographic products, generally, represented by maps and cartograms. Efforts in order to test new technologies in fisheries are being conducted in the national territory and other countries most of the time in the context of higher education institutions and yet little explored in government institutions, that monitor or supervise the fishing activities.

These geo-technology are already inserted some time in other fields of human activity, como mineração, transports etc., although, use in fishing activity, except for the Global Position System (GPS), still barely visible, but tends to increase due to the importance it has taken in recent years because, in a GIS environment, for example, the accumulation of geographical information—in local or global levels and the possibility of the relationship of spatial data and attribute of these geometries, with speed and agility in processing of cartographic product generation, strengthens the processes of decision making and planning by the user, since a GIS allows grouping, automatically collecting and analyzing spatial information, in order to be more optimized and faster than was possible with traditional research techniques ? especially in analog format.

However, when using geo-processing tools, such as GIS, One can not ignore the data from field activities, much less, One can not ignore the data from field activities, much less, the different that are interrelated with the form of appropriation of space and rationalized use of natural resources, so that there is overlapping of scales with regard to regulation of the use of these natural resources. Currently, except for the tactile maps and models, most cartographic products (maps, charts, plans etc.) can be represented in computational media, which allows the user the ability to view/temporary demonstration of a specific geographical information, where a certain space is presented according to the will and the need to be viewed by the user [24] .

These types of map can demonstrate the dynamism of actions on the Earth's surface, for example, of fishing activities, which according to their mobility characteristics, seasonality and instability etc., may represent the flows of fisheries and fishermen, and movements beyond the temporal, climatic and ecological changes which influence the fisheries productivity. In all studies the environmental analysis of this volatility can be improved/enhanced with the use of remote sensing and GIS programs that enable the generation—in real time of spatial geo-information linked to the phenomenon analyzed, with the possibility of different interpolation and simultaneous information presentation through symbols and cartographic conventions internationally accepted. Although, for Cruz and Menezes [24] this representation is not only restricted to computerized presentation, because the display can also be demonstrated by copies, assuming that in this case the preview feature of paper maps, where information is made permanent, making it difficult to update the cartographic product, except for the construction of a new map.

The main requirement for any phenomenon can be represented on a map is the association of spatial or geographical distribution. In other words, must be known and clearly defined its occurrence on the earth's surface. This is the link between the phenomenon and the map. Thus, any phenomenon that is spatially distributed is likely to have represented their occurrence on the earth's surface through a map. A well characterized phenomenon is known as geo-referenced [24] .

Therefore, as shown in the citation, regardless of the form of presentation of the cartographic product one must know the phenomena inherent to the objects to be represented, as when considering the use of fisheries resources again, for example, it must be emphasized that the precise knowledge of the ecology of the majority of all species of commercial and non-commercial fish is still unknown, mainly due to the great diversity of species, or the fact that some species are more sought—commercially (then others) follows the search by knowing only the main ecological characteristics of the most profitable species.

Accordingly, in relation to the scale, one can generally define it as a relationship between the size of the object represented on the map—starting from their geometry and attributes and their size in real space. When one speaks in drafting a cartographic product, it is important to mention that the scale will always be present at any level of mapping exercises, being considered a determinant factor in defining the physical space degree of detail of a representation or identification. In analog terms, on printed maps, the scales remain unchanged only changing when reproducing copies or preparation of new maps. In computational terms, however, the issue of scale can be a complex factor, once the scale maps show the relative independence of the digital basis [25] [26] , i.e. in most of the software geo-technology, "the functions towards and away from (zoom in and zoom out) providing a sense of independence of scale, since they can generate a series of views on continuous scales" [24] , this means, the possibility of unlimited zoom can generate "illusory maps" in the interpretation of its contents. Speaking of the choice of scale, Fitz [26] informs that:

In the case of maps stored in digital files, this situation tends to be relegated to the background, because in principle, the scale can be easily changed to any values. Nevertheless, this can create a lot of problems. One should be very careful when dealing with this type of structure, because what actually matches the reality is the source of the information generated. Thus, a map created in digital media, originally conceived on the scale 1:50,000, will never have better accuracy than allowed for this scale (Remember This) [26] .

In such cases, as seen in fishing activity, one can speak of "multiscale" within the computing environment but at the time of transfer to the analog model map scale will be one, according to the scale of the base map source and due to its permanent feature when printed, but varied when taken into account differing scales found in the cartographic database that generated the final cartographic product—is based on a vector or raster in a computerized environment. It is complex, however, establish the threshold for each scale, because the concept of large, medium and small is very subjective, and this association to a numeric value is set to establish a reference to the relative size of objects presented in real space, where it is possible to classify them according to global, regional and local characteristics, but also in a very subjective way, generating controversy upon its association to numerical scales and the final representation of the environment [27] .

As a suggestion in choosing the correct scale work in environmental studies, and here be inserting fishing activities, Silva [27] analyzes the different work schedules are determined according to the aspects that define both the location and extent of events and entities, about the properties and relationships of these components, which generate certain situations studied, due to the "geographic scale factors and geo-economic". For this author [27] , commonsense recommendations and reflection on the application of the recommended use of this form of analysis procedures, where one can distinguish four levels of causality in a geo-environmental scale:

1) Local level: Mainly refers to situations that occur on municipal scale, with emphasis on the surveys which take into account the environmental resources available physical, biotic and socioeconomic). In these respects, the scales of representation at this level of causality are of 1:50,000, for the treatment of municipal area as a whole, and 1:10,000 for the analysis of processes and phenomena that take place in specific local areas, in case of fisheries the representation would be in fishery well defined territorially, in the catchment area of a municipality in question;

2) Intermunicipal level: At this level of causality is considered the dependence of events and processes that exist in the inter municipal space, the interaction between municipalities and their political-administrative boundaries. Silva [27] emphasizes the importance of this approach for studies directed research in geoprocessing, of small watersheds, which typically cover the area of more than one municipality. The geographical scales of data processing for that level of environmental analysis are 1:50,000 and 1:100,000 [27] , in the Brazilian territory, these scales cover with relative detail the processes that unfold in inter municipal level, and fishing, facilitate the analysis that can be made between conflicts of fishermen from different municipalities that operate in the same region;

3) Regional or national level: For Silva [27] one should understand that scale level as the area encompassing the "judicial power coupled to a nation which could be the director of research strictly within its territory". However, this scale level exceeds often the national territory of fishermen or historical-political boundaries, only one nation such as the Amazon region, observed by the author or in studies that analyze the delineation of watersheds that does not adhere to political boundaries but the environmental issues. In this particular case, Silva [27] explains that the analysis can be supranational, regional or respecting the concepts of the nation and national territory. At this level of causality the author shows scales of 1:100,000 and 1:250,000.

4) Global level: For Silva [27] this level of detail refers to the economic-demographic system on a global scale, "for actions or phenomena that affect the world's population, in particular by warn for problems related to depletion of non-renewable natural resources". One realizes that by trying to synthesize the complex reality, this level of causality is extremely criticized, although advances in geo-technology have enabled further analysis on a global level. The author suggests that the scales of treatment at the global level are of 1:1,000,000 or less and has direct military-strategic and geopolitical connotation. Based on these levels of causation Figure 3 seeks to simplify how you can work with cartographic products, in a computational environment according to "multi-scale".

Thus, from the analysis of the levels of causality proposed by Silva [27] and simplified in Figure 3, geo-in- formation can be represented in different ways in different scales, it is observed that the that depend on the demand and mode of investigation of the end user, enabling different levels of detail and various forms of interpretation. Therefore, when choosing the working scale, professionals who use geo-processing should use the structures analysis compatible with the environmental complexity that they face, according to Cruz and Menezes [24] the larger cartographically scale represent a level higher than that of smaller scales detailing, approaching in turn a smaller geographical area.

Thus, the scale size, or the level of causality will depend on the type of study that the researcher is doing, which will cover a specific area according to the object or phenomenon studied. In the case of fishing activity, the local level is appropriate for studies of fish species that maintains a small performance area; hence the maps will be prepared on a large scale that is reflective of the variety of details found in the cartographic product. For studies that reflect the migratory species, regional and even global, is the most appropriate because it encompasses all moments of life of fish, from birth to maturity and reproduction.

However, it is currently visible efforts to conceptualize and better understand the activities in computer environment using geo-processing tools. In this sense, appears in recent years to the proposal of the so called "digital territory, which can be considered as forms of computer representations of geographic space" where those areas exist in reality are represented on the computer screen and manipulated to the various scales that geo-processing tools offer, as seen before.

With this idea, which has been widespread in recent years, the scale takes "multiscale" sense, appropriate to the user's work, where the base map used is diverse and widespread is common, starting from the accumulation of knowledge that the user has the represented space and the contributions of different types of knowledge for the generation of the final cartographic product.

In this type of cartographic representation, yet to be defined categorically, actual territories, contained in geographic space, tend to be represented in GIS environments where the scale variability would be assured by the accuracy of data collected in real space and the elements of spatial reference that the user intends to show, in this way eliminating, secondary information, which may be omitted. In this example, the spatial information, existing in a Geographical Database (GDB), are arranged in registration form, where the geometry is guaranteed and attribute data are subject to editing, with the addition of new information or subtraction of unnecessary information. Such geo-technology can already be viewed as today WebGis tools available on the internet with specific themes, according to the convenience of its administrator.

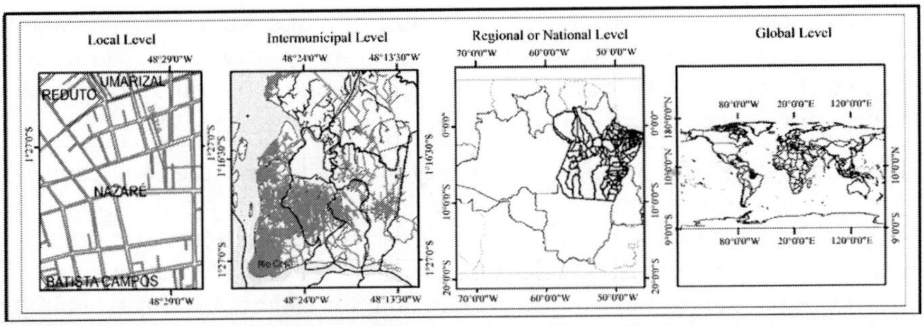

Figure 3. Levels of causality, according to the research object and different scales. (source: Organized by the authors starting with Silva [27] and Cruz and Menezes [24]).

It is important to inform that the potential for spatial relationships in these environments are feasible and can generate new information from various data interpolation with creating future scenarios or the discovery of other existing attributes in the present time. The ideas that threaten to digital territories are innovative propositions, directed to the geo-processing environments, where faults which must be tested and presented to the scientific community in the coming years are still observed. For fisheries, the concept of digital territories can mean a change in a change in the land use model known, in addition to environmental and ecological characteristics, may also be aggregated socio-cultural information that will encourage the analysis of the manager/user of fishery

4. CONCLUSIONS

From the conceptual discussion on fishing territory it can be seen that the use of geo-technologies is of fundamental importance for the (re)cognition of certain defined spaces at the present time. Furthermore with respect to monitoring and environmental surveillance applied by environmental agencies, it is noticed that these techniques are key allies to the fight against illicit activities or predatory activities which are processed by the Brazilian territory.

Generally speaking, fisheries studies, when considering its superficial coverage of a horizontal character, are similar to studies that analyze phenomena and objects in continental ecosystems. However, when presenting the aspects of vertical character, related to the water column in an aquatic environment, the complexities in working with fishery studies are added, where resources of different continental perspective can be defined with greater precision. In addition, this complexity tends to increase from the moment in which the fish are considered dynamic, such as, mobility, non territorial limitation, species in different aquatic environments, ecological seasonality etc.

The complexity of working with fishery studies shows the importance of this type of study for geo-processing, where the possibility of representing each element can be treated as a separate attribute, connected to a particular spatial geometry. In this way, when considering element by element, spatial relations appear as relevant, for example, to investigate the crossing of data on the ecology of species in watersheds and/or various aquatic environments (rivers, streams, lakes etc.).

The scale factor, before limiting element in the analysis of the printed map, becomes flexible when the cartographic product is disposed on a computer. Yet, one wants to encourage confusing generalizations that omit important elements in landscape studies, if one intends to pay attention to the need to generate effective and temporally agile tools and rapid development, which assist in analysis of ideal geographic space.

In this perspective, the development of GDB or the Geographic Information Systems (GIS), incremented by the emergence of remote sensing images of high spatial resolution with different spectral bands, are indispensable tools for planning and land use in fishing nowadays, especially when related to multitemporal analysis and thermal and climatic information, as necessary to identify different species of fisheries. Thus, the possibility of coupling of complex information, with a very large number of attributes, calls into question the proposal of analysis of digital territories as possible to be applied in activities such as fisheries, mainly due to the large amount of information generated by type of fish species. Nevertheless, the government's indecision regarding the use of geo-technologies fishing is visible, and has not benefited the activity of fewer fishermen who subsist it a fact that it tends to change, in view of the visibility for these tools which they had in recent years.

REFERENCES

1. Villar, F. (1945) A missão do cruzador "José Bonifácio": Os pescadores na defesa nacional—A nacionalização da pesca e a organização dos seus serviços (1919-1923). Rio de Janeiro, Gráfica Laemmert.

2. Berkes, F., et al. (2006) Gestão da pesca de pequena escala: Diretrizes e métodos alternativos. FURG, Rio Grande do Sul.
3. Almeida Pinto, W.H., et al. (2007) Critérios para setorização e espacialização de dados pesqueiros na região de Parin- tins-AM e Santarém-PA. Simpósio Brasileiro de Sensoriamento Remoto, 13, Florianópolis, INPE, 3019-3026.
4. Botelho, A.Z., Calado, H. and Costa, A.C. (2010) GIS-Based Marine Biodiversity Mapping for Assessment of Coastal and Marine Priority Áreas for Conservation. In: Calado, H., and Gil, A., Eds., Geographic Technologies Applied to Marine Spatial Planning and Integrate Coastial Zone Management, Universidade dos Açores/Centro de Informação Geográfica e Planeamento Territorial (CIGPT), Ponta Delgada-Portugal, 31-37.
5. Garcia, L.M., et al. (2010) Geographical Information Systems (GIS) as a Tool for the Environmental Evaluation and Management of Coastal Area of Tazacorte, La Palma, Canary Islands, Spain. In: Calado, H. and Gil, A., Eds., Geographic Technologies Applied to Marine Spatial Planning and Integrate Coastial Zone Management. Universidade dos Açores/Centro de Informação Geográfica e Planeamento Territorial (CIGPT), Ponta Delgada-Portugal, 53-60.
6. Lozano-Rivera, P., Garcia-Valencia, C. And Rodrígues, A.L. (2010) Geographic Information Technologies for Integrated Coastal Zone Management in Colombia: A National Experience. In: Calado, H. and Gil, A., Eds., Geographic Technologies Applied to Marine Spatial Planning and Integrate Coastal Zone Management. Universidade dos Açores/ Centro de Informação Geográfica e Planeamento Territorial (CIGPT), Ponta Delgada-Portugal, 86-93.
7. Pitt, J.M. and Shailer, M.L. (2010) GIS Applications in Marine Resources Management: Examples of Spatial Management Measures from Bermuda. In: Calado, H. and Gil, A. Eds., Geographic Technologies Applied to Marine Spatial Planning and Integrate Coastial Zone Management. Universidade dos Açores/Centro de Informação Geográfica e Planeamento Territorial (CIGPT), Ponta Delgada-Portugal, 103-111.
8. Zagaglia, C.R., Brichta, M. And Cabral, D.H.G.S.L.B. (2007) Mapas de gestão pesqueira como base para o Programa Nacional de Rastreamento de Embarcações Pesqueiras por Satélite e demais atividades do IBAMA. Simpósio Brasil- eiro de Sensoriamento Remoto, 13, Florianópolis, 4339-4341.
9. Câmara, G. and Carvalho, M.S. (2004) Análise espacial de dados geográficos. Embrapa, Brasília.
10. Jensen, J.R. (2009) Sensoriamento remoto do ambiente: Uma perspectiva em recursos terrestres. Trad.: José Carlos Epiphanio (coord.). São José dos Campos, Parêntese.
11. Ruffino, M.L. (2005) Gestão do uso dos recursos pesqueiros na Amazônia. IBAMA, Manaus.
12. Câmara, G. and Monteiro, A.M. (2001) Introdução da ciência da Geoinformação. São José dos Campos, INPE, 07-41.
13. Cruz, M.N. and Almeida, O.T. (2009) Estratégias para a co-gestão dos recursos pesqueiros no estuário amazônico: O caso dos acordos de pesca em Abaetetuba-PA. Simpósio Nacional de Geografia Política, Território e Poder, UFPR, Cu- ritiba.
14. Raffestin, C. (1993) Por uma geografia do poder. Ática, São Paulo.
15. D'almeida, B.G. (2006) Os acordos de pesca na Amazônia: Uma perspectiva diferenciada de gestão das águas. En- contro Preparatório do Conselho Nacional de Pesquisa e Pós-Graduação em Direito, 15. CONPEDI, Recife.
16. Cardoso, E.S. (2001) Pescadores artesanais: Natureza, território, movimento social. Tese de Doutorado em Geografia, Universidade de São Paulo, São Paulo.
17. Mcgrath, D.G. (1993) Manejo comunitário dos lagos de várzea do Baixo Amazonas. In: Gonçalves, L.G., Leitão, W. and Mello, A.F., Eds., Povos das águas: Realidades e perspectivas na Amazônia, MCT/CNPq/MPEG, Belém, 389- 402.
18. Mcgrath, D.G. and Câmara, E.P.L. (1995) A viabilidade da Reserva de Lago como unidade de manejo sustentável dos recursos da várzea. Boletim do Museu Paraense Emílio Goeldi, MPEG, Belém, 87-132.
19. Maldonado, S.C. (1993) Mestre & Mares: Espaço e indivisão na pesca marítima. Annablume, São Paulo.

20. Begossi, A. (2001) Mapping Spots: Fishing Areas or Territories among Islanders of the Atlantic Forest (Brazil). Regional Environmental Change, 2, 1-12. http://dx.doi.org/10.1007/s101130100022
21. Begossi, A. (2004) Ecologia de pescadores da Mata Atlântica e da Amazônia. Hucitec, São Paulo.
22. Begossi, A. (2006) Temporal Stability in Fishing Spots: Conservation and Co-Management in Brazilian Artisanal Coastal Fisheries. http://www.ecologyandsociety. org/ vol11/ iss1/art5/
23. Guedes, E.B. (2010) Território e territorialidade de pescadores no Marajó: O exemplo das localidades Céu e Cajuúna, Soure (PA). In: da Silva, J.M.P. and Silva, C.N., Eds., Pesca e territorialidades: Contribuições para a análise espacial da atividade pesqueira, GAPTA/UFPA, Belém, 24-45.
24. Cruz, C.B.M. and Menezes, P.M.L. (2009) A cartografia no Ordenamento territorial do espaço geográfico brasileiro. In: Almeida, F.J. and Soares, L.D.A., Eds., Ordenamento Territorial, Coletânea de textos com diferentes abordagens no contexto brasileiro, Bertrand Brasil, Rio de Janeiro, 195-225.
25. Fitz, P.R. (2008) Geoprocessamento sem complicação. Oficina de Textos, São Paulo.
26. Fitz, P.R. (2008) Cartografia básica. Oficina de Textos, São Paulo.
27. Silva, J.X. (2001) Geoprocessamento: Para a análise ambiental. Edição do Autor, Rio de Janeiro.

NOTES

[1]According to Câmara and Monteiro [12] , the fundamental object of science is the study of geo-information and the implementation of different forms of computational representation of geographic space.

CHAPTER 14

Recreational fisheries in the USA: economics, management strategies, and ecological threats

Robert M. Hughes

Amnis Opes Institute and Department of Fisheries and Wildlife, Oregon State University, 2895 SE Glenn, Corvallis, OR, USA

ABSTRACT

Recreational fishing has a long history in the USA, developing from the subsistence fisheries of Native Americans together with a partial subsistence fishery of later immigrants. Marine, diadromous, and aquatic taxa are targeted, including both vertebrates and macroinvertebrates. This paper defines recreational fishing, describes the main fishing techniques, identifies target taxa by region, summarizes the socio-economic values of recreational fishing, and discusses management strategies and major ecological threats.

Keywords: Angling Sport fishing, Outdoor recreation, Ecosystem services

1. INTRODUCTION

Recreational or sport fishing or angling is defined as fishing for pleasure, as opposed to commercial fishing for income or subsistence fishing for survival. Angling is typically conducted with a rod, reel and line with a baited hook, lure or fly attached. Some recreational fishing is conducted with a spear, net, or bow and arrows. In addition to finfish, recreational fishers collect crustaceans by net or trap, mollusks by hand, rake or shovel, frogs by spear or lures, and turtles by net. Fishing may occur from the shore of the water body, by wading in shallow waters, or from watercraft ranging in size from large multi-passenger live-aboard ocean-going ships to single-passenger kayaks or float tubes. Most of the catch is consumed as food and it may be an important dietary component for some families, especially in rural areas where fishers have ready access to the potential catch. Increasingly, anglers are required by management regulations to practice catch-and-release fishing to conserve a fishery. Numerous finfish and shellfish species are recreationally fished in coastal waters (Appendices 1, 2, 3, 4) and many finfish species are recreationally fished in inland (fresh) waters (Appendices 5, 6).

2. FISHERY CATCH AND ECONOMICS

Recreational fishing is an economically and culturally important activity in the USA (Table 1). Based on national census data, an estimated 33 million anglers in 2011 participated in over 443,000 fishing trips and generated over $40 billion in retail sales. Because of economic multiplier effects, these expenditures produced an estimated $115 billion economic impact and over 800,000 jobs. Based on marine survey data, an estimated 12 million marine anglers took about 85,000 fishing trips in 2012 and spent nearly $31 billion, which had an $82 billion economic impact and provided 500,000 jobs. Although there are more freshwater anglers, marine anglers have a relatively greater economic impact because of the need for larger and more expensive gear and boats. In northern latitude USA states where recreational fishing is economically important, 14–43 % of the population fishes, producing 10,000–38,000 jobs and a $1.1–4.3 billion economic impact (Table 2). In two river basins in Idaho and Wyoming with high-quality catch-and-release trout fisheries, 341–851 jobs and $12–29 million in county income were created. Although this paper focuses on recreational fisheries, it is useful to compare them with the USA commercial seafood fisheries (including harvesting, processing, distributing, and sales); in 2012, that industry provided 1.3 million jobs and a $239 billion economic impact (NMFS Web: http://www.st.nmfs.noaa.gov/ Assets/economics/

documents/feus/2012/FEUS2012_National Overview.pdf, accessed June 2014).

Table 1 Estimated economic impacts of USA recreational fisheries in 2011

Data source	No. of anglers (million)	No. of trips (thousand)	Direct sales ($ billion)	Economic impact ($ billion)	No. of jobs (thousand)
USFWS Web (US census)	33.1 (marine and freshwater)	443 (freshwater)	41.8	115 (ASA Web)	800 (ASA Web)
NMFS Web (marine surveys)	12	85	31	82	500

Table 2 Estimated economic impacts of state recreational fisheries in 2011

State	No. of anglers (million)	Wages ($ billion)	Economic impact ($ billion)	No. of jobs (thousand)
Alaska	0.5	0.3	1.1	10
Michigan	1.7	1.4	4.3	38
Minnesota	1.6	1.3	4.1	36

Fishery catch data are collected from multiple sources and in multiple ways. Individual recreational catches of marine species are estimated through use of coastal telephone surveys, vessel telephone surveys, angler dockside surveys at ports, and state and regional logbooks (NMFS Web: http://www.st.nmfs.noaa.gov/Assets/commercial/fus/fus12/03_recreational2012.pdf). The data gathered include number of trips, angler residency, species composition and catch rates, and species weights and lengths. The data are combined through use of models to produce catch, effort, and catch per unit effort metrics, which are reported by fish species, fishing method, state, and management region. Economic data for inland fisheries are obtained mostly from national census surveys of householders and state angler surveys. Canada relies on mail surveys of anglers every 5 years to estimate species-specific catch rates [1]. Oregon requires salmon, steelhead, and sturgeon anglers to purchase and complete a punch card for recording the species, number, location, and date of each of those species harvested (DFW Web: http://www.dfw.state.or.us/resources/fishing/sportcatch.asp, accessed June 2014).

3. FISHERIES MANAGEMENT

3.1 Management agencies

In the USA, recreational fisheries are managed at both state and federal levels. State fish and wildlife agencies manage inland (freshwater) and near-coastal (within 5 km of the shoreline) fisheries. These agencies set and enforce fishing dates and times, fishing gear, and catch limits on fish size and number. The National Marine Fisheries Service manages marine fisheries outside the state management limits and regulates anadromous and marine species listed by the USA government as threatened or endangered. The US Fish and Wildlife Service regulates fisheries on listed freshwater species. The following fish species are listed as endangered or threatened by the federal government, thereby curtailing their fisheries: Alabama sturgeon Scaphirhynchus suttkusi, pallid sturgeon S. albus, Gulf sturgeon Acipenser oxyrinchus, shortnose sturgeon A. brevirostrum, bull trout Salvelinus confluentus, Apache trout Oncorhynchus apache, gila trout O. gilae; greenback cutthroat trout O. clarkiistomias, Lahontan cutthroat trout O. clarkii henshawi, Paiute cutthroat trout O.clarkii seleniris, black abalone Haliotis chracherodii, and white abalone H. sorenseni. Ecologically significant units of chum salmon O. keta, Chinook salmon O. tshawytscha, coho salmon O. kisutch, sockeye salmon O. nerka, steelhead (anadromous rainbow trout O.mykiss); golden trout O. mykiss aguabonita, and Pacific eulachon Thaleichthys pacificus are listed as threatened or endangered. The number of federally listed threatened and endangered fish species is strongly correlated with economic and population growth [2].

Although they do not manage fisheries, other federal resource management agencies have fisheries programs. Because their land and water management affects fisheries, the US Forest Service, Bureau of Land Management, National Park Service, US Geological Survey, Army Corps of Engineers, and US Environmental Protection Agency (USEPA) employ numerous fishery biologists. The USEPA has implemented a rigorous nationwide ecological monitoring and assessment program of surface waters (lakes, reservoirs, streams, rivers, near-coastal marine, wetlands). See Shapiro et al. [3] for an overview of the program, Paulsen et al. [4] and USEPA [5] for examples of stream and river assessments, and USEPA [6] for an example of a national lakes assessment.

3.2 Regulating recreational versus commercial fishing

Most conflicts involving relative take by recreational versus commercial fisheries occur in marine ecosystems. Total and relative catch limits in marine systems are set regionally by one of eight regional Fishery Management Councils established by the Fishery Conservation and Management Act of 1976: North Pacific, Pacific, Western Pacific, New England, Mid-Atlantic, South Atlantic, Gulf of Mexico, Caribbean. The voting members of each Council include one National Marine Fisheries Service representative, a representative from each State fishery agency in the Council region, private citizens nominated by State governors, and a representative from tribal or territorial governments in some regions (USFC Web: http://www.fisherycouncils.org, accessed June 2014). Because each region supports different types of fisheries and different levels of commercial versus recreational fishing, there are no nationwide rules for decision-making. Differing levels and types of fishing are also affected by whether a species or stock is listed as overfished, vulnerable, threatened, or endangered, by its relative economic value to one type of fishery or another, and by previous take levels. For example, proposed options for allocating catch quotas between commercial and recreational fisheries on red snapper ranged from 51 and 49 % to 0 and 100 %, respectively, with total catch baselines of 4–5 million kg (GMFMC Web: http://www.gulfcouncil.org/docs/amendments/Amendment%2028%20-%20Allocation%20PH% 20 Draft%20March%202014.pdf, accessed June 2014). In addition, if catches exceed established limits, a fishery may be closed for the season or it may be reduced in the subsequent year (GMFMC Web: http://www .gulfcouncil.org, accessed June 2014).

3.3 Management funding

Funding for state fishery agencies comes mostly from fishing license sales. In addition, federal excise taxes on fishing gear and boat fuels are dispersed to the states for fishery improvement projects (research, hatcheries, improved access, habitat rehabilitation). The enabling legislation for those taxes is the Dingell-Johnson Act of 1950 (Federal Aid in Sport Fish Restoration Act) and the Wallop-Breaux amendments of 1984. Between 1955 and 2006, those taxes yielded $36 to $212 million (in 2009 dollars) annually to the states (ASC Web: http://asafishing.org/uploads/Benefits_to_Business_2011_Technical_Report.pdf, accessed April 2014).

3.4 The AFS and recreational fisheries management

The American Fisheries Society (AFS) is the premier society for professionals interested in fish, fisheries, and their ecosystems. Through its efforts, the US government established the US Commission for Fish and Fisheries in 1871, with the requirement that it be led by a fish scientist. The 6,000–9,000 AFS members are employed by federal and state agencies, universities, private contracting firms, and nongovernmental agencies.

Although most AFS members reside in the USA and Canada, there are chapters in Mexico and Puerto Rico and members are spread across the globe in 62 nations. As part of its mission, the AFS publishes books, a Fisheries magazine, and five scientific journals (Transactions of the American Fisheries Society; North American Journal of Fisheries Management; Marine and Coastal Fisheries: Dynamics, Management, and Ecosystem Science; Journal of Aquatic Animal Health; North American Journal of Aquaculture).

Perhaps the most popular AFS book is Inland Fisheries Management in North America [7]. That book's chapters describe techniques for managing stream, river, lake, and reservoir fisheries from individual stocks or populations to ecologically significant units or entire species through use of population dynamics [8]. The stock concept has been useful, both for maintaining healthy populations and for rehabilitating threatened populations of marine and freshwater fish. Three key fishery management techniques are regulating harvest [9], using hatchery fish to supplement populations [10], and enhancing the physical and chemical habitat [11]. Typically all three approaches are used in concert. Of course, quantitative data are needed before implementing such techniques, which require rigorous monitoring and assessment programs [12, 13, 14], including multistock and multispecies assessments [15].

4. THREATS TO RECREATIONAL FISHERIES

4.1 Historical and current threats

Significant historical and current threats to USA recreational fisheries include intensified land use, physical habitat and hydrological modification, chemical contaminants, eutrophication and hypoxia, overfishing, and introductions of invasive non-native species (including hatchery fish) and diseases. In a study of over 9,000 European river sites, Schinegger et al. [16] reported disrupted connectivity (from dams and road crossings) in 85 % of the catchments and 35 % of the segments; 47 % of the sites were altered by multiple factors and 90 % of lowland rivers had altered water quality, hydrology, connectivity, and morphology. Similarly, in a survey of 1,900 USA sites, USEPA [5] determined that 20–40 % of rivers and streams in the USA were in poor condition because of excess nutrients and disturbed riparian zones. In other words, multiple stressors frequently limit fish assemblages.

4.2 Land use

Land use, particularly intensive silviculture and agriculture, urbanization, and mining fundamentally alter the condition of surface waters, whether they occur in the USA or elsewhere. Mebane et al. [17] found that fish assemblage condition in large Oregon rivers declined with increased catchment agriculture and decreased catchment forest. Snyder et al. [18] reported poor conditions in fish assemblages once urban land use exceeded 7 % in West Virginia catchments, especially with steeper channel slopes. Wang et al. [19] demonstrated that the percent of impervious area that was connected to streams negatively affected fish assemblages and trout densities in Minnesota and Wisconsin streams. Stranko et al. [20] reported brook trout almost never occurred in Maryland streams once catchment impervious cover exceeded 4 %. Trautwein et al. [21] reported that fish assemblage condition in Austrian streams declined with increased levels of catchment agriculture and urbanization and decreased levels of forest. In a series of western USA case studies, Woody et al. [22] summarized the negative effects of metal mining on salmonids. Daniel et al. (unpublished data, 2014) found that mineral and coal mining at the catchment scale had a greater effect on fish assemblages, including game species, than did that mining at the local scale and compared with catchment-scale urbanization or agriculture.

Catchment conditions may have a greater influence on fish assemblages than site conditions, whether they occur in the USA or elsewhere. Roth et al. [23] and Wang et al. [24] determined that catchment agriculture had a greater effect on fish assemblages in Michigan and Wisconsin streams, respectively, than did local riparian vegetation. Van Sickle et al. [25] found that the riparian conditions along Oregon stream networks explained more variability in fish assemblages than did conditions of the entire catchment or at the site. Sály et al. [26] demonstrated that variation in fish species and abundances in Rumanian streams was explained more by catchment and catchment-site shared variables than by site variables alone. For French streams, Marzin et al. [27] reported that anthropogenic variables at network riparian, catchment, shared site-riparian, or shared site-catchment scales explained more variation than site-scale anthropogenic variables. However, Macedo et al. [28] determined that site variables in Brazilian streams explained more variability in fish species richness than did catchment land use variables. For additional reading on landscape-scale effects on fish assemblages, see the books by Hughes et al. [29] and Yeakley et al. [30].

4.3 Hydromorphological modification

Physical and hydrological modification of fish habitat varies from major dams that fundamentally alter flows and convert rivers to lakes, to local changes in substrate and riparian vegetation. Stanford et al. [31] considered flow regulation the most pervasive change in large rivers, and Poff et al. [32] argued that flow regimes and flood pulses were master variables governing the condition of rivers. Based on studies of 27 large American rivers, Hughes et al. [33] reported that flow and channel alterations resulting from large dams were key factors disturbing fish assemblages. Carlisle et al. [34] determined that flows had been altered in 2,500 USA stream sites and that reduced flow magnitudes were better predictors of fish assemblage condition and impairment than site chemical or physical factors. Sedell and Froggatt [35] described how channelization and large wood removal from the Willamette River, Oregon, separated the channel from its floodplain and removed vast amounts of salmon spawning and rearing habitat. Similarly, in the eastern USA before European settlement, low gradient streams were anastomosing with extensive wetlands (hydromorphologically complex), rather than the single incised channels existing now [36].

4.4 Substrate and riparian modification

Lower fish assemblage condition was associated with excess fine sediments and reduced channel complexity at stream sites in the Oregon and Washington Coast Range [37]. Bryce et al. [38] concluded that surficial fine sediments <5 % were needed to maintain the habitat potential for trout and other sediment sensitive aquatic vertebrates in western USA mountain streams. Habitat simplification and loss of large wood debris in lakes of the midwestern and northeastern USA were associated with reduced game fish populations [39, 40, 41, 42]. In a study of northeastern USA lakes, Kaufmann et al. [43] found that the richness of intolerant fish species was positively correlated with greater physical habitat quality, whereas the richness of tolerant fish species declined.

4.5 Chemical contaminants

Chemical contamination of USA waters has been reduced significantly since implementation of the Clean Water Act of 1972. Nonetheless, mercury concentrations exceeded the USEPA 300 ppb fish tissue consumption criterion at nearly half of USA lakes [44]. In a survey of 600 western USA streams and rivers, Peterson et al. [45] found that large piscivorous game fish exceeded that mercury criterion in 57 % of the assessed stream length.

As a result of airborne chemical pollutants, even high elevation and high latitude lakes in national parks contain persistent pollutants that may place their fish and fisheries at risk [46].

4.6 Eutrophication and hypoxia

Despite effective nutrient removal from urban sewage, eutrophication remains a pervasive problem in USA waters, largely as a result of agricultural runoff. USEPA [5] reported that excess phosphorus was associated with poor fish assemblage condition in over 20 % of the USA stream and river length. USEPA [6] found that 50 % of USA lakes and reservoirs were eutrophic or hypereutrophic, with the highest proportions of eutrophic and hypereutrophic lakes in the agricultural central USA (36 and 24 %, respectively). When this nutrient rich runoff reaches near-coastal marine waters, it produces ocean hypoxia. For example, a 20,000 km^2 hypoxic area exists in the Gulf of Mexico as a result of nitrogen and phosphorus from the Mississippi River Basin [47]. Along the USA East Coast, agricultural runoff has led to coastal marine eutrophication [48]. Nationally, Dodds et al. [49] estimated that eutrophication creates over $2 billion in damages annually.

4.7 Overfishing

Recreational overfishing has been a historical issue for many highly valued species (e.g., GMFMC Web: http://www.gulfcouncil.org, accessed June 2014) but its impacts have been reduced by fishery management agencies, and recreational fishing typically has markedly less impact than commercial fishing. Nonetheless, as any angler can attest, additional fishing does not improve one's chances of catching fish or the fishing experience.

4.8 Non-native fish

By sampling 1,000 western USA stream and river sites (representing 90,000 stream kilometers), Lomnicky et al. [50] estimated that 52 % of the stream length contained non-native aquatic vertebrates (83 % of large river length). Three different trout were the most commonly occurring non-native species. Sanderson et al. [51] estimated that the effects of non-native species on Pacific salmonids equalled or exceeded that of hatcheries, harvest, hydropower and habitat degradation. In addition to non-native fish, hatchery fish stray onto spawning grounds, diluting the genetic pool of wild salmonids, and increase feeding competition in freshwater [52, 53, 54, 55, 56], in estuaries, and at sea [57, 58, 59, 60]. Such changes limit the recovery of listed wild salmonids.

4.9 Developing threats to USA recreational fisheries

Significant developing threats to USA recreational fisheries include endocrine disrupters, nanoparticles, and climate change. All three are pervasive and threaten freshwater and marine recreational fisheries.

4.10 Endocrine disrupting chemicals

For years, we have known about the presence of endocrine disrupting chemicals in aquatic environments, as well as their physiological effects on fish and other aquatic vertebrates [61, 62, 63]. Recently, their potential population-level effects in natural environments have been modelled quantitatively [64], as have their management implications [65]. Because of the number of such chemicals in the environment and their developmental and immunological toxicity at very low levels, they are considered a ticking time bomb [66].

4.11 Nanoparticles

The physico-chemical effects of nanoparticles on fish populations are at an early stage of understanding [67], but their rapidly increasing use and widespread distribution are troubling. Shaw and Handy [68] reported that nanocopper was twice as toxic as dissolved copper and that nanoparticles altered respiration and caused gill, liver, intestine and brain tissue pathology. Cedervall et al. [69] found that nanopolystyrene passed up the algae-zooplankton-fish food chain and altered lipid metabolism and halved food consumption rates of fish.

4.12 Climate change

Climate change is predicted to have substantial impacts on USA marine and freshwater recreational fisheries. By 2100, ten states are predicted to lose all their cold and cool freshwater fisheries and in 17 states those fisheries will be halved [70]. Predicted national economic losses range from $80 to $320 million, depending on the predictive model used and the degree to which warm water fishery gains offset coldwater fishery losses. Jones et al. [71] predicted that the USA would lose 50 % of its coldwater fisheries habitat by 2100, with only western and northeastern mountainous areas supporting coldwater fisheries under the highest emission scenario. Under that scenario, coldwater fishing days would decline by 6.4 million, resulting in economic losses of $81 million to $6.4 billion depending on the emission scenario and discount rate. By altering ocean pH, climate change is a serious potential threat to the zooplankton food base of marine recreational fishes and the ability of marine and estuarine mollusks and corals to fix calcium carbonate in their shells (AFS Web: http://fisheries.org/docs/policy_statements/policy_33f.pdf, accessed April 2014). Comte and Grenouillet [72] reported that French stream fish, including numerous species fished recreationally, have been unable to shift their ranges to higher elevations to keep pace with temperature changes from 1980 to 2009. Nonetheless, Tedesco et al. [73] noted that current and historical anthropogenic pressures account for more species extinctions than does predicted climate change.

5. CONCLUSIONS

Individuals and governments around the world value the conservation and sustainable use of recreational fisheries. In the USA and Europe that concern has led, respectively, to the Clean Water Act of 1972 with the goal of swimmable and fishable waters, and the Water Framework Directive with the goal of good water body condition and fisheries by 2015. Although we understand well how to manage sustainable recreational marine and freshwater fisheries, major historical, current and developing threats result from other demands on our lands and waters. Those threats either singly or together have markedly reduced once-substantial recreational fisheries on Great Lakes lake trout and lake sturgeon [74], Pacific salmon [75, 76], Atlantic salmon (WWF Web: http://awsassets.panda.org/downloads/salmon2.pdf), and Atlantic cod (IUCN Web: http://www.seafish.org/media/ Publications/FS17_201003_IUCNRedList.pdf). Continued human population and economic growth drive those threats and thus fundamentally conflict with healthy and sustainable fisheries on those and many other species [2, 77, 78, 79].

ACKNOWLEDGMENTS

Yadan Liu invited me to present an earlier draft of this paper at the Third International Recreational Fisheries Forum in Xiamen, Fujian Province, China. Ian Winfield assisted manuscript preparation by sharing a draft of his recreational fishery manuscript with me and Sabrina Lovell provided valuable information regarding marine fisheries. Yumi Terashima invited me to submit this review manuscript to Fisheries Science and Tadashi Tokai suggested several helpful improvements. As its President, the American Fisheries Society supported my airfare to present this paper at the March 2014 annual meeting of the Japanese Society of Fishery Science in Hakodate, Hokkaido, Japan.

APPENDIX 1

Common recreational Atlantic Coast marine finfish fisheries.

Amberjack, pompano Seriola, Carangidae

Atlantic cod Gadus, Gadidae

Drum (black, red) Pogonias, Sciaenops, Sciaenidae

Black sea bass Centropristis, Serranidae

Bluefish Pomatomus, Pomatomidae

Bonefish Albula, Albulidae

Bonito Sarda, Scombridae

Dolphinfish Coryphaena, Coryphaenidae

Marlin (blue, white) Makaira, Tetrapturus, Istiophoridae

Plaice Hippoglossoides, Pleuronectidae

Red snapper Lutjanus, Lutjanidae

Sailfish Istiophorus, Istiophoridae

Snook Centropomus, Centropomidae

Spotted sea trout, weakfish Cynoscion, Sciaenidae

Striped bass Morone, Moronidae

Summer flounder Paralichthys, Paralichthyidae

Swordfish Xiphias, Xiphiidae

Tarpon Megalops, Megalopidae

Tautog Tautoga, Labridae

Wahoo Acanthocybium, Scombridae

Yellowfin tuna Thunnus, Scombridae.

APPENDIX 2

Common recreational Atlantic Coast marine shellfish fisheries.

Clam (hardshell, quahog, razor, softshell, surf) Mercenaria, Veneridae; Ensis, Pharidae; Mya, Myidae; Spisula, Mactridae

Conch Lobatus, Strombidae

Crab (blue, stone) Callinectes, Portunidae; Menippe, Menippidae

Lobster (American, spiny) Homarus, Nephropidae; Panulirus, Palinuridae

Mussel Mytilus, Mytilidae

Oyster Crassostrea, Ostreidae

Scallop (bay, calico) Argopecten, Pectinidae

Whelk Busycon, Buccinidae.

APPENDIX 3

Common recreational Pacific Coast marine finfish fisheries.

Cabezon Scorpaenichthys, Cottidae

Dolphinfish Coryphaena, Coryphaenidae

Greenling Heagrammos, Hexagrammidae

Halibut Hippoglossus, Pleuronectidae

Lingcod Ophiodon, Hexagrammidae

Pacific herring Clupea, Clupeidae

Redtail surfperch Amphistichus, Embiotocidae

Rockfish (black, black-and-yellow, blue, brown, calico, china, copper, gopher, grass, kelp, olive, treefish) Sebastes, Sebastidae

Starry flounder Platichthys, Pleuronectidae

Tuna (albacore, yellowfin, yellowtail) Thunnus, Scombridae.

APPENDIX 4

Common recreational Pacific Coast marine shellfish fisheries.

Clams (butter, littleneck, gaper, razor, softshell) Saxidomus,Prothothaca, Veneridae; Treusus, Mactridae; Siliqua, Pharidae; Mya, Myidae;

Crab (Dungeness, red rock) Metacarcinus, Cancer, Cancridae

Mussel Mytilus, Mytilidae

Squid Loglio, Logliginidae.

APPENDIX 5

Common recreational inland coldwater fisheries.

American shad Alosa, Clupeidae

Salmon (Atlantic, coho, Chinook, sockeye) Salmo, Oncorhynchus, Salmonidae

Sturgeon (lake, white) Acipenser, Acipenseridae

Trout (brook, brown, cutthroat, lake, rainbow) Salvelinus, Salmo, Oncorhynchus, Salmonidae.

APPENDIX 6

Common recreational inland warmwater fisheries.

Black bass (largemouth, redeye, smallmouth, spotted) Micropterus, Centrarchidae

Bullhead (black, brown, yellow) Ameiurus, Ictaluridae

Catfish (blue, channel, flathead) Ictalurus, Pylodictus, Ictaluridae

Crappie (black, white) Pomoxis, Centrarchidae

Pike (northern, muskellunge) Esox, Esocidae

Rock bass (Roanoak, rock, shadow) Ambloplites, Centrarchidae

Sunfish (bluegill, longear, orangespotted, pumpkinseed, redbreast, redear, spotted) Lepomis, Centrarchidae

Temperate bass (striped, yellow, white, white perch) Morone, Moronidae

Walleye, sauger Sander, Percidae

Yellow perch Perca, Percidae.

REFERENCES

1. Brownscombe JW, Bower SD, Bowden W, Nowell L, Midwood JD, Johnson N, Cooke SJ (2014) Canadian recreational fisheries: 35 years of social, biological, and economic dynamics from a national survey. Fisheries 39:251–260
2. Limburg KE, Hughes RM, Jackson DC, Czech B (2011) Population increase, economic growth, and fish conservation: collision course or savvy stewardship? Fisheries 36:27–35
3. Shapiro MH, Holdsworth S, Paulsen SG (2008) The need to assess the condition of aquatic resources in the US. J N Am Benthol Soc 27:808–811
4. Paulsen SG, Mayio A, Peck DV, Stoddard JL, Tarquinio E, Holdsworth SM, Van Sickle J, Yuan LL, Hawkins CP, Herlihy A, Kaufmann PR, Barbour MT, Larsen DP, Olsen AR (2008) Condition of stream ecosystems in the US: an overview of the first national assessment. J N Am Benthol Soc 27:812–821
5. USEPA (US Environmental Protection Agency) (2013) National rivers and streams assessment 2008-2009: a collaborative survey. EPA/841/D-13/001.Office of Wetlands, Oceans and Watersheds and Office of Research and Development, Washington, DC
6. USEPA (US Environmental Protection Agency) (2009) National lakes assessment: a collaborative survey of the Nation's lakes. EPA/841/R-09/001.Office of Water and Office of Research and Development, Washington, DC
7. Hubert WA, Quist MC (eds) (2010) Inland fisheries management in North America. American Fisheries Society, Bethesda
8. Allen MS, Hightower JE (2010) Fish population dynamics: mortality, growth, and recruitment. In: Hubert WA, Quist MC (eds) Inland fisheries management in North America. American Fisheries Society, Bethesda, pp 43–80
9. Isermann DA, Paukert CP (2010) Regulating harvest. In: Hubert WA, Quist MC (eds) Inland fisheries management in North America. American Fisheries Society, Bethesda, pp 185–212
10. Trushenski J, Flagg T, Kohler C (2010) Use of hatchery fish for conservation, restoration, and enhancement of fisheries. In: Hubert WA, Quist MC (eds) Inland fisheries management in North America. American Fisheries Society, Bethesda, pp 261–294
11. Pegg MA, Chick JH (2010) Habitat improvement in altered systems. In: Hubert WA, Quist MC (eds) Inland fisheries management in North America. American Fisheries Society, Bethesda, pp 295–394
12. Bonar S, Hubert W, Willis D (eds) (2009) Standard methods for sampling North American freshwater fishes. American Fisheries Society, Bethesda
13. Kwak TJ, Freeman MC (2010) Assessment and management of ecological integrity. In: Hubert WA, Quist MC (eds) Inland fisheries management in North America. American Fisheries Society, Bethesda, pp 353–394
14. Pope KL, Lochmann SE, Young MK (2010) Methods for assessing fish populations. In: Hubert WA, Quist MC (eds) Inland fisheries management in North America. American Fisheries Society, Bethesda, pp 325–352
15. Chipps SR, Graeb BDS (2010) Ecology and management of lake food webs. In: Hubert WA, Quist MC (eds) Inland fisheries management in North America. American Fisheries Society, Bethesda, pp 395–424
16. Schinegger R, Trautwein C, Melcher A, Schmutz S (2012) Multiple human pressures and their spatial patterns in European running waters. Water Environ J 26:261–273
17. Mebane CA, Maret TR, Hughes RM (2003) An index of biological integrity (IBI) for Pacific Northwest rivers. Trans Am Fish Soc 132:239–261
18. Snyder CD, Young JA, Villella R, Lemarie DP (2003) Influences of upland and riparian land use patterns on stream biotic integrity. Landscape Ecol 18:647–664
19. Wang L, Lyons J, Kanehl P (2003) Impacts of urban land cover on trout streams in Wisconsin and Minnesota. Trans Am Fish Soc 132:825–839

20. Stranko SA, Hilderbrand RH, Morgan RP II, Staley MW, Becker AJ, Roseberry-Lincoln A, Perry ES, Jacobson PT (2008) Brook trout declines with land cover and temperature changes in Maryland. N Am J Fish Manag 28:1223–1232
21. Trautwein C, Schinegger R, Schmutz S (2012) Cumulative effects of land use on fish metrics in different types of running waters in Austria. Aquat Sci 74:329–341
22. Woody CA, Hughes RM, Wagner EJ, Quinn TP, Roulsen LH, Martin LM, Griswold K (2010) The US general mining law of 1872: change is overdue. Fisheries 35:321–331
23. Roth NE, Allan JD, Erickson DL (1996) Landscape influences on stream biotic integrity assessed at multiple spatial scales. Landscape Ecol 11:141–156
24. Wang L, Lyons J, Kanehl P, Gatti R (1997) Influences of watershed land use on habitat quality and biotic integrity in Wisconsin streams. Fisheries 22(6):6–12
25. Van Sickle J, Baker J, Herlihy A, Bayley P, Gregory S, Haggerty P, Ashkenas L, Li J (2004) Projecting the biological condition of streams under alternative scenarios of human land use. Ecol Appl 14:368–380
26. Sály P, Takács P, Kiss I, Bíró P, Erős T (2011) The relative influence of spatial context and catchment- and site-scale environmental factors on stream fish assemblages in a human-modified landscape. Ecol Freshwat Fish 20:251–262
27. Marzin A, Verdonschot PFM, Pont D (2013) The relative influence of catchment, riparian corridor, and reach-scale anthropogenic pressures on fish and macroinvertebrate assemblages in French rivers. Hydrobiol 704:375–388
28. Macedo DR, Hughes RM, Ligeiro R, Ferreira WR, Castro M, Junqueira NT, Silva DRO, Firmiano KR, Kauffman PR, Pompeu PS, Callisto M (2014) The relative influence of catchment and site variables on fish and macroinvertebrate assemblage richness in cerrado biome streams. Landscape Ecol 29:1001–1016. doi:10.1007/s10980-014-0036-9
29. Hughes RM, Wang L, Seelbach PW (eds) (2006) Landscape influences on stream habitat and biological assemblages. American Fisheries Society, Bethesda
30. Yeakley JA, Maas-Hebner KG, Hughes RM (eds) (2014) Wild salmonids in the urbanizing Pacific Northwest. Springer, New York
31. Stanford JA, Ward JV, Liss WJ, Frissell CA, Williams RN, Lichatowich JA, Coutant CC (1996) A general protocol for restoration of regulated rivers. Regul Rivers Res Manag 12:391–413
32. Poff NL, Allan JD, Bain MB, Karr JR, Prestegaard KL, Richter BD, Sparks RE, Stromberg JC (1997) The natural flow regime. Bioscience 47:769–784
33. Hughes RM, Rinne JN, Calamusso B (2005) Historical changes in large river fish assemblages of the Americas: a synthesis. In: Rinne JN, Hughes RM, Calamusso B (eds) Historical changes in large river fish assemblages of the Americas. American Fisheries Society, Bethesda, pp 603–612
34. Carlisle DM, Wolock DM, Meador MR (2011) Alteration of streamflow magnitudes and potential ecological consequences: a multiregional assessment. Front Ecol Environ 9:264–270
35. Sedell JR, Froggatt JL (1984) Importance of streamside forests to large rivers: the isolation of the Willamette River, Oregon, USA, from its floodplain by snagging and streamside forest removal. Verh Intl Verein Limnol 22:1828–1834
36. Walter RC, Merritts DJ (2008) Natural streams and the legacy of water-powered mills. Science 319:299–304
37. Kaufmann PR, Hughes RM (2006) Geomorphic and anthropogenic influences on fish and amphibians in Pacific Northwest coastal streams. In: Hughes RM, Wang L, Seelbach PW (eds) Landscape influences on stream habitat and biological assemblages. American Fisheries Society, Bethesda, pp 429–455
38. Bryce SA, Lomnicky GA, Kaufmann PR, McAllister LS, Ernst TL (2008) Development of biologically based sediment criteria in mountain streams of the western United States. N Am J Fish Manag 28:1714–1724
39. Jennings MJ, Bozek MA, Hatzenbeler GR, Emmons EE, Staggs MD (1999) Cumulative effects of incremental shoreline habitat modification on fish assemblages in north temperate lakes. N Am J Fish Manag 19:18–27

40. Sass GG, Kitchell JE, Carpenter SR, Hrabik TR, Marburg AE, Turner MG (2006) Fish community and food web responses to a whole-lake removal of coarse woody habitat. Fisheries 31:321–330
41. Whittier TR, Paulsen SG, Larsen DP, Peterson SA, Herlihy AT, Kaufmann PR (2002) Indicators of ecological stress and their extent in the population of northeastern lakes: a regional-scale assessment. Bioscience 52:235–247
42. Wagner T, Jubar AK, Bremigan MT (2006) Can habitat alteration and spring angling explain largemouth bass nest success? Trans Am Fish Soc 135:843–852
43. Kaufmann PR, Hughes RM, Whittier TR, Bryce SA, Paulsen SG (2014) Relevance of lake physical habitat assessment indices to fish and riparian birds. Lake Reserv Manag 30:177–191
44. Stahl LL, Snyder BD, Olsen AR, Pitt JL (2009) Contaminants in fish tissue from US lakes and reservoirs: a national probabilistic study. Environ Monitor Assess 150:3–19
45. Peterson SA, Van Sickle J, Herlihy AT, Hughes RM (2007) Mercury concentration in fish from streams and rivers throughout the western United States. Environ Sci Tech 41:58–65
46. Landers DH, Simonich SM, Jaffe D, Geiser L, Campbell DH, Schwindt A, Schreck C, Kent M, Hafner W, Taylor HE, Hageman K, Usenko S, Ackerman L, Schrlau J, Rose N, Blett T, Erway MM (2010) The Western Airborne Contaminant Assessment Project (WACAP): an interdisciplinary evaluation of the impacts of airborne contaminants in western U.S. National Parks. Environ Sci Tech 44:855–859
47. Rabalais NN, Turner RE, Scavia D (2002) Beyond science into policy: Gulf of Mexico hypoxia and the Mississippi River. Bioscience 52:129–142
48. Boyer EW, Goodale CL, Jaworski NA, Howarth RW (2002) Anthropogenic nitrogen sources and relationships to riverine nitrogen export in the northeastern USA. Biogeochemistry 57:37–169
49. Dodds WK, Bouska WW, Eitzmann JL, Pilger TJ, Pitts KL, Riley AJ, Schloesser JT, Thornbrugh DJ (2008) Eutrophication of U.S. freshwaters: analysis of potential economic damages. Environ Sci Tech 43:12–19
50. Lomnicky GA, Whittier TR, Hughes RM, Peck DV (2007) Distribution of nonnative aquatic vertebrates in western U.S. streams and rivers. N Am J Fish Manag 27:1082–1093
51. Sanderson BL, Barnas KA, Wargo Rub M (2009) Nonindigenous species of the Pacific Northwest: an overlooked risk to endangered salmon. Bioscience 59(3):245–256
52. Nickelson T (2003) The influence of hatchery coho salmon (Oncorhynchus kisutch) on the productivity of wild coho salmon populations in Oregon coastal basins. Can J Fish Aquat Sci 60:1050–1056
53. Nickelson TE, Solazzi MF, Johnson SL (1986) Use of hatchery coho salmon (Oncorhynchus kisutch) presmolts to rebuild wild populations in Oregon coastal streams. Can J Fish Aquat Sci 43:2443–2449
54. Pearsons TN (2008) Misconception, reality, and uncertainty about ecological interactions and risks between hatchery and wild salmonids. Fisheries 33:278–290
55. Naman SW, Sharpe CS (2012) Predation by hatchery yearling salmonids on wild subyearling salmonids in the freshwater environment: a review of studies, two cases histories, and implications for management. Environ Biol Fish 94:21–28
56. Tatara CP, Berejikian BA (2012) Mechanisms influencing competition between hatchery and wild juvenile anadromous Pacific salmonids in fresh water and their relative competitive abilities. Environ Biol Fish 94:7–19
57. Daly EA, Broduer RD, Fisher JP, Weitkamp LA, Teel DJ, Beckman BR (2012) Spatial and trophic overlap of marked and unmarked Columbia River Basin spring Chinook salmon during early marine residence with implications for competition between hatchery and naturally produced fish. Environ Biol Fish 94:117–134
58. Kaeriyama M, Seo H, Kudo H, Nagata M (2012) Perspectives on wild and hatchery salmon interactions at sea, potential climate effects on Japanese chum salmon, and the need for sustainable salmon fishery management reform in Japan. Environ Biol Fish 94:165–177

59. Naish KA, Taylor JE III, Levin PS, Quinn TP, Winton JR, Huppert D, Hilborn R (2008) An evaluation of the effects of conservation and fishery enhancement hatcheries on wild populations of salmon. Adv Mar Biol 53:61–194
60. Ruggerone GT, Agler BA, Nielsen JL (2012) Evidence for competition at sea between North Sound chum salmon and Asian hatchery chum salmon. Environ Biol Fish 94:149–163
61. Colborn T, vom Saal FS, Soto AM (1993) Development effects of endocrine-disrupting chemicals in wildlife and humans. Environ Health Persp 101:378–384
62. Tyler CR, Jobling S, Sumpter JP (1998) Endocrine disruption in wildlife: a critical review of the evidence. Crit Rev Toxicol 28:319–361
63. Hayes TB, Case P, Chui S, Chung D, Haeffele C, Haston K, Lee L, Mai VP, Marjuoa Y, Parker J, Tsui M (2006) Pesticide mixtures, endocrine disruption, and amphibian declines: are we underestimating the impact? Environ Health Persp 114(S-1):40–50
64. Brown AR, Riddle AM, Winfield IJ, Fletcher JM, James JB (2005) Predicting the effects of endocrine disrupting chemicals on healthy and disease impacted populations of perch (Perca fluviatilis). Ecol Model 189:377–395
65. Crane M, Gross M, Matthiessen P, Ankley GT, Axford S, Bjerregaard P, Brown R, Chapman P, Dorgeloh M, Galay-Burgos M, Green J, Hazlerigg C, Janssen J, Lorenzen K, Parrott J, Rufli H, Schäfers C, Seki M, Stolzenberg H-C, van der Hoeven N, Vethaak D, Winfield IJ, Zok S, Wheeler J (2010) Multi-criteria decision analysis of test endpoints for detecting the effects of endocrine active substances in fish full life cycle tests. Integr Environ Assess Manag 6:378–389
66. Colborn T, Dumanosk D, Myers JP (1996) Our stolen future. Dutton, New York
67. Handy RD, Owen R, Valsami-Jones E (2008) The ecotoxicology of nanoparticles and nanomaterials: current status, knowledge gaps, challenges, and future needs. Ecotoxicol 17:315–325
68. Shaw BJ, Handy RD (2011) Physiological effects of nanoparticles on fish: a comparison of nanometals versus metal ions. Environ Int 37:1083–1097
69. Cedervall T, Hansson L-A, Lard M, Frohm B, Linse S (2012) Food chain transport of nanoparticles affects behaviour and fat metabolism in fish. PLoS One 7(2):e32254. doi:10.1371/journal.pone.0032254
70. Michaels G, O'Neal K, Humphrey J, Bell K, Camacho R, Funk R (1995) Ecological impacts from climate change: an economic analysis of freshwater recreational fishing EPA-230-R-95-004. Office of Policy Planning and Evaluation. USEPA, Washington
71. Jones R, Travers C, Rodgers C, Lazar B, English E, Lipton J, Vogel J, Strzepek K, Martinich J (2013) Climate change impacts on freshwater recreational fishing in the United States. Mitig Adapt Strateg Glob Change 18:731–758
72. Comte L, Grenouillet G (2013) Do stream fish track climate change? Assessing distribution shifts in recent decades. Ecography 36:1–11
73. Tedesco PA, Oberdorff T, Cornu J-F, Beauchard O, Brosse S, Dürr HH, Grenouillet G, Leprieur F, Tisseuil C, Zaiss R, Hugueny B (2013) A scenario for impacts of water availability loss due to climate change on riverine fish extinction rates. J Appl Ecol 50:1105–1115
74. Regier HA, Hughes RM, Gannon JE (2013) The lake sturgeon as survivor and integrative indicator of changes in stressed aquatic ecosystems in the Laurentian Basin. In: Dempsey D, Auer N (eds) The great lake sturgeon. Michigan State University Press, Lansing, pp 27–57
75. Nehlsen WJ, Williams JE, Lichatowich JA (1991) Pacific salmon at the crossroads: stocks at risk from California, Oregon, Idaho, and Washington. Fisheries 16(2):4–21
76. Gustafson RG, Waples RS, Myers JM, Weitkamp LA, Bryant GJ, Johnson OW, Hard JJ (2007) Pacific salmon extinctions: quantifying lost and remaining diversity. Cons Biol 21:1009–1020
77. Clausen R, York R (2008) Global biodiversity decline of marine and freshwater fish: a cross-national analysis of economic, demographic, and ecological influences. Soc Sci Res 37:1310–1320

78. Leprieur F, Beauchard O, Blanchet S, Oberdorff T, Brosse S (2008) Fish invasions in the world's river systems: when natural processes are blurred by human activities. PLoS Biol 6(2):e28
79. Czech B (2013) Supply shock: economic growth at the crossroads and the steady state solution. New Society Publishers, Gabriola Island

CHAPTER 15

First Record in Brazil of Epistylis sp. (Ciliophora) Adhered to Argulus sp. (Argulidae), a Parasite of Hoplias aimara (Erithrinidae)

Lincoln Lima Corrêa[1]*, Marcos Sidney Brito Oliveira[2], Luiza Prestes[3], Marcos Tavares-Dias[4]

[1]*Universidade Federal do Oeste do Pará (UFOPA), Instituto de Ciências e Tecnologia das Águas (ICTA), Santarém, Brasil*

[2]*Programa de Pós-Graduação em Recursos Aquáticos Continentais Amazônicos (PPGRACAM), Universidade Federal do Oeste do Pará (UFOPA), Instituto de Ciências e Tecnologia das Águas (ICTA), Santarém, Brasil*

[3]*Programa de Pós-Graduação em Ecologia Aquática e Pesca (PPGEAP), Universidade Federal do Pará (UFPA), Belém, Brasil*

[4]*Embrapa Amapá, Rodovia Juscelino Kubitschek, Macapá, Brasil*

ABSTRACT

This paper records the first occurrence of Epistylis sp. on the body surface of Argulus sp. parasitizing Hoplias aimara from the upper Araguari River, in the eastern Amazon region, in the north of Brazil. Of the 16 specimens of H. aimara examined, 93.7% had their pelvic, caudal and pectoral fins and tegument infested by Argulus sp. (n = 73), which in turn were infested by

Epistylis sp. The specimens of Epistylis sp. from the body surface of Argulus sp. were analyzed using Scanning Electron Microscopy (SEM). The present study also identified a widening of the geographic distribution of these two species of ectoparasites to the eastern Amazon region of Brazil.

Keywords: Parasite Ecology, Opportunistic Parasite, Rio Araguari, Eastern Amazon

1. INTRODUCTION

Hoplias aimara Valenciennes, 1847 is a Characiforme from the Erythrinidae family which is distributed from Central America to Argentina, and is popularly known as giant lungfish, lungfish or Aymara. In Brazil, this species is distributed in the Trombeta, Xingu, Tapajós, Jari, Tocantins and Araguari Rivers, as well as the coastal rivers of Surinam and French Guyana [1] [2]. This fish lives in a variety of habitats, from lotic environments such as large rivers and waterfalls to lentic environments such as lakes and reservoirs [3]. However, little is known about the parasites that the fish hosts.

Among Branchiura Thorell, 1818 crustaceans, species of the genus *Argulus* Müller, 1785 are highly adapted to ectoparasitism in different fish species [4] - [6]. Some *Argulus* spp. are endemically distributed in the region of the Brazilian Amazon [6]. These ectoparasite crustaceans can reduce the weight, growth rates, breathing capacity, and muscle and fat percentage of fish, and lead to tissue traumatization. The severity of parasitic infection mainly depends on the infestation intensity. When it is high, there may be fish mortality, causing damage to fishing and fish farming [4] [7] - [9].

Species of *Epistylis* Ehrenberg, 1830 are colony forming ciliates of whom several species are free-living (freshwater and marine) or live in epibiosis, colonizing the body surface of insects and aquatic plants, mollusks, amphibians, reptiles, rotifers and fish. In fish, the main organ colonized is the gills and in Brazil, *Epistylis* sp. is an emerging parasite [10] [11] . The death of thousands of Centrarchidae in the USA has been attributed to heavy infestation by *Epistylis* sp., which coincide with ulcers and lesions. As this is not an obligate parasite, these injuries can be gateways to secondary infections, so these mortalities are most often associated with secondary infections caused by bacteria [12] [13] . *Epistylis* spp. commonly colonizes planktonic copepods [14] - [17]. However, colonization studies of ectoparasite crustaceans are rare [11] .

This study describes the first record of *Epistylis* sp. in Argulus sp. collected from *Hoplias aimara,* a fish from the upper Araguari River, in the state of Amapá, in the north of Brazil.

2. MATERIALS AND METHODS

A total of 16 specimens of *H. aimara* were collected from a stretch of the Araguari River, in the state of Amapá, Brazil (Figure 1), which is the largest river to be truly located in Amapá, being 617 km long. Its source is found in the Parque Nacional Montanhas do Tumucumaque (Tumucumaque Mountains National Park), in the Guiana Shield [18] . It passes through the municipalities of Porto Grande and Ferreira Gomes, in the Floresta Nacional do Amapá (Amapá National Forest) (FLONA), the Floresta Estadual do Amapá (Amapá State Forest) (FLOTA) and the Reserva Biológica do Lago Piratuba (Lake Piratuba Biological Reserve) (REBIO), and there are two hydroelectric power plants along its course.

The collection of *H. aimara* took place in July 2014, using a hand line and gillnet with different mesh sizes. Immediately after capture, the fish were macroscopically examined for the presence of ectoparasite crustaceans. Then, each fish was weighed (g) and its standard length measured (cm). The crustacean specimens collected were fixed in 70% ethanol before being transferred to 70% glycerinated alcohol after 48 hours (10%) and identified in accordance with [8] . The prevalence values, mean abundance and mean intensity were calculated in accordance with [19] . These procedures were carried out in accordance with the principles of the Colégio Brasileiro de Experimento Animal (the Brazilian College of Animal Experimentation) (COBEA).

For analysis with a scanning electron microscope (SEM), five specimens of *Argulus* sp. previously fixed in 70% ethanol were transferred to a 2.5% glutaraldehyde solution in 0.15 M phosphate buffer (pH 7.3) for 24 h, and post-fixed in 1% osmium tetroxide in the same buffer for 2 h. Dehydration was performed with an increasing sequence of ethanol solutions, and critical point drying was carried out with a CPD 020 (Union Balzer) with liquid CO_2. The specimens were coated with a 20 nm layer of gold using the MED 010 apparatus (Balzer Union). Analysis was performed by SEM (SEM 515 Philips).

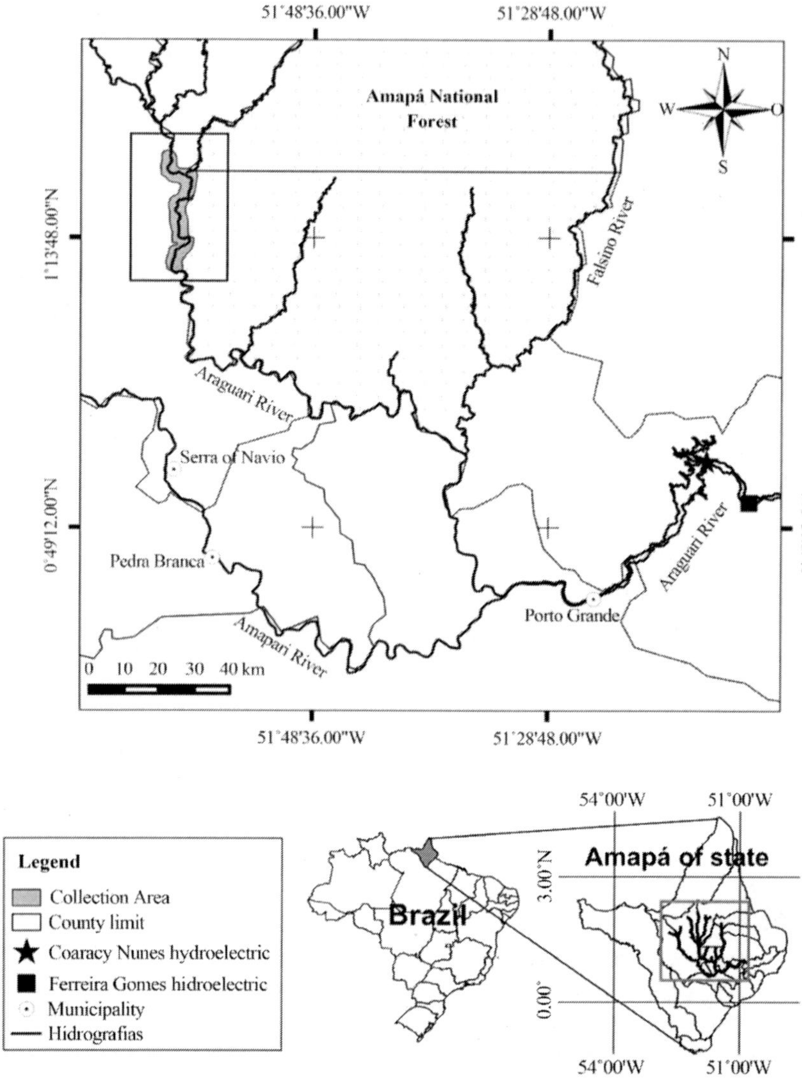

Figure 1. Geographic location of Hoplias aimara collection area in a stretch of the upper Araguari River, in the state of Amapá, in the north of Brazil.

3. RESULTS

A total of 16 *H. aimara* fish with a body weight of 1619.9 ± 1054.7 g and a standard length of 40.7 ± 9.3 cm were collected. From these hosts (n = 15) a total of 73 specimens of *Argulus* sp. were collected from the pelvic, caudal and pectoral fins and tegument, with a prevalence of 93.7%, average intensity of 4.9 host and mean abundance of 4.6. Of these specimens of *Argulus* sp. five were infested by Epistylis sp., which were adhered in large quantity to the ventral region of the argulidae (Figure 2).

The specimens of *Epistylis* sp. established themselves on *Argulus* sp., which in turn serves as a substrate and means of transport, with no lysis or mechanical injury to the infested *Argulus* sp. observed. The Argulus sp. were infested by *Epistylis* sp. in various parts of the bodies, with a stalk bearing several zooids (Figure 2A and Figure 2B). This stalk is relatively long and may be divided to form colonies which can be either longer or shorter or more or less ramified (Figure 2B). The vaulted epistomial disc is slightly elevated above the peristomial lips and slanted. The macronuclei are long (Figure 2D).

4. DISCUSSION

Hoplias aimara was infested by *Argulus* sp., which is a new species found parasitizing this host from the Araguari River basin in the state of Amapá (eastern Amazon region, Brazil). Different species of *Argulus* have been found infesting Brazilian fish, especially species of Anostomidae, Pimelodidae, Cichlidae and Serrasalmidae [6] .

These ectoparasites are often found parasitizing the lentic fish environment, as natant-free forms can find their host more easily in such an environment [4] [7] [20] . However, in *H. aimara* we found a high prevalence of *Argulus* sp., indicating that the infestations also occur in fish from lotic environments, as this host fish also inhabits rapids environments [3] , such as the Rio Araguari, the basin of this study. Additionally, the levels of infestation of *Argulus* sp. in H. aimara were similar to those of *Argulus foliaceus* described for Cyprinus carpio and *Abramis brama* [4] and greater than *Argulus pestifer* in *Pseudoplatystoma faciatum* [7]

Species of *Epistylis* are not obligate ectoparasites, but live in epibiosis with shellfish, aquatic plants, rotifers, aquatic insects, amphibians, reptiles, free crustaceans and parasitic crustaceans [10] [11] as has been described for the occurrence of *Epistylis* sp. in *Ergasilus chelangulatus* from *Pimelodus maculatus* [11] . In this manner, the results of the present study suggest that Epistylis sp. also colonizes *H. aimara*, the fish from which specimens of *Argulus* sp. were collected. Species of *Epistylis* have been reported in different species of freshwater fish in Brazil, such as *Colisa lalia, Ctenopharyngodon idella, Oreochromis niloticus, Piaractus mesopotamicus, Pseudoplatystoma corruscans, Pseudoplatystoma reticulatum, Leiarius marmoratus, Hemisorubim platyrhynchos* and *Zungaro jahu* [10] [21] [22] , which were possibly also infected by crustacean species, suggesting that crustaceans are important for the development and maintenance of the life of *Epistylis* sp. [11] [14] . Furthermore, it is suggested that the infective forms seek new hosts for the fixation and development of new colonies, such as zooplankton copepods, which can act as vectors or host reservoirs of *Epistylis* spp., as it is common to find these ciliates fixed on the body surface of copepods [14] - [16] thereby increasing the dissemination of this ciliate.

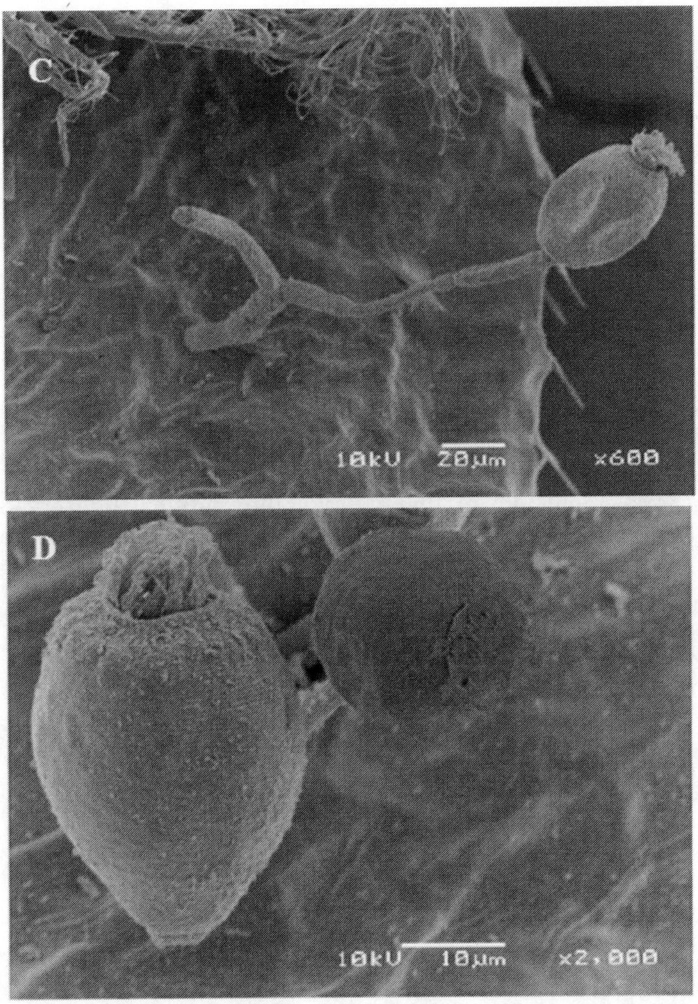

Figure 2. A. Scanning Electron Microscopy (SEM) detailing Epistylis sp. adhered to the anterior region of Argulus sp. B. Extensive and ramified colonies of Epistylis sp. adhered to Argulus sp. C. Epistylis sp. D. Vaulted epistomial disc, slightly elevated above peristomial lips.

While *Epistylis* spp. are considered commensal, this protozoan causes damage such as bodily injury, locomotor disorders, low fertility, reduced competitiveness, increased susceptibility to predation, increased energy demand, faster sinking rates and a decrease in the survival rate to hosts, as well as creating the possibility of bacterial infections [12] - [14] [16] . However, in *Argulus* sp. parasitizing *H. Aymara* no injury was found. At the same time, the fixation of Epistylis spp. to fish can result in the proliferation of the parasite, causing depigmentation and injuries to the integument with bleeding issues and the excessive production of mucus, triggering anorexia and itching [21] . Thus, it is suggested that *Epistylis* spp. is an opportunistic parasite that uses *Argulus* sp. as a substrate and a means of infecting other animals such as *H. aimara*, which have a greater area for their proliferation.

This first report of *Epistylis* sp. on the body surface of *Argulus* sp., in addition to being the first record of *Argulus* sp. parasitizing *H. aimara*, extends the geographical distribution of these two species of ectoparasites to the eastern Amazon region in the extreme north of Brazil.

ACKNOWLEDGEMENTS

The authors would like to thank Brazil International Conservation (CI), Walmart Brazil, the Instituto Chico Mendes de Conservação da Biodiversidade (ICMBio), and the Biologia e Etnoecologia da Ictiofauna na Floresta Nacional do Amapá (the Biology and Ethno-ecology of the Ichthyofauna of the Amapa National Forest) (FLONA) project for its support. We would also like to thank the fishermen for their help with specimen collection and the evaluators for suggestions that improved the quality of this manuscript, as well as the National Council for Research and Technological Development (CNPq, Brazil), for the productivity scholarship for Dr. Marcos Tavares-Dias.

REFERENCES

1. Planquette, P., Keith, P. and Le Bail, P.-Y. (1996) Atlas des poissons d'eau douce de Guyane: Tome 1. Collection du Patrimoine Naturel, Vol. 22. Service du Patrimoine Naturel/Institut d'écologie et de Gestion de la Biodiversite?/ Muse?um National D'histoire Naturelle/Institut National de la Recherche Agronomique/Conseil Supe?rieur de la Pe?che, Paris.
2. Froese, R. and Rauly, D. (2016) FishBase. World Wide Web Electronic Publication. www.fishbase.org
3. Oyakawa, O. and Mattox G. (2009) Revision of the Neotropical Trahiras of the Hoplias lacerdae Species Group (Ostariophysi: Characiformes: Erythrinidae) with Descriptions of Two New Species. Neotropical Ichthyology, 7, 117-140. http://dx.doi.org/10.1590/S1679-62252009000200001
4. Walker, P.D., Harris, J.E., Van der Velde, G. and Wendelaar Bonga, S.E. (2008) Effect of Host Weight on the Distribution of Argulus foliaceus (L.) (Crustacea, Branchiura) within a Fish Community. Acta Parasitologica, 53, 165-172. http://dx.doi.org/10.2478/s11686-008-0020-0
5. Moller, O.S. (2009) Branchiura (Crustacea)—Survey of Historical Literature and Taxonomy. Arthropod Systematics & Phylogeny, 67, 41-55.

6. Tavares-Dias, M., Dias-Júnior, M.B.F., Florentino, A.C., Silva, L.M.A. and Cunha, A.C. (2015) Distribution Pattern of Crustacean Ectoparasites of Freshwater Fish from Brazil. Veterinary Parasitology, 24, 136-147. http://dx.doi.org/10.1590/s1984-29612015036
7. Malta, J.C.O. (1984) Os peixes de um lago de várzea da Amaz?nia central (Lago Janauacá, Rio Solimões) e suas relações com os crustáceos ectoparasitos (Branchiura: Argulidae). Acta Amazonica, 14, 355-372. http://dx.doi.org/10.1590/s1984-29612015036
8. Thatcher, V.E. (2006) Amazon Fish Parasites. 2nd Edition, Pensoft Publishers, Moscow.
9. Lima, F.S., Casali, G.P. and Takemoto, R.M. (2013) Crustacea. In: Pavanelli, G.C., Takemoto, R.M. and Eiras, J.C., Org., Parasitologia de peixes de água doce do Brasil, Eduem, Maringá, 371-397.
10. Pádua, S.B., Ishikawa, M.M., Ventura, A.S., Jer?nimo, G.T., Martins, M.L. and Tavares, L.E.R. (2013) Brazilian Catfish Parasitized by Epistylis sp. (Ciliophora, Epistylididae), with Description of Parasite Intensity Score. Parasitology Research, 11, 443-446. http://dx.doi.org/10.1007/s00436-012-3069-5
11. Azevedo, R.K., Brandão, H., Abdallah, V.D. and Silva, R.J. (2014) First Record of an Epibiont Protozoan Epistylis sp. (Ciliophora, Peritrichia) Attached to Ergasilus chelangulatus (Ergasilidae) in Brazil. Brazilian Journal of Biology, 74, 460-463. http://dx.doi.org/10. 1590/1519-6984.10112
12. Rogers, W.A. (1971) Disease in Fish Due to the Protozoan Epistylis (Ciliata: Peritricha) in the Southeastern US. Proceedings of the Southeastern Association of Game Fish Commisioners, 25th Annual Conference, Charleston, 493-496.
13. Hazen, T.C., Raker, M.L., Esch, G.W. and Fliermans, C.B. (1978) Ultrastructure of Red-Sore Lesions on Largemouth Bass (Micropterus salmoides): Association of the Ciliate Epistylis sp. and the Bacterium Aeromonas hydrophila. Journal of Protozoology, 25, 351-355. http://dx.doi.org/10.1111/j.1550-7408.1978.tb03901.x
14. Xu, Z. and Burns, C.W. (1991) Effects of the Epizoic Ciliate, Epistylis Duphniae, on Growth, Reproduction and Mortality of Boeckella triarticulata (Thomson) (Copepoda: Calanoida). Hydrobiologia, 209, 183-189. http://dx.doi.org/10.1007/BF00015341
15. Visse, M. (2007) Detrimental Effect of Peritrich Ciliates (Epistylis sp.) as Epibionts on the Survival of the Copepod Acartia bifilosa. Proceedings of the Estonian Academy of Sciences, Biology and Ecology, 56, 173-178.
16. Bozkurt, A. and Genc, E. (2009) First Record of the Epibiont Protozoan Epistylis sp. on Zooplankton (Copepoda, Cladocera, and Rotifera) from the Asi River, Turkey. Crustaceana, 82, 515-530.http://dx.doi.org/10.1163/156854009X407687
17. Rajabunizal, K. and Ramanibai, R. (2011) Presence of an Epibiont Epistylis niagarae (Protozoa, Ciliophora) on Mesocyclops aspericornis in Velachery Lake Chennai India. Journal of Biological Sciences, 11, 189-195. http://dx.doi.org/10.3923/jbs.2011.189.195
18. Cunha, A.C., Vilhena, J.E.S., Santos, E.S., Saraiva, J.M.B., Kuhn, P.A.F., Brito, D.C., Souza, E.B., Rocha, E.P., Cunha, H.F.A., Brito, A.U., Júnior, A.C.P.B., Paca, V.H.M. and Santos, P.V.C.J. (2014) Evento Extremo de chuva-vazão na bacia hidrográfica do Rio Araguari, Amapá, Brasil. Revista Brasileira de Meteorologia, 29, 95-110. http://dx.doi.org/10.1590/0102-778620130051
19. Bush, A.O., Lafferty, K.D., Lotz, J.M. and Shostack, A.W. (1997) Parasitology Meets Ecology on Its Own Terms: Margolis et al. Revisited. The Journal of Parasitology, 83, 575-583. http://dx.doi.org/10.2307/3284227
20. Maman, M., Hamel, C. and Van Damme, P.A. (2014) Ectoparasites (Crustacea: Branchiura) of Pseudoplatystoma fasciatum (surubí) and P. tigrinum (Chuncuina) in Bolivian Whitewater Floodplains. Ecología en Bolivia, 39, 9-20.
21. Pavanelli, G.C., Eiras, J.C. and Takemoto, R.M. (2008) Doença de peixes, profilaxia, diagnóstico e tratamento. 3th Edition, Eduem, Maringá, 311 pp.
22. Adriano, E.A., Ceccarelli, P.S., Silva, M.R.M. and Maia, A.A.M. (2012) Prevalência, Distribuição geográfica e sazonal de protozoários e mixozoários parasitos de jaú (Zungaro jahu) no Pantanal Matogrossense. Pesquisa Veterinária Brasileira, 32, 1341-1344. http://dx.doi.org/10.1590/S0100-736X2012001200020

CHAPTER 16

Japanese chum salmon stock enhancement: current perspective and future challenges

Shuichi Kitada

Graduate School of Marine Biosciences, Tokyo University of Marine Science and Technology, Tokyo 108-8477, Japan

ABSTRACT

This study reviews the present status of the Japanese chum salmon *Oncorhynchus keta* stock enhancement program and considers the ecological sustainability of wild populations while providing fishery production, exemplified by the hatchery-based Kitami region set net fishery. The return rate and the number of returns have been historically high in the Sea of Okhotsk, but have decreased in other regions since 2005. Natural spawning of chum salmon occurred in at least 160 rivers in Hokkaido. The genetic diversity of Japanese chum salmon was similar to or higher than that of other Pacific Rim populations. Numbers of alleles were high at microsatellite loci, but the loss of rare haplotypes was observed in all populations. The estimated N_e/N ratio for the Kitami region was >0.15 % including hatchery and wild fish under the present high fishing pressure. Four regional populations were inferred in Hokkaido, however, genetic differentiation was weak and some river-populations were nested. Substantial changes in run timing were observed, but it has recovered gradually owing to the recent practice of escapement. Our analyses highlight the importance of juvenile quality and the vital roles of escapements in enhanced and non-enhanced rivers. New research is needed to minimize the genetic risks associated with hatchery programs.

Keywords: Integrated hatchery program, Escapement, Effective population, size, Genetic diversity, Population structure

1. INTRODUCTION

The hatchery stock enhancement program in Japan was initiated in the 1880s to increase Japanese chum salmon *Oncorhynchus keta* fishery production and has continued for over 120 years. The numbers of chum salmon returning to Japan have increased remarkably since the late 1970s. Hatchery technology innovation is considered the major cause for the increase [1, 2], and the favorable ocean and climate conditions have supported this increase [3, 4]. The closure of high seas fisheries and improved river water quality may have boosted the recent abundance of chum salmon in Hokkaido [5], the principal salmon production area in Japan. Although past management of Japanese chum salmon focused on producing and releasing hatchery-reared fish (hereafter hatchery fish), Japanese salmon scientists and hatchery managers have become aware of the importance of conserving wild fish. Counting surveys of wild fish recently conducted in Hokkaido have found that natural spawning occurs in many rivers [6]. However, the majority of chum salmon returning to Hokkaido are hatchery-released fish; therefore, the ecological sustainability of the chum salmon stock is controversial [7, 8, 9]. While traditional hatcheries have concentrated on fish production, the goals of modern hatcheries are to conserve wild populations and/or provide harvest with minimal risk to the wild population [10, 11]. There are strong ecological and economic incentives in Hokkaido to reform salmon fishery management by focusing on the coexistence of hatchery and wild fish [2, 12, 13, 14]. Further discussion on the use of hatcheries from the viewpoint of conservation may be useful.

This study reviews the present status of the Japanese chum salmon stock enhancement program and considers the role of hatcheries in the ecological sustainability of wild populations, focusing on the hatchery-based set net fishery in the Kitami region, Hokkaido, the largest chum salmon fishery in Japan. Onsite surveys of the chum salmon stock enhancement program were conducted, and statistics and information were collected from the Salmon and Freshwater Fisheries Research Institute (SFFRI) at Eniwa, the Hokkaido Federation of Fisheries

Cooperative Association (HFFCA) at Sapporo, and the Kitami Salmon Enhancement Program Association (KSEPA) at Abashiri. After conducting a brief review of the chum salmon stock enhancement program and the results of wild fish counting surveys in Hokkaido, the ratio of effective population size to the number of returning adult fish was estimated for the Kitami region by combining hatchery broodstock and hatchery/wild escapements. Genetic diversity indices of Japanese chum salmon populations were compared with those of other populations in the Pacific Rim. Fine-scale population structures of Japanese chum salmon populations were estimated based on publicly available mitochondrial (mt) DNA and microsatellite (ms) DNA data. Our analyses highlight the importance of juvenile fish quality and the vital role of hatchery-wild integrated reproduction through escapements to natural streams.

2. HATCHERY RELEASES AND RETURN RATES TO HOKKAIDO

The chum salmon hatchery program in Hokkaido is one of the largest salmon hatchery programs in the world [2]. The number of juveniles released from Hokkaido has been stable since the 1990s, averaging 1,030 ± 22 (standard deviation) million per annum during the last decade. There are currently 150 private and 10 national hatcheries in Hokkaido, and the majority of the chum salmon released (87 %) are produced by private hatcheries, with the residual portion (13 %) being produced and released by national hatcheries [2]. Hatchery fish are released in 140 rivers, of which 60 rivers use juveniles produced from parent fish collected in adjacent rivers (transplants). In addition, transplants are reared in 80 net pens and released from the rearing sites [2]. Such transplants are allowed within the geographic region in order to minimize genetic disturbance. All juveniles released from national hatcheries have otolith thermal marks. Fish from private hatcheries are marked only when scientific research is conducted. The otolith-marked chum salmon are currently <15 % of the total number released [2].

The number of chum salmon returning to Hokkaido has increased remarkably since the late 1970s, and the average number was 49 ± 10 million (range 38–61 million) during 2008–2012. This number has been decreasing since 2005, and as of 2012 (39 million) was ~64 % of the historical maximum of 61 million in 2004 (Fig. 1a). The average return rate was ~5 % for the brood years after 1996 [2]. In this article, the return rate was simply calculated by the number of returns (i.e., coastal commercial catches plus the number of adults caught for broodstock in rivers) divided by the number of hatchery fish released in the previous 4 years with the assumption that the number of naturally spawned chum salmon was negligible if any [2]. Returning salmon are caught by set net fisheries in the coastal waters off Hokkaido. Approximately 1,000 salmon set nets are operated from late August to late December, with landings peaking in late September and early October, although the fishing season differs among regions [14]. Commercial fishers pay for ~7 % of the landings every year to support the hatchery program. Commercial and recreational fishing for adult chum salmon are prohibited in all rivers and in coastal waters near the river mouths where hatchery broodstock are collected. Therefore, chum salmon that have escaped capture by commercial set nets are only captured at weirs at enhanced rivers [2]. Surplus fish caught in weirs are sold to support hatchery operations.

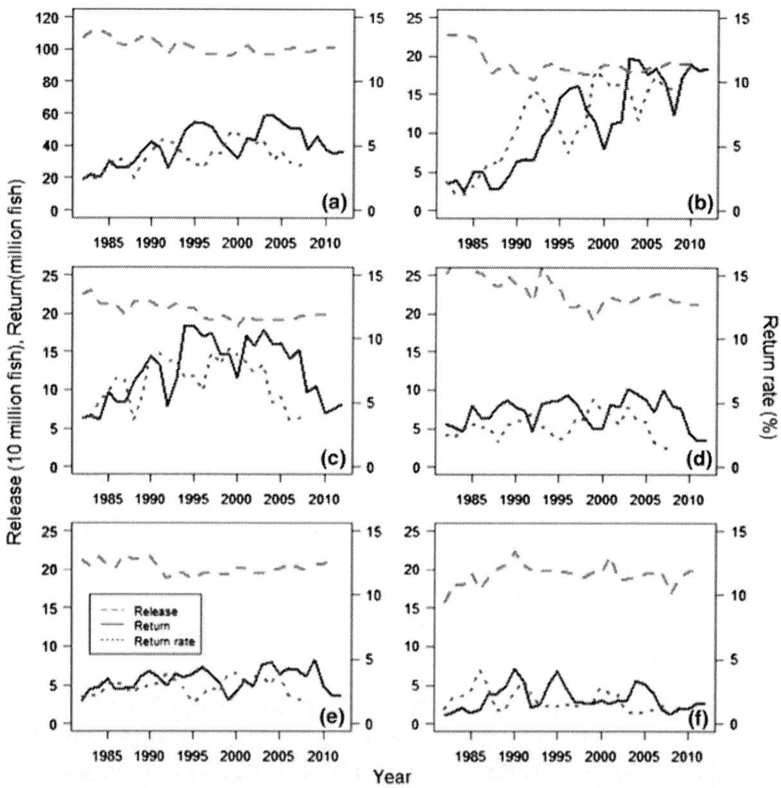

Figure. 1 Numbers of releases and returns with return rates of chum salmon in **A** total, and in the regions of the **B** Sea of Okhotsk coast (Kitami region), **C** Nemuro Strait, **D** eastern Pacific, **E** western Pacific, and **F** Sea of Japan, Hokkaido. The return rate was estimated as (number returned 4 years after release/number released)

The Hokkaido Prefectural Government established the salmon enhancement plan in 1999, which prescribes the number of adults to be captured for broodstock, number of eggs to be taken in each river, number of fry to be reared in each hatchery, and the number of fry to be released in each river or net pen location [2]. In the enhancement plan, Hokkaido is divided into five geographical regions (Sea of Okhotsk coast, Nemuro Strait, eastern Pacific coast, western Pacific coast, and Sea of Japan coast) [14], each of which is further subdivided into 2–4 sub-regions, resulting in 14 management regions [2]. The hatcheries are operated by the nine regional salmon enhancement program associations, including KSEPA, which are supported by the Hokkaido Salmon Propagation Association, Sapporo. The salmon enhancement plan is temporally divided into three periods to manage the coastal set net fishery and broodstock collection in rivers [2]. These periods for the coastal fishery include early- (September), middle- (October), and late-periods (November and later). However, this is a period for fishery management, and not for identifying biological differences in returning fish. Here, the term "early run" is tentatively defined for fish that return during September–November, and "late run" for fish that return during December–January. The term "wild fish" is defined as naturally-produced individuals in the wild, even though they could include hatchery descendants [15, 16, 17].

A decreasing trend in returns has been observed in all regions except the Sea of Okhotsk. The recent return rates and the number of fish returning have been historically high in the Sea of Okhotsk and relatively low in the Sea of Japan region (Fig. 1). Release sizes have been increasing in all regions, and at most hatcheries they exceed the target size to be reared, i.e., 5 cm in fork length and 1 g in weight [2]. Despite the large release sizes, declining return rates have been observed in the Nemuro Strait, eastern Pacific, and western Pacific regions (Fig. 1). A plausible cause for the regional difference in return rates is the recent changes in the coastal environment around release time, but there remains an insufficient understanding of the causes [2].

The average number of chum salmon returning to the Sea of Okhotsk coast was 1,369,000 during 1959–1970. The average number of returning fish for the past 5 years (2008–2012) was 16,998,000 (12.4 times larger), most of which were hatchery fish. Hatchery innovation techniques are considered the main reason for the remarkable increase, which mimics the wild recruitment process and allows release timing to match favorable conditions [13] in river and coastal areas. In addition, KSEPA, an agency that enhances chum salmon populations in the Sea of Okhotsk region, has improved seed production techniques during the past decade to produce healthy juveniles. The quality of released juveniles may have contributed to the increased number of returning fish and return rates after 2000 (Fig. 1b). It may be useful to evaluate the present management system in the

Kitami region and consider chum salmon sustainability, which depends on both hatchery and wild fish.

3. NATURAL SPAWNING IN NON-ENHANCED AND ENHANCED RIVERS

About 1,500 river systems occur in Hokkaido, of which 140 (9.3 %) rivers currently receive hatchery-releases of chum salmon [6]. The rivers are categorized into four types in relation to hatchery operations: (1) no enhancement to date, (2) currently no enhancement, (3) enhanced without broodstock collection, and (4) enhanced with broodstock collection for seed production [6, 14]. The rivers are relatively short, so hatchery fish are released in most large rivers with lengths >50 km, and naturally spawning chum salmon populations remain in many small (short) rivers. A total of 322 rivers (length > 8 km) were selected for counting surveys throughout Hokkaido, because water levels in smaller rivers may be inadequate to allow the passage of chum salmon [6]. Therefore, the number of rivers possible for chum salmon runs is estimated to be at least 322 in Hokkaido. Among the rivers, the number of enhanced rivers (categories 3 and 4) was 140 (43.4 %), of which 76 (2008) (23.8 %) and 84 (2009) (26.0 %) had weirs to collect hatchery broodstock (category 4).

Miyakoshi et al. [6] conducted foot surveys of 238 rivers during 2008 and 2009 and found that natural spawning occurred in 59 (31.4 %) and 50 (37.6 %) of the non-enhanced rivers, respectively. Furthermore, chum salmon were seen to ascend upstream in 191 and 175 rivers, including rivers from which broodstock are collected (76 and 84 rivers) and rivers <8 km (11 and eight rivers), respectively. These observations show that natural spawning of chum salmon occurs in 30–40 % of non-enhanced rivers (Fig. 2a) and in 70–80 % of 238 rivers surveyed, including categories 1–4.

The escapement of hatchery fish into enhanced rivers with hatcheries creates wild fish. Escapement during hatchery operations supplements early-run wild fish, and escapement creates late-run wild fish after hatchery operations. All weirs in the enhanced rivers of the Kitami region (Fig. 2b, c) have been remodeled recently to provide passage of fish for spawning (KSEPA, pers. comm., Fig. 2d). KSEPA recorded the number of escapements after 2005 in some enhanced rivers with hatcheries, although the number was low (Fig. 3). Weirs are removed around the end of November, when hatchery operations are finished. Returning hatchery and wild fish freely ascend the stream and spawn in the wild on enhanced rivers, and supplement early and late-run wild fish, respectively. The natural run in enhanced rivers without hatcheries and non-enhanced rivers creates wild fish throughout the run season. Thus, regardless of enhancement, wild fish can be created in all river systems. Therefore, it is important to evaluate how the sustainability of the chum salmon populations is impacted by integrated reproduction in hatcheries and natural streams.

Figure. 2 A A non-enhanced river where natural spawning occurs (Nikura River, a tributary of the Saromabetsu River, where counting surveys of natural spawning fish are conducted), **B**, **C** a weir to collect hatchery broodstock (Tokoro River) and **D** a remodeled fish pool to adjust escapement upstream (Shari River)

4. INTEGRATED HATCHERY AND WILD REPRODUCTION

Huge numbers of hatchery fish return to hatchery-enhanced rivers, mainly in September–November. In 2012, 18,460,000 fish returned to the Kitami region. Among them, ~16 million fish were caught by set nets in coastal areas, and 2 million fish were caught at weirs on enhanced rivers, of which 114,000 fish were sent to hatcheries as parent fish, which provided an estimate that 4,140 hatchery fish escaped upstream for wild stock conservation. The surplus hatchery fish caught in weirs (1,886,000) were sold to support hatchery operations. The exploitation rates were very high at 86.7 % (set nets) and 94.3 % (weirs). This calculation indicates that 0.62 % of returning hatchery fish were used for hatchery broodstock and 0.02 % escaped from weirs to upstream rivers.

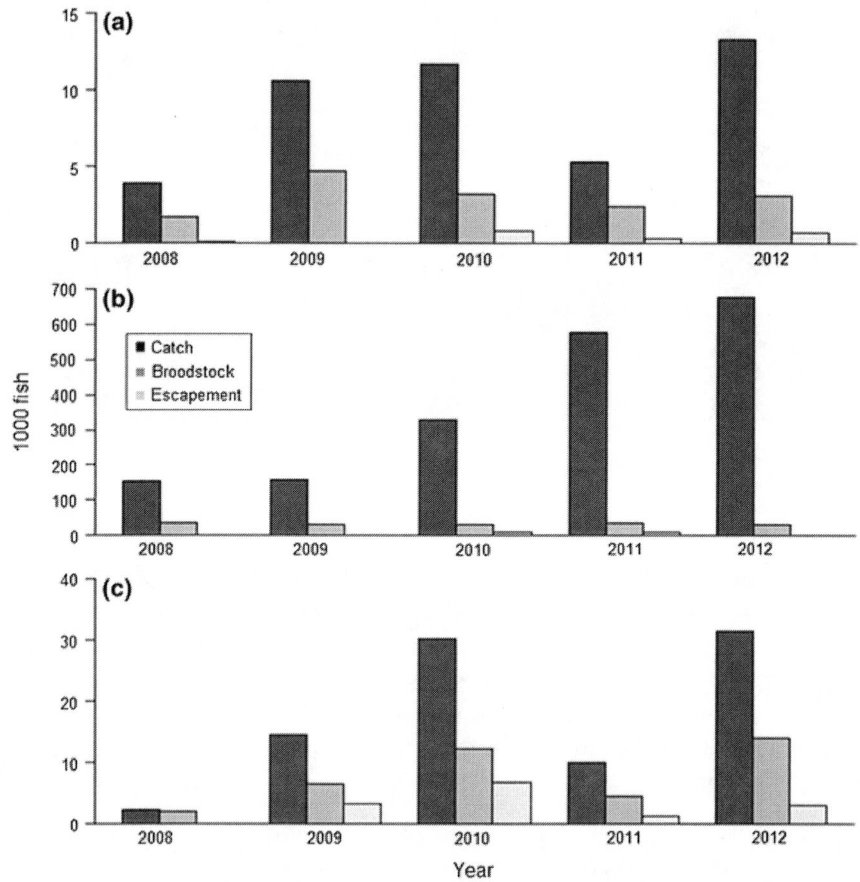

Figure. 3 Numbers of fish caught at the weir, hatchery broodstock, and escapement of chum salmon in the **A** Iwaobetsu, **B** Abashiri, and **C** Horonai Rivers of the Kitami region

Substantial numbers of wild fish return to some Kitami-region rivers [6, 18]. The largest portion of this natural return occurs during December and early January, after returns of hatchery fish and the hatchery-focused fishery are completed in late November. The management system believes that fish returning during this time frame represent the historical natural return and are not subject to significant harvest. Escapements of these fish have been relatively consistent for the last several years, and current wild salmon escapement was estimated at 20,191 in the Kitami region (SFFRI, unpubl. data, 2013).

The percentage of hatchery breeders among all returned fish is very small (0.62 %). However, this is not surprising because the effective population size (N_e) is very small compared with the census population size (N) [19]. It is generally accepted that the effective population size of highly fecund marine teleosts is between two and five orders of magnitude smaller than the census spawning size [20, 21, 22]. For example, the N_e/N ratio was 10^{-5} for North Sea cod *Gadus morhua* [22] and New Zealand snapper *Pagrus auratus* [23], 10^{-3} for red drum *Sciaenops ocellatus* [24], and 10^{-4} for Japanese Spanish mackerel *Scomberomorus niphonius* in the Seto Inland Sea [17]. This may be explained by a biased sex ratio caused by dominant males monopolizing females, a disproportionate contribution of genotypes through egg production, an increase in larval viability with increasing female size, overlapping generations, or high variance in individual reproductive success [22]. In addition, the chance spatial and temporal match or mismatch of spawning with suitable oceanic conditions [20, 25] could result in the majority of surviving recruits being spawned by a minority of the population [17, 22].

The N_e/N ratio for salmonids should be larger than that of marine fish because of lower fecundity with larger eggs. The N_e/N ratio was estimated to be 4.3 % for a natural population of Chinook salmon *O. tshawytscha* [26]. The N_e/N ratio is on the order of 0.2 for most natural stocks of salmonid fish, where N is the number of spawning adults [27]. No N_e/N ratio estimate for a natural and a hatchery population of chum salmon was found in the literature, so 20 % ($\sim 10^{-1}$) was used as a tentative estimate of N_e/N for both hatchery and wild spawning chum salmon. The number of spawning adults in hatcheries was 114,000, which was assumed to be hatchery-released fish. The rough estimate of the N_e/N ratio for the Kitami hatchery program was ~0.12 % (=114,000 × 0.2/18,460,000 × 100), and ~0.03 % (=24,331 × 0.2/18,460,000 × 100) for the estimated escapement (24,331 = 4,140 hatchery + 20,191 wild fish). The total N_e estimate was 27,666 (=(114,000 + 24,331) × 0.2) and the N_e/N ratio was ~0.15 % (10^{-3}) after combining the hatchery broodstock and escapement. Current wild salmon escapement (20,191) was ~0.1 % of the 18,460,000 fish returned to the Kitami region in September–November, which was an underestimate because it did not account for naturally spawning salmon in major enhanced rivers (SFFRI, pers. comm.). Therefore, the N_e/N ratio estimate (~0.15 %) should be an underestimate. In addition, the N_e/N ratio might be much higher in hatcheries, as reported for rainbow trout *O.*

mykiss (~90 %) [26]. A large number of adults (114,000) were used for artificial fertilization and the sex ratio was adjusted to ~0.5. However, even if this was the case, the N_e/N ratio of the chum salmon in the Kitami region was very small (~10^{-2}). It may be much smaller than those of Chinook salmon, coho salmon *O. kisutch*, steelhead trout *O. mykiss*, and Atlantic salmon *Salmo salar*, reflecting differences in early life history and the large census population size of chum salmon. To estimate the N_e/N ratio correctly, the effective population size of returned fish and the effective number of breeders in hatcheries needs to be estimated based on genetic monitoring [28].

5. GENETIC DIVERSITY

Preceding studies on chum salmon using different genetic markers have provided valuable information on population sustainability. Sato et al. [29] examined the genetic population structure of chum salmon in the Pacific Rim using the mtDNA control region for 2,154 individuals from 47 rivers in 15 populations from Japan, Korea, Russia, and North America, and found that haplotype diversity was highest in the Japanese populations. Beacham et al. [30] surveyed genetic variation at 14 microsatellite loci in 3,602 individuals from 26 Japanese chum salmon populations, one population from Kamchatka, and three populations from North America, and found that the Japanese populations are more genetically diverse at each of the 14 microsatellites surveyed than that of North American populations in the Yukon River and British Columbia. Seeb et al. [31] surveyed the allelic variation of 10,458 chum salmon at 60 loci of single-nucleotide polymorphisms (SNPs) in 114 locations from Japan (16 sites), Korea, Russia, and North America, and showed that neither marker type suggested that the Japanese populations are inherently less diverse than those from North America, as suggested by the current SNP panel. Genetic diversity indices of Japanese chum salmon for the three markers were generally higher or similar in comparison with Korean, Russian, and North American chum salmon populations (Fig. 4).

6. POPULATION STRUCTURE

Analyses of the mtDNA control region and SNPs coincidently show that the Japanese and Korean populations form a single, distinct cluster [29, 31].

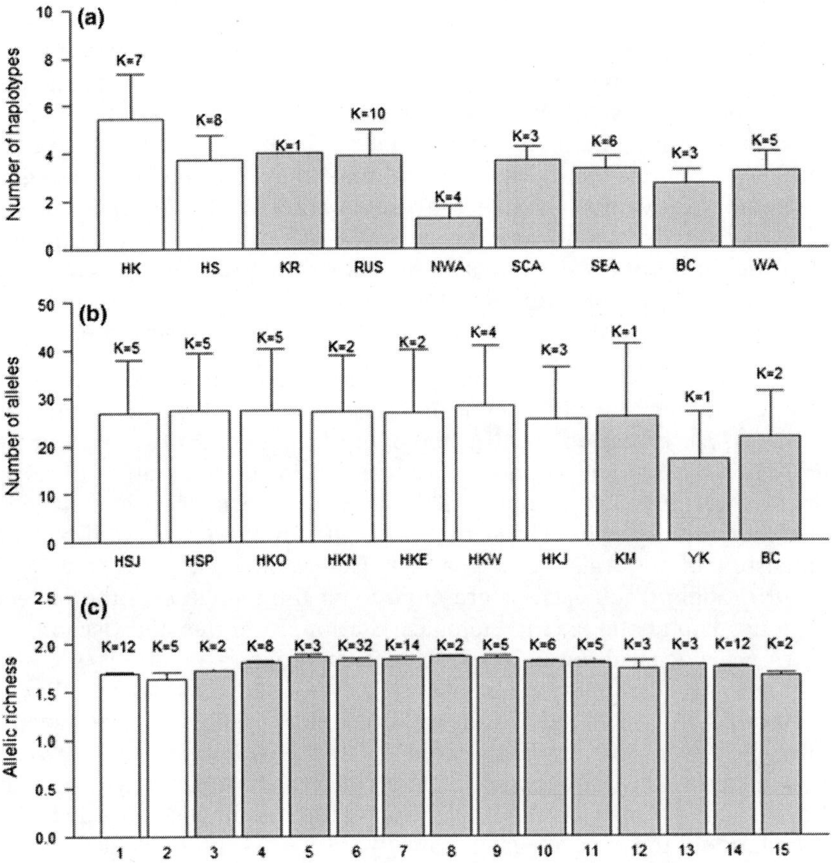

Figure. 4 Genetic diversity of chum salmon in the Pacific Rim. *K* refers to the number of rivers where samples were taken (*white*: Japan, *grey*: other countries). **A** Number of haplotypes for the mitochondrial DNA control region (data from [29], *n* = 2,154). *Error bars* show standard deviations. *HK* Hokkaido, *HS* Honshu, *KR* Korea, *RUS* Russia, *NWA* northwest Alaska and Alaskan peninsula, *SCA* south- Alaska, *SEA* southeast Alaska, *BC* British Columbia, *WA* Washington. **B** Number of alleles for 14 microsatellite DNA loci (data from [30], *n* = 3,602). *HSJ* Honshu, Sea of Japan coast), *HSP* Honshu, Pacific coast, *HKO* Hokkaido, Sea of Okhotsk coast, *HKN* Hokkaido, Nemuro Strait, *HKE* Hokkaido, eastern Pacific coast, *HKW* Hokkaido, western Pacific coast, *HKJ* Hokkaido, Sea of Japan coast, *KM* Kamchatka, *YK* Yukon, *BC* British Columbia. **C** Number of alleles for 60 single-nucleotide polymorphic loci (data from [31], *n* = 10,458). *1*, Hokkaido; *2*, Honshu; *3*, Amur; *4*, Kamchatka; *5*, Alaska; *6*, western Alaska; *7*, Yukon; *8*, Yukon2; *9*, Bristol Bay; *10*, north Alaskan peninsula; *11*, southern Alaskan peninsula; *12*, Kodiak Island; *13*, Susitna River; *14*, Alaska Bay; *15*, Washington

Analyses of msDNA [30] found five clusters of populations in Hokkaido (Sea of Okhotsk coast, Nemuro Strait, eastern Pacific coast, western Pacific coast, and Sea of Japan coast) and two in Honshu (Pacific coast and Sea of Japan coast). Honshu Pacific coast had the most distinct cluster. KSEPA and Japanese salmon scientists suggest that the December–January return (late run) represents the native return and that the hatchery-dominated September–November return represents transplants. An analysis of the mtDNA control region from four sites in Hokkaido (n = 323) [32] showed that the Yurappu River chum salmon has two populations: the late native-population spawning in December and the early mixed-population introduced from other river populations in October and November.

To infer the fine population structure including early and late-run populations, genetic differentiation was estimated using a common scale in terms of F_{ST} based on mtDNA haplotype frequencies of 15 samples (n = 273) [29, 32, 33] and msDNA allele frequencies of 26 samples (n = 2,670) at 14 loci [30]. The samples [29, 30, 33] were collected in hatcheries (hatchery-released fish), and the Yurappu sample of Yokotani et al. [32] consisted of naturally-spawning fish (Kaeriyama M. pers. comm.). The msDNA allele frequencies were cited from the web site of the Molecular Genetics Lab at the Pacific Biological Station, Fisheries and Ocean Canada (http://www.pac.dfo-mpo.gc.ca/science/facilities-installations/pbs-sbp/mgl-lgm/data-donnees/index-eng.html). We used an empirical Bayesian method [34] with the software POPDIF 1.0 (http://popgene.co-site.jp/software_eng.html) to avoid a biased estimation of pairwise F_{ST} (e.g., negative F_{ST} values), which is a concern for species with high gene flow such as marine fish. The maximum likelihood global F_{ST} over populations, the rate of gene flow (=1/F_{ST} − 1), and the means of the posterior distributions of pairwise F_{ST} were estimated based on haplotype and allele frequencies.

Global F_{ST} was estimated as 0.0433 ± 0.0130 for the mtDNA haplotype frequencies. The rate of gene flow was 22.1 ± 6.9, suggesting that 22 females in terms of the effective population size migrate between rivers per generation. The unweighted pair group method with arithmetic mean (UPGMA) dendrogram of pairwise F_{ST} showed a distinct cluster in the samples taken in September and October. Honshu Pacific coast samples also formed a cluster. In contrast, samples taken in November and December were included in a different cluster for except Tokushibetsu, collected in September (Fig. 5). The November samples were distinct from the October samples at Tokoro and Yurappu, suggesting that it was influenced by transplantations from the Chitose hatchery. The pairwise F_{ST} value [95 % credible interval] between October and November was small at 0.0160 [0.0013, 0.0482] for Tokoro and 0.0125 [0.0017, 0.0352] for Yurappu, respectively. F_{ST} between November and December for Yurappu was 0.0290 [0.0050, 0.0708] and between October and December was

0.0529 [0.0158, 0.1072]. The genetic difference was greater for the later run. However, the posterior F_{ST} distributions overlapped, clearly showing that early- and late-run populations were not genetically distinct with high gene flow (Fig. 6). The difference between Yurappu (December) and Tokushibetsu (September) was also similar at 0.0349 [0.0084, 0.0750]. It was unclear why the early sample from Tokushibetsu was close to the late one from Yurappu.

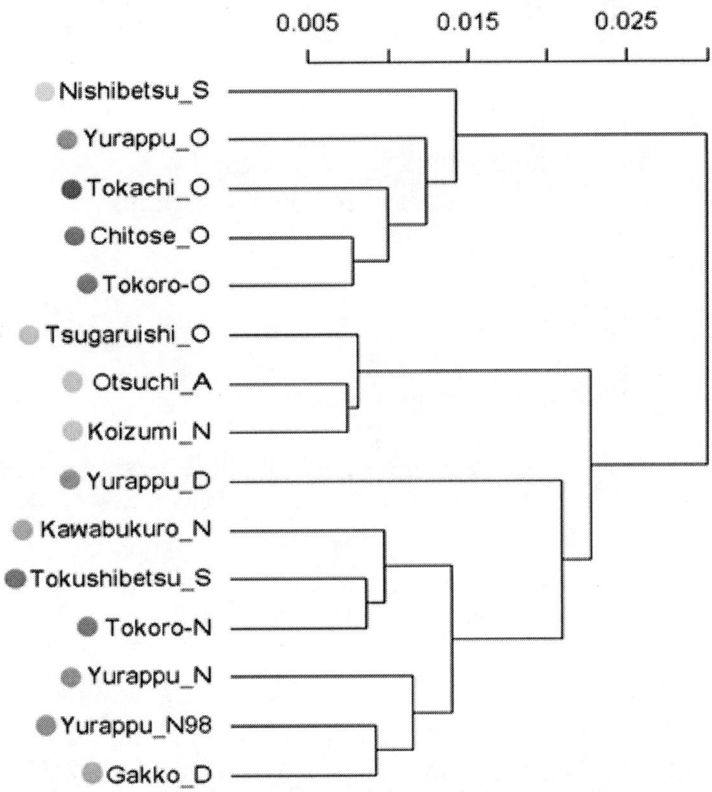

Figure. 5 Genetic differentiation of chum salmon based on the posterior mean of pairwise F_{ST}, inferred from mitochondrial DNA control region haplotype frequencies of 15 geographical samples (n = 273) (data from Sato et al. [29, 33], Yokotani et al. [32]). Hokkaido, Sea of Okhotsk Coast (*red circle*); Hokkaido, Nemuro Strait (*yellow circle*); Hokkaido, eastern Pacific coast (*brown circle*); Hokkaido, western Pacific coast (*dark green circle*); Hokkaido, Sea of Japan coast (*purple circle*); Honshu, Pacific coast (*light green circle*); Honshu, Sea of Japan coast (*blue circle*). *S*, September; *O*, October; *N*, November; *D*, December; *A*, April

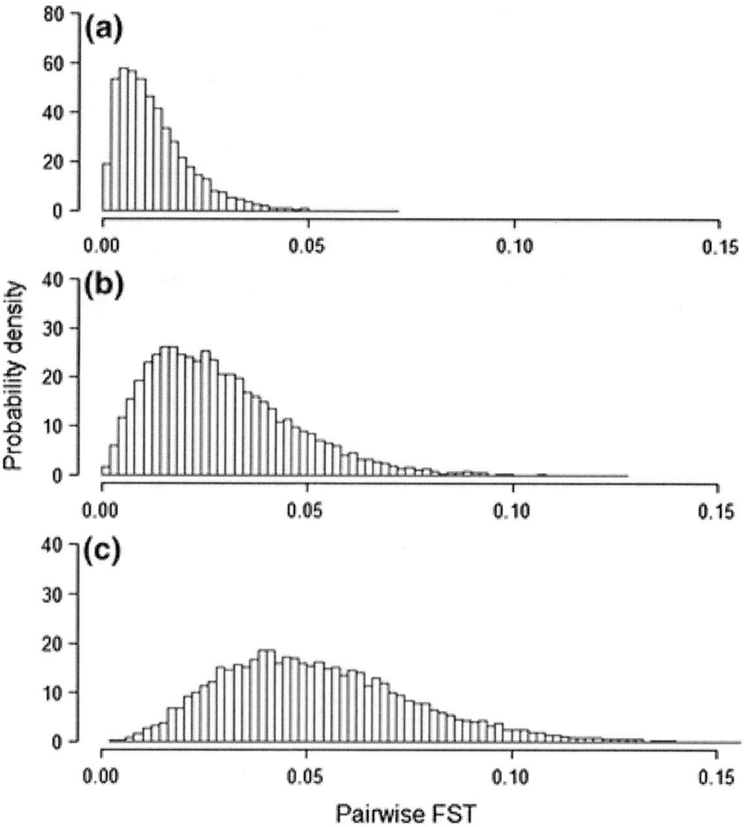

Figure. 6 Posterior distributions of pairwise F_{ST} between early- and late-run samples of chum salmon in the Yurappu River inferred from the mitochondrial DNA control region haplotype frequencies; **A** October vs. November, **B** November vs. December, **C** October vs. December

The global F_{ST} was much smaller (0.0069 ± 0.0002) than that from the mtDNA haplotype frequencies based on the msDNA allele frequencies. The rate of gene flow was 155.3 ± 4.7. The UPGMA dendrogram of pairwise F_{ST} showed three clusters. Honshu Pacific coast samples formed a distinct cluster, coinciding with the results from the mtDNA analysis. Five geographical samples (except from the eastern Pacific coast of Hokkaido) were located closely in the large middle cluster, but some were nested. The Honshu Sea of Japan coast formed one cluster with Teshio and Chitose, but Kawabukuro was close to the western Pacific coast of Hokkaido, and Uono was separated from the cluster. Hokkaido samples from the Sea of Okhotsk, Nemuro Strait, the western Pacific coast, and the Sea of Japan coast were generally located nearby. In contrast, samples from the Pacific coast of Hokkaido (Kushiro and Tokachi) were not close and Horonai was different from other Sea of Okhotsk samples (Fig. 7), though the F_{ST} value between Horonai and Tokoro was very small at 0.0036 [0.0011, 0.0082]. Our results generally coincided with the results of Beacham et al. [30], who found seven regional populations (five on Hokkaido and two on Honshu) within Japan, and learned that the Honshu Pacific coast populations were the most distinct based on Cavalli–Sforza and Edwards chord distance. However, the genetic differentiation was very small and population structure was not rigid in terms of the posterior mean of F_{ST}. The samples were all hatchery fish, as mentioned above. The nested population structure in Hokkaido may have been influenced by the transplantation of hatchery fish. No sample set was available from the same site including both early and late runs, so it was not possible to detect the genetic difference between the two runs. Beacham et al. [30] analyzed the original temporal samples from three populations within 1 year: (1) Gakko River, October and December samples; (2) Yurappu River, September and November samples; and (3) Chitose River, October and December samples. Their F_{ST} values were 0.015 ± 0.002 (Gakko), 0.008 ± 0.002 (Yurappu), and 0.010 ± 0.001 (Chitose). Beacham et al. [30] suggested that transplantation history could influence the observed temporal variation within these populations.

In the Yurap River, our F_{ST} value based on mtDNA between October and December (0.0529 [0.0158, 0.1072]) was 6.6 times larger than that between September and November based on msDNA (0.008 ± 0.002) [30]. Our estimate of the rate of global gene flow for msDNA (155.3 ± 4.7) was seven times larger than that of mtDNA (22.1 ± 6.9), suggesting that females return to natal rivers and spawning sites, but males stray into non-natal rivers to extend the chance of reproduction. Straying of males should serve to enhance genetic diversity in populations and adaptability to environmental changes. It is unclear if straying reflects a failure to return home or a decision to spawn elsewhere, but in either case, straying has been essential for the persistence and distribution of salmon [35]. Chinook salmon stray more (mean population range 0.11–34.6 %) than coho salmon (0.08–0.94 %), ocean-type Chinook (5.2–18.6 %) stray more than stream-type Chinook (0.11–10 %), and Chinook salmon (0.90–54.9 %) stray more than steelhead (0.30–2.3 %) [36].These patterns are largely the result of species-specific behavioral and endocrine factors during juvenile life stages [36]. The level of straying in hatchery chum salmon has not been assessed in Japan [2]. Further research is needed on chum salmon and pink salmon *O. gorbuscha*, which exhibit similar juvenile life histories (individuals typically migrate soon after emergence) [36]. The distributions of the mean pairwise F_{ST} (0.0037 ± 0.0007) and the mean pairwise rate of gene flow (278 ± 49), estimated from msDNA allele frequencies of 26 samples (n = 2,670), suggested that straying is typical in chum salmon (Fig. 8). The global F_{ST} values for masu salmon *O. masou* were 0.016–0.032 within the Uryu River system in Hokkaido [37], which were estimated from genotypes of 10 msDNA loci using the same software POPDIF. These point estimates were 2.3–4.6 times larger than our estimate of chum salmon (0.0069), which reflect the smaller rate of straying of masu salmon.

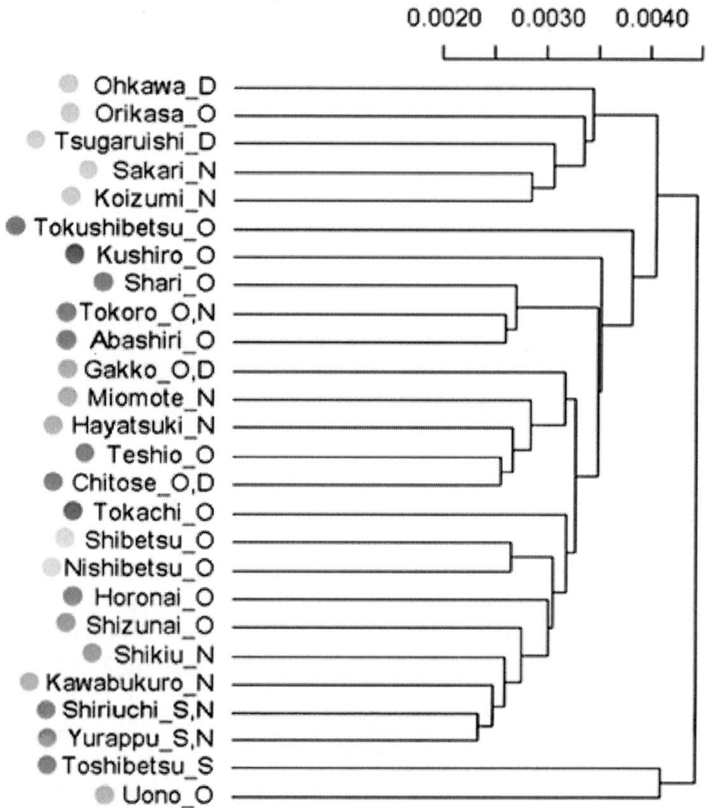

Figure. 7 Genetic differentiation of chum salmon based on the posterior mean of pairwise F_{ST}inferred from microsatellite DNA allele frequencies of 26 geographical samples (n = 2,670) at 14 loci [30] (data source: http://www.pac.dfo-mpo.gc.ca/science/facilities-installations/pbs-sbp/mgl-lgm/data-donnees/index-eng.html). Hokkaido, Sea of Okhotsk coast (*red circle*); Hokkaido, Nemuro Strait (*yellow circle*); Hokkaido, eastern Pacific coast (*brown circle*); Hokkaido, western Pacific coast (*dark green circle*); Hokkaido, Sea of Japan coast (*purple circle*); Honshu, Pacific coast (*light green circle*); Honshu, Sea of Japan Coast (*blue circle*). *S*, September; *O*, October; *N*, November; *D*, December

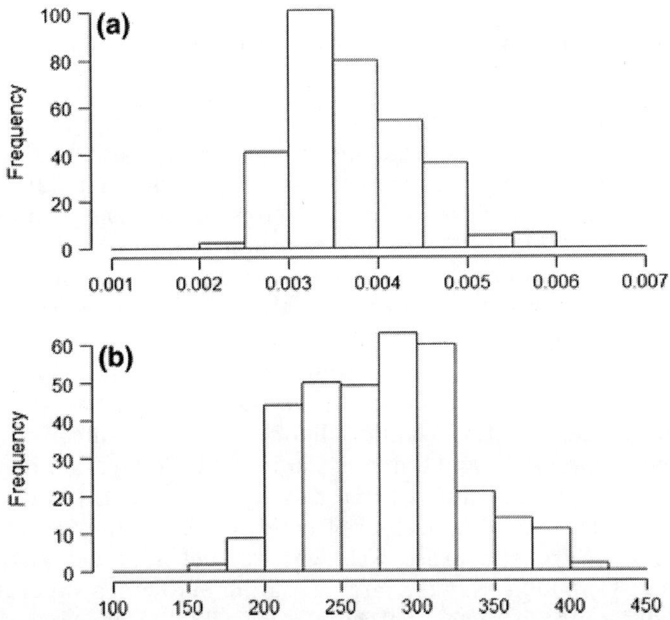

Figure. 8 Distributions of **A** pairwise F_{ST} and **B** the rate of gene flow (= the effective population size per generation) of chum salmon between rivers in Japan, inferred from microsatellite DNA allele frequencies in 26 geographical samples (n = 2,670) at 14 loci [19]

7. GENETIC EFFECTS OF HATCHERY RELEASE AND ROLES OF INTEGRATED HATCHERY MANAGEMENT

Japanese populations had a similar number of haplotypes as in the Pacific Rim (Fig. 4a). However, the number of haplotypes ranged from two to seven in a population, and the loss of rare haplotypes was observed in all populations. Similar reductions in the numbers of haplotypes were observed in other foreign populations. In contrast, the number of microsatellite alleles was very high at 26.8 ± 11.3 (range 4.8–73.9) [30]. A large number of parents have been used since the mid-1970s when the stock enhancement program became effective. Currently, an annual total of 1.27 million fish (3.2 % of returned fish in 2012) are collected for hatchery broodstock from about 75 rivers in Hokkaido [2]. The large number of parent fish used for artificial fertilization may have contributed to maintaining the high number of microsatellite alleles (26.8 ± 11.3) [30]. However, it is unclear if long-term hatchery release is responsible for the loss of rare haplotypes, as the reproductive ecology of females might also be a factor.

Fish generally remain in a chum salmon hatchery (from eggs to release) for ~6 months. Survival rates in hatcheries are very high at >90 %. In contrast, survival rates should be very low in the wild, particularly during the early life stages. The rearing process in hatcheries should relax natural selection until release. In addition, eggs and sperm are taken artificially, so parent fish do not experience natural spawning. Carrying out this artificial fertilization process for more than 10 generations might reduce the natural spawning behavior performance of hatchery and wild fish. Such artificial selection could cause a reduction in the fitness of hatchery fish in the wild. Evidence for a reduction in reproductive success (RS) of hatchery fish has accumulated for Atlantic salmon [38, 39], steelhead trout [40, 41, 42], coho salmon [43], and Chinook salmon [44]. A Bayesian meta-analysis for the steelhead data showed that the average level of relative RS (RRS) implied the reproductive decline of hatchery fish, but RRS varies greatly depending on genetic cross, sex, year of release, and environmental conditions [45]. Reflecting the large variation in RRS in a super population, no RS difference between hatchery and wild fish has been found for steelhead trout [46], brown trout *Salmo trutta* [47], or coho salmon [48]. If the ratio of hatchery fish with lower RS is substantial in the stocked population, the total number of progeny recruited in the stocked population could be reduced in the next generation [49]. The carryover effect of reduced RS in the next generation, as estimated in a steelhead population [50], should amplify the reduction in recruitment.

The mechanisms responsible for reduced RS of hatchery fish are unknown [40]. The most likely cause is thought to be selection pressures associated with domestication in captivity [51]. Björnsson et al. [52] suggested that it is important not to lose sight of the fact that when handling expression data in endocrinology, it is the circulating hormones that stimulate target tissues. Gene expression mirrors the physiological activity of an organism. Highly significant differences in transcription levels have been found between farmed and wild Atlantic salmon [53]. A significant correlation between the domesticated/wild-type gene expression ratio and the transgenic/wild-type gene expression ratio for coho salmon was found [54]. Transgenesis of the growth hormone gene causes inferior reproductive behavior and reduces fertilization success of Atlantic salmon [55]. These findings provide an important suggestion that enhanced growth in captivity causes a reduction in reproductive performance of hatchery fish. Actually, hatchery-reared steelhead are almost ubiquitously released as yearlings, whereas wild steelhead typically spend two or more years in freshwater before migrating to sea [56]. It is crucial to discover the essential genetic cascades related to survival and reproductive traits, and to identify the main trigger genes, which will determine the mechanisms of the fitness reduction caused by captive rearing [57]. The results will contribute to the development of juvenile production techniques that help to conserve natural populations.

The duration of hatchery rearing for chum salmon (~6 months) is about half that for steelhead. The growth of hatchery-reared chum salmon is also similar to wild fish and is not enhanced in the hatchery, as in steelhead and farmed Atlantic salmon. The effect of domestication might be smaller in chum salmon than in other salmon, except in pink salmon. No significant difference in RS between hatchery and wild chum salmon has been found [58]. The RS of wild and hatchery chum salmon has not been evaluated in Japan, so it is unknown if hatchery fish have lower RS in the wild and if releases reduce the RS of the population. It should be noted that DNA parentage assignment for the precise estimation of RS requires a complete sampling of the parent and progenies for hatchery and wild fish [59]. The number of returns and return rates in the Kitami region have continued to increase in the last two decades (Fig. 1b), showing that reductions in the survival rate of hatchery-released fish are unlikely. However, a consistent decreasing trend in the number of returns and the return rate since about 2005 in other regions in Hokkaido clearly indicates a survival decline of hatchery-released fish. Mizuno et al. [60] suggested the need for improving juvenile quality in the areas where recent returns have been decreasing in order to recover the chum salmon stocks.

Efforts to enhance the early-run population have been encouraged since the early 1980s, and resulted in a change in run timing [2], as found in coho salmon [48]. Miyakoshi et al. [2] demonstrated the long-term change (1970–2010) in the distribution of adult chum salmon captured in enhanced rivers in Hokkaido. There were both early- and late-run populations of chum salmon in Hokkaido until the early 1980s, and the numbers of fish in the two runs were almost the same (see Fig. 8 in Miyakoshi et al. [2]). The late-run population had almost disappeared by the late 1990s. The late-run population has continued to decrease because adults captured late in the season for several years in the 1980s were not used for hatchery fertilization, but instead were simply sold. Hatchery managers have recently become aware that such extreme intentional selection is undesirable and have made an effort to escape the late-run fish, as mentioned above. The late-run population has been recovering gradually, suggesting the effectiveness of such escapement in enhanced rivers (Fig. 8 in Miyakoshi et al. [2]). The run-timing distribution for 1970–1975 should reflect that of the natural (original) populations of chum salmon in Hokkaido, because that period was the beginning of the increase in returned fish due to success of the hatchery program. The very small population differentiation of chum salmon was evaluated by neutral genetic markers. The F_{ST} analysis evaluates differences in allele or haplotype frequencies and might not reflect the genetic characteristics of the indigenous demes, but the present run-timing distribution suggests that indigenous individuals remain in many rivers even with hatcheries. Therefore, escapement in enhanced and non-enhanced rivers throughout the run season should enable the recovery of the natural (original) genetic component. Further efforts for the escapement of both early- and late-run fish are key to the recovery and maintenance of natural populations of chum salmon.

The escapement of hatchery [H] fish into the wild [W] population allows for the reproduction between genetic crosses of [HH], [HW], and [WW], which creates three types of wild fish, denoted as W [HH], W [HW], and W [WW]. There are also three types of hatchery fish, such as H [HH], H [HW] and H [WW]. Substantial introgression of hatchery-released fish into a wild population has been observed, even after several generations [17]. Considering long-term intensive hatchery release, almost all chum salmon returning to the Kitami region may be hatchery descendants. The increasing trend in return rate may suggest that hatchery-wild integrated reproduction acts to compensate for domestication selection caused by captive rearing and artificial fertilization in hatcheries. The Sea of Okhotsk region is considered as a genetic management unit, and the genetic difference between hatchery (early-run) and wild fish (late-run) is very small (F_{ST} = 0.008–0.010) [30]. This coincides with the concept of the genetically-integrated hatchery program [10]. In the successful salmon stock enhancement in southeast Alaska, hatcheries are located away from significant wild stocks [61], regarded as a genetically-segregated hatchery program [10]. However, the contribution of hatchery-origin chum salmon is 78 % in southeast Alaska [61], showing that any hatchery program is an integrated one within enhanced regions. Under integrated management, the idea of zoning management for enhanced and non-enhanced rivers in Hokkaido [13, 14] is useful both for conservation and for the production of chum salmon. Spawning habitats suitable for the reproduction of chum salmon are estimated to comprise 891 km in the Kitami region, including the Shiretoko area (Urabe H, SFFRI, unpublished data). It is crucial to allow escapements throughout the run season in order to sustain the habitat capacity of enhanced and non-enhanced rivers. Establishing river-specific escapement goals in enhanced rivers is needed. Marking and genetic monitoring will contribute to an evaluation of the effectiveness of the integrated hatchery program. Furthermore, the restoration of rivers, including dams [13, 14], increases spawning habitat capacity.

8. CONCLUDING REMARKS

The number of returns and return rates of hatchery fish have continued to increase in the past two decades in the Sea of Okhotsk region, supported by the favorable ocean and climate conditions. The increasing trend in return rates suggests that a survival reduction of hatchery-released fish in the Kitami chum salmon population does not occur. On the other hand, a consistent decreasing trend in the number of returns and the return rate since about 2005 in other regions in Hokkaido clearly indicates a survival decline of hatchery-released fish. However, it is unknown if hatchery fish have lower RS in the wild and whether releases reduce the RS of the stocked population. It is crucial to determine the mechanisms of the fitness reduction caused by captive rearing. New approaches are needed to discover the essential genetic cascades related to the survival and reproductive traits and to identify the main trigger genes, which will contribute to developing juvenile production techniques aimed at conserving natural populations.

The genetic diversity of Japanese chum salmon is higher or similar compared with other populations in the Pacific Rim. However, the number of haplotypes ranged from two to seven in a population, and a loss of rare haplotypes was observed in all populations. In contrast, the number of microsatellite alleles was very high at 26.8 ± 11.3 (range 4.8–73.9). A tentative estimate of N_e/N ratio for the Kitami hatchery program was >0.15 % under the present high fishing pressure. The integrated hatchery-wild reproduction system, realized by a large number of broodstock in hatcheries and escapement into enhanced and non-enhanced rivers, maintains the present genetic diversity. However, it is unclear if long-term hatchery release is responsible for the loss of rare haplotypes, which might be caused by the reproductive ecology of females. Six regional populations (four on Hokkaido and two on Honshu) are assumed to occur within Japan, which suggests the management units of Japanese chum salmon. However, genetic differentiation is very low, and the population structure is not rigid. The nested population structure in Hokkaido might reflect the transplantation history of hatchery fish. The large gene flow suggests that straying, particularly by males, is a chum salmon strategy to maintain genetic diversity. Early and late-run populations are not genetically distinct in terms of F_{ST}.

Intensive hatchery enhancement for the early run populations has changed the distribution of run timing for the chum salmon populations in Hokkaido. Recent escapement practices of late-run fish have gradually recovered the late-run populations. Hatchery releases serve to maintain the genetic composition of the early run, whereas escapement practices recover the late run, which contributes to maintaining the genetic characteristics of the natural (original) populations. Further efforts for escapement of both early- and late-run fish are the key to recovering and maintaining natural populations of chum salmon. Surplus spawning habitat capacity exists in the Kitami region including the Shiretoko area. The challenge is to extend present escapement practices to other enhanced rivers with allowable escapements for habitat capacity. The restoration of rivers and dams increases spawning habitat capacity. A region-wide integrated hatchery management plan with river-specific escapement goals in enhanced rivers will recover the natural populations and contribute to sustainable chum salmon fisheries in the future. New research is needed to evaluate the effectiveness of the integrated hatchery program, which will contribute to the minimization of ecological and genetic risks for hatchery fish.

ACKNOWLEDGMENTS

I am indebted to Y. Miyakoshi (SFFRI), who provided data for Figs. 1 and 3 with valuable information. I thank M. Kaeriyama (Hokkaido University) and M. Nagata (SFFRI) for constructive comments on the earlier version of the manuscript. I also thank H. Urabe and Y. Sasaki (SFFRI), K. Fukuda (HFFCA), N. Masukawa, and O. Ishizuka (KSEPA) for providing useful materials. This study was supported by a JSPS Grant-in-Aid for Scientific Research (B) 25280006.

REFERENCES

1. Kaeriyama M (1999) Hatchery programmes and stock management of salmonid populations in Japan. In: Howell BR, Moksness E, Svåsand T (eds) Stock enhancement and sea ranching. Blackwell, Oxford, pp 153–167
2. Miyakoshi Y, Nagata M, Kitada S, Kaeriyama M (2013) Historical and current hatchery programs and management of chum salmon in Hokkaido, northern Japan. Rev Fish Sci 21:469–479
3. Beamish RJ, Bouillon DR (1993) Pacific salmon production trends in relation to climate. Can J Fish Aquat Sci 50:1002–1016
4. Kaeriyama M, Seo H, Kudo H (2009) Trends in run size and carrying capacity of Pacific salmon in the North Pacific Ocean. N Pac Anadr Fish Comm Bull 5:293–302
5. Morita K, Saito T, Miyakoshi Y, Fukuwaka M, Nagasawa T, Kaeriyama M (2006) A review of Pacific salmon hatchery programmes on Hokkaido Island, Japan. ICES J Mar Sci 63:1353–1363
6. Miyakoshi Y, Urave H, Saneyoshi H, Aoyama T, Sakamoto H, Ando D, Kasugai K, Mishima Y, Takada M, Nagata M (2012) The occurrence and run timing of naturally spawning chum in northern Japan. Environ Biol Fish 94:197–206
7. Hilborn R (1992) Hatcheries and the future of salmon in the northwest. Fisheries 17:5–8
8. Waples RS (1999) Dispelling some myths about hatcheries. Fisheries 24:12–21

9. Brannon EL, Amend DF, Cronin MA, Lannan JE, LaPatra S, McNeil WJ, Noble RE, Smith CE, Talbot AJ, Wedemeyer GA, Westers H (2004) The controversy about salmon hatcheries. Fisheries 29:12–31
10. Mobrand L, Barr J, Blankenship HL, Campton DE, Evelyn TTP, Flagg TA, Mahnken CVW, Seeb LW, Seidel RR, Smoker WW (2005) Hatchery reform in Washington state: principles and emerging issues. Fisheries 30:11–23
11. Waples RS, Drake J (2004) Risk/benefit considerations for marine stock enhancement: a Pacific salmon perspective. In: Leber KM, Kitada S, Svåsand T, Blankenship HL (eds) Stock enhancement and sea ranching, 2nd edn. Blackwell, Oxford, pp 260–306
12. Kaeriyama M, Edpalina RR (2004) Evaluation of the biological interaction between wild and hatchery population for sustainable fisheries management of Pacific salmon. In: Leber KM, Kitada S, Svåsand T, Blankenship HL (eds) Stock enhancement and sea ranching, 2nd edn. Blackwell, Oxford, pp 247–259
13. Kaeriyama M, Seo H, Kudo H, Nagata M (2012) Perspectives on wild and hatchery salmon interactions at sea, potential climate effects of Japanese chum salmon at the need for sustainable salmon fisheries management reform in Japan. Environ Biol Fish 94:165–177
14. Nagata M, Miyakoshi Y, Urabe H, Fujiwara M, Sasaki Y, Kasugai K, Torao M, Ando D, Kaeriyama M (2012) An overview of salmon enhancement and the need to manage and monitor natural spawning in Hokkaido, Japan. Environ Biol Fish 94:311–323
15. Kitada S, Kishino H (2006) Lessons learned from Japanese marine finfish stock enhancement programmes. Fish Res 80:101–112
16. Kitada S, Shishidou H, Sugaya T, Kitakado T, Hamasaki K, Kishino H (2009) Genetic effects of the long-term stock enhancement programs. Aquaculture 290:69–79
17. Nakajima K, Kitada S, Habara Y, Sano S, Yokoyama E, Sugaya T, Iwamoto A, Kishino H, Hamasaki K (2014) Genetic effects of marine stock enhancement: a case study based on the highly piscivorous Japanese Spanish mackerel. Can J Fish Aquat Sci. doi:10.1139/cjfas-2013-0418.21
18. Morita K, Takahashi S, Ohkuma K, Nagasawa T (2013) Estimation of the proportion of wild chum salmon *Oncorhynchus keta* in Japanese hatchery rivers (in Japanese with English abstract). Nippon Suisan Gakkaishi 79:206–213
19. Frankham R (1995) Effective population size/adult population size ratios in wildlife: a review. Genet Res 66:95–107
20. Hedgecock D (1994) Does variance in reproductive success limit effective population sizes of marine organisms? In: Beaumont AR (ed) Genetics and evolution of aquatic organisms. Chapman and Hall, London, pp 122–134
21. Grant WS, Waples RS (2000) Spatial and temporal scales of genetic variability in marine and anadromous species: implications for fisheries oceanography. In: Harrison PJ, Parsons TR (eds) Fisheries oceanography. Blackwell Science, Cambridge, pp 61–93
22. Hutchinson WF, van Oosterhout C, Rogers SI, Carvalho GR (2003) Temporal analysis of archived samples indicates marked genetic changes in declining North Sea cod (*Gadus morhua*). Proc Royal Soc B-Biol Sci 270:2125–2132
23. Hauser L, Adcock GJ, Smith PJ, Bernal Ramirez JH, Carvalho GR (2002) Loss of microsatellite diversity and low effective population size in an overexploited population of New Zealand snapper (*Pagrus auratus*). Proc Natl Acad Sci USA 99:11742–11747
24. Turner TF, Wares JP, Gold JR (2002) Genetic effective size is three orders of magnitude smaller than adult census size in an abundant, estuarine-dependant marine fish (*Sciaenops ocellatus*). Genetics 162:1329–1339
25. Cushing DH (1995) Population production and regulation in the sea: a fisheries perspective. Cambridge University Press, Cambridge
26. Bartley D, Bagley M, Gall G, Bently B (1992) Use of linkage disequilibrium data to estimate effective size of hatchery and natural fish populations. Cons Biol 6:365–375
27. Allendorf FW, Bayles D, Bottom D, Currens KP, Frissell CA, Hankin D, Lichatowich JA, Nehlsen W, Trotter PC, Williams TH (1997) Prioritizing Pacific salmon stocks for conservation. Cons Biol 11:140–152
28. Schwartz MK, Luikart G, Waples RS (2006) Genetic monitoring as a promising tool for conservation and management. Trends Ecol Evol 22:25–33

29. Sato S, Kojima H, Ando J, Ando H, Wilmot RL, Seeb LW, Efremov V, LeClair L, Buchholz W, Jin DH, Urawa S, Kaeriyama M, Urano A, Abe S (2004) Genetic population structure of chum salmon in the Pacific Rim inferred from mitochondrial DNA sequence variation. Environ Biol Fish 69:37–50
30. Beacham TD, Sato S, Urawa S, Lei KD, Wetklo M (2008) Population structure and stock identification of chum salmon *Oncorhynchus keta* from Japan determined by microsatellite DNA variation. Fish Sci 74:983–994
31. Seeb LW, Templin WD, Sato S, Abe S, Warheit K, Park JY, Seeb JE (2011) Single nucleotide polymorphisms across a species' range: implications for conservation studies of Pacific salmon. Molec Ecol Res 11(Suppl 1):195–217
32. Yokotani R, Azuma N, Kudo H, Abe S, Kaeriyama M (2009) Genetic differentiation between early- and late-run populations of chum salmon (*Oncorhynchus keta*) naturally spawned in the Yurappu River inferred from mitochondrial DNA analysis. Fish Genet Breed Sci 39:1–8
33. Sato S, Ando J, Ando H, Urawa S, Urano A, Abe S (2001) Genetic variation among Japanese populations of chum salmon inferred from the nucleotide sequences of the mitochondrial DNA control region. Zool Sci 18:99–106
34. Kitada S, Kitakado T, Kishino H (2007) Empirical Bayes inference of F_{ST} and its distribution in the genome. Genetics 177:861–873
35. Quinn TP (2005) The behavior and ecology of Pacific salmon and trout. University of Washington Press, Seattle
36. Westley PAH, Quinn TP, Dittman AH (2013) Rates of straying by hatchery-produced Pacific salmon (*Oncorhynchus* spp.) and steelhead (*Oncorhynchus mykiss*) differ among species, life history types, and populations. Can J Fish Aquat Sci 70:735–746
37. Kitanishi S, Yamamoto Y, Edo K, Higashi S (2012) Influences of habitat fragmentation by damming on the genetic structure of masu salmon populations in Hokkaido, Japan. Conserv Genet 13:1017–1026
38. Fleming I, Hindar K, Mjølnerød IB, Jonsson B, Balstad T, Lamberg A (2000) Lifetime success and interactions of farm salmon invading a native population. Proc R Soc Lond B 267:1517–1523
39. McGinnity P, Prodöhl P, Ferguson A, Hynes R, Ó Maoiléidigh N, Baker N, Cotter D, O'Hea B, Cooke D, Rogan G, Taggart J, Cross T (2003) Fitness reduction and potential extinction of wild populations of Atlantic salmon *Salmo salar* as a result of interactions with escaped farm salmon. Proc R Soc Lond B 270:2443–2450
40. Araki H, Cooper B, Blouin MS (2007) Genetic effects of captive breeding cause a rapid, cumulative fitness decline in the wild. Science 318:100–103
41. Berntson EA, Carmichael RW, Flesher MW, Ward EJ, Moran P (2011) Diminished reproductive success of steelhead from a hatchery supplementation program (Little Sheep Creek, Imnaha Basin, Oregon). Trans Am Fish Soc 140:685–698
42. Christie MR, Marine ML, French RA, Blouin MS (2012) Genetic adaptation to captivity can occur in a single generation. PNAS 109:238–242
43. Thérault V, Moyer G, Jackson LS, Blouin MS, Banks MA (2011) Reduced reproductive success of hatchery coho salmon in the wild: insights into most likely mechanisms. Molec Ecol 20:1860–1869
44. Anderson JH, Faulds PL, Atlas WI, Quinn TP (2013) Reproductive success of captively bred and naturally spawned Chinook salmon colonizing newly accessible habitat. Evol Appl 6:165–179
45. Kitada S, Kishino H, Hamasaki K (2011) Bias and significance of relative reproductive success estimates based on steelhead trout (*Oncorhynchus mykiss*) data: a Bayesian metaanalysis. Can J Fish Aquat Sci 68:1827–1835
46. Araki H, Ardren WR, Olsen E, Cooper B, Blouin MS (2007) Reproductive success of captive-bred steelhead trout in the wild: evaluation of three hatchery programs in the Hood River. Conserv Biol 21:181–190
47. Dannewitz J, Petersson E, Dahl J, Prestegaard T, Löf AC, Järvi T (2004) Reproductive success of hatchery-produced and wild-born brown trout in an experimental stream. J Appl Ecol 41(2):355–364

48. Ford MJ, Fuss H, Boelts B, LaHood E, Hard J, Miller J (2006) Changes in run timing and natural smolt production in a naturally spawning coho salmon (*Oncorhynchus kisutch*) population after 60 years of intensive hatchery supplementation. Can J Fish Aquat Sci 63:2343–2355
49. Chilcote MW, Goodson KW, Falcy MR (2011) Reduced recruitment performance in natural populations of anadromous salmonids associated with hatchery-reared fish. Can J Fish Aquat Sci 68:511–522
50. Araki H, Cooper B, Blouin MS (2009) Carry-over effect of captive breeding reduces reproductive fitness of wild-born descendants in the wild. Biol Lett 5:621–624
51. Araki H, Berejikian BA, Ford MJ, Blouin MS (2008) Fitness of hatchery-reared salmonids in the wild. Evol Appl 1:342–355
52. Björnsson BT, Stefansson SO, McCormick SD (2011) Environmental endocrinology of salmon smoltification. Gen Comp Endocrinol 170:290–298
53. Roberge C, Normandeau E, Einum S, Guderley H, Bernatchez L (2008) Genetic consequences of interbreeding between farmed and wild Atlantic salmon: insights from the transcriptome. Mol Ecol 17:314–324
54. Devlin RH, Sakhrani D, Tymchuk WE, Rise ML, Goh B (2009) Domestication and growth hormone transgenesis cause similar changes in gene expression in coho salmon (*Oncorhynchus kisutch*). Proc Natl Acad Sci USA 106(9):3047–3052
55. Moreau DTR, Conway C, Fleming IA (2011) Reproductive performance of alternative male phenotypes of growth hormone transgenic Atlantic salmon (*Salmo salar*). Evol Appl 4:736–748
56. Berejikian BA, Larsen DA, Swanson P, Moore ME, Tatara CP, Gale WL, Pasley CR, Beckman BR (2012) Development of natural growth regimes for hatchery-reared steelhead to reduce residualism, fitness loss, and negative ecological interactions. Environ Biol Fish 94:29–44
57. Nakamichi R, Kishino H, Kitada S (2013) A novel method to identify key factors of the gene regulatory network behind salmonids reproductive behavior using directed graphical modeling. N Pac Anadr Fish Comm Tec Rep 9:50–53
58. Berejikian BA, Van Doornik DM, Scheurer JA, Bush R (2009) Reproductive behavior and relative reproductive success of natural- and hatchery-origin Hood Canal summer chum salmon (*Oncorhynchus keta*). Can J Fish Aquat Sci 66:781–789
59. Marshall TC, Slate J, Kruuk LEB, Pemberton JM (1998) Statistical confidence for likelihood-based paternity inference in natural populations. Molec Ecol 7:639–655
60. Mizuno S, Nakajima M, Naito K, Koyama T, Saneyoshi H, Kobayashi M, Koide N, Ueda H (2010) Physiological impacts of high rearing density on chum salmon *Oncorhynchus keta* fry. Aquaculture Sci 58:387–399
61. Heard WR (2012) Overview of salmon stock enhancement in southeast Alaska and compatibility with maintenance of hatchery and wild stocks. Environ Biol Fish 94:273–283

CHAPTER 17

Effect of global warming on the life history and population dynamics of Japanese chum salmon

Masahide Kaeriyama, Hyunju Seo, Yu-xue Qin

Hokkaido University, N15W8, Kita-ku, Sapporo, Hokkaido 060-0815, Japan

ABSTRACT

We have reviewed the effects of long-term climatic/oceanic conditions on the growth, survival, production dynamics, and distribution of Hokkaido chum salmon *Oncorhynchus keta* in Japan during the period 1945–2005 using path analysis, back-calculation, and scale analyses, and applied a prediction method based on the SRES-A1B scenario of the intergovernmental panel on climate change. The populations of Hokkaido chum salmon were found to have had high growth rates at age 1 year since the late 1980s. Path analysis indicated that the growth at age 1 year in the Okhotsk Sea was directly affected by warm sea surface temperature associated with global warming, with the increased growth at age 1 year resulting in higher rates of survival and large population sizes. Predictions on the global warming effects on the chum salmon were (1) decreased carrying capacity and distribution area, (2) occurrence of a strong density-dependent effect, and (3) loss of migration route to the Sea of Okhotsk, especially for Hokkaido chum salmon. We have also outlined the future challenges of establishing a sustainable conservation management scheme for salmon that include adaptive management and precautionary principles, as well as conservation of natural spawning populations and recovery of natural river ecosystems in Japan despite the warming climate.

Keywords: Climate change, Global warming, Chum, salmon, Life history, Population, dynamics, Risk management

1. INTRODUCTION

The Pacific salmon (*Oncorhynchus* spp.) is a diadromous fish that begins its life in freshwater but subsequently migrates between seawater and freshwater during its life cycle. Salmon that migrate to the sea at an early stage of development have a large area of ocean distribution and high biomass, suggesting that such fish evolved the ability to migrate seaward during the Ice Age as a means to utilize the high biological productivity of the ocean and retained the ability to return to freshwater for reproduction [1, 2, 3]. Of all the salmon species, the chum salmon *O. keta* and the pink salmon *O. gorbuscha* show the highest degree of evolution, spending a large part of their life histories in the sea [4, 5, 6]. In the study reported here, we focused mainly on the chum salmon, discussing its life history and dynamics of biomass with respect to global warming and climate change.

The number of chum salmon returning to Japanese waters significantly increased up to the late 1970s but thereafter decreased and stabilized at around 40–50 million, after peaking at over 80 million in the late 1990s (Fig. 1). Since 2000 there has been a significant decrease in their number, primarily due to the collapse of the diatom–euphausiid food web by coccolith blooms in the eastern Bering Sea ecosystem caused by the 1997/1998 Super El Niño [7, 8]. The disappearance of euphausiids caused a significant population decrease in marine biomass, including that of the Japanese salmon, the Bristol Bay sockeye salmon *O. nerka*, and seabirds, such as the short-tailed shearwater *Puffinus tenuirostris* [7, 9].

2. LONG-TERM CLIMATE CHANGE AND BIOMASS DYNAMICS OF SALMON

Salmon biomass is often associated with long-term climate change. The North Pacific Ocean has undergone a climate change within a cycle of >10 years, as measured by atmosphere and ocean coordinates, and these parameters have been indexed using the Pacific decadal oscillation (PDO). The PDO index is defined as the leading principal component of the North

Pacific monthly sea surface temperature (SST) variability (poleward of 20 N for the period extending from 1900 up to the present); a positive value indicates that the SST is lower than the average at the central area of the North Pacific and the SST is higher at the eastern area of the North Pacific and near the equator [10]. According to the PDO, the amount of harvested salmon from the northern Pacific increases over time when the PDO is positive and decreases when the PDO is negative (Fig. 2) [11].

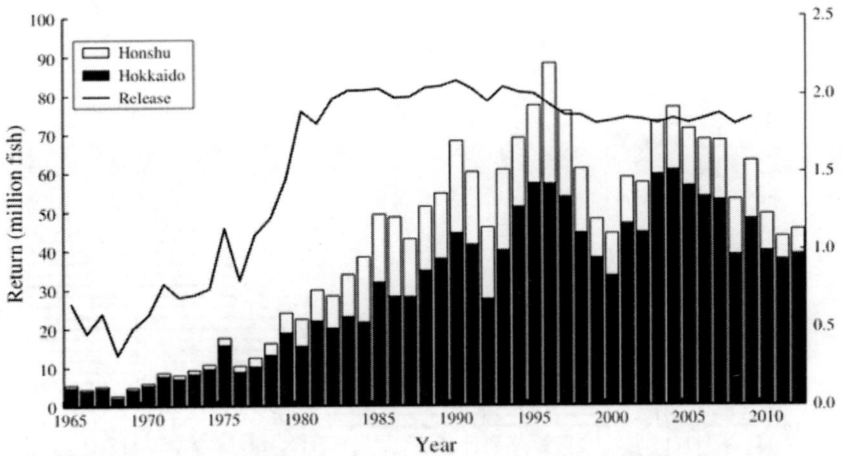

Figure. 1 Annual changes in the numbers of returning adult and released juvenile chum salmon in Japan

The PDO is closely related to the strength of the Aleutian Low in the winter, with its value being positive when the pressure is strong and negative when the pressure is weak. On the eastern side of the northern Pacific, including the Bering Sea, westerlies intensify when the strength of the Aleutian Low increases, thereby resulting in strengthening of the Subarctic Current and subsequently of the Alaska Current and Alaska Coast Current. Strengthening of the coastal current causes increases in eddy currents, churns the coastal water through synergic interactions with the tide, and carries a large amount of nutrients (e.g., nitrogen and phosphorus) up from the sea bottom to the surface layer. These nutrients are then distributed to the Gulf of Alaska through the Alaska Current and the Alaskan Gyre. At the same time, winter storms become active, vertical mixing of seawater becomes greater, and nutrients at the bottom layer diffuse to the surface layer through a strengthened Aleutian Low in the eastern Bering Sea. Furthermore, a counter-clockwise eddy of low pressure brings in the wet and warm air and warm water from the south, increasing the water temperature at the surface layer of the Gulf of Alaska and the eastern Bering Sea. The Aleutian Low has a significant impact on biological productivity in the Gulf of Alaska and the Bering Sea. Storms in the winter thus improve the carrying capacity of salmon [12, 13, 14].

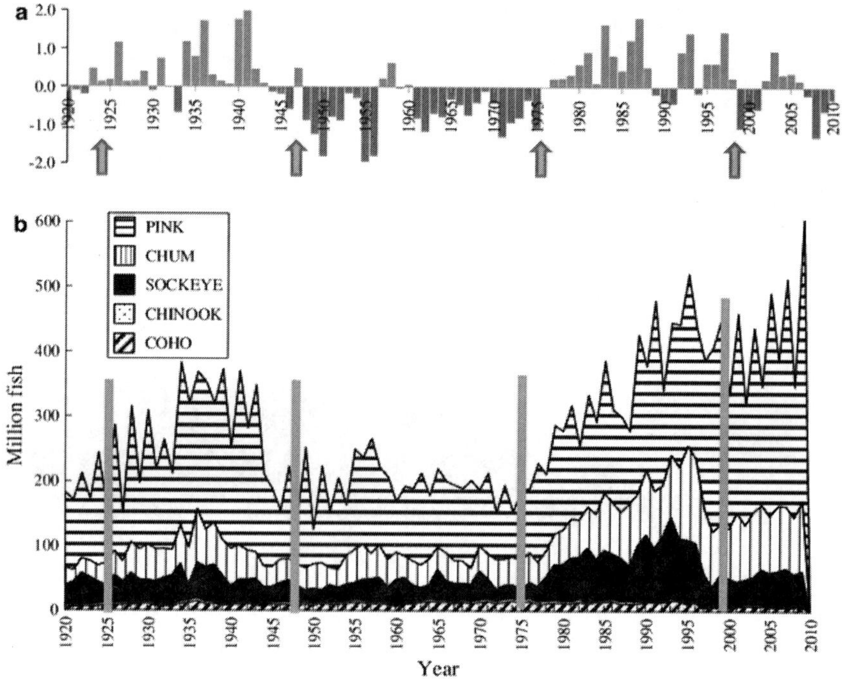

Figure. 2 Annual changes in the Pacific decadal oscillation (PDO) (**a**) and catch of Pacific salmon (**b**) in the North Pacific Ocean. *Arrows, bars* Years of climate regime shifts (Kaeriyama et al. [11])

The mid- to long-term shift in the PDO from positive to negative or vice versa is called a 'climate regime shift' [15]. The PDO showed a strong tendency to be positive between 1975/1976 and the late 1990s, and the salmon population increased during this time. However, it has tended to be negative since 1997/1998, the year when the 'Super El Niño', the greatest El Nino of the 20th century, developed and initiated the next regime shift [7, 9, 16]. The overall amount of harvested salmon in the entire North Pacific has remained at a high level due to the significant increase in pink salmon and chum salmon in Russia (Fig. 3), but the carrying capacity has definitely passed its peak. The number of sockeye salmon has decreased since this peak, and a similar trend has been observed in the Japanese chum salmon [11]. The same trends were also observed in salmon in the

U.S. states of Washington and Oregon [17]. Overall, the population of southern chum salmon has decreased since 1997/1998, shifting away from the 'good-old-days' that extended from the late 1970s until the shift in climate regime in the late 1990s.

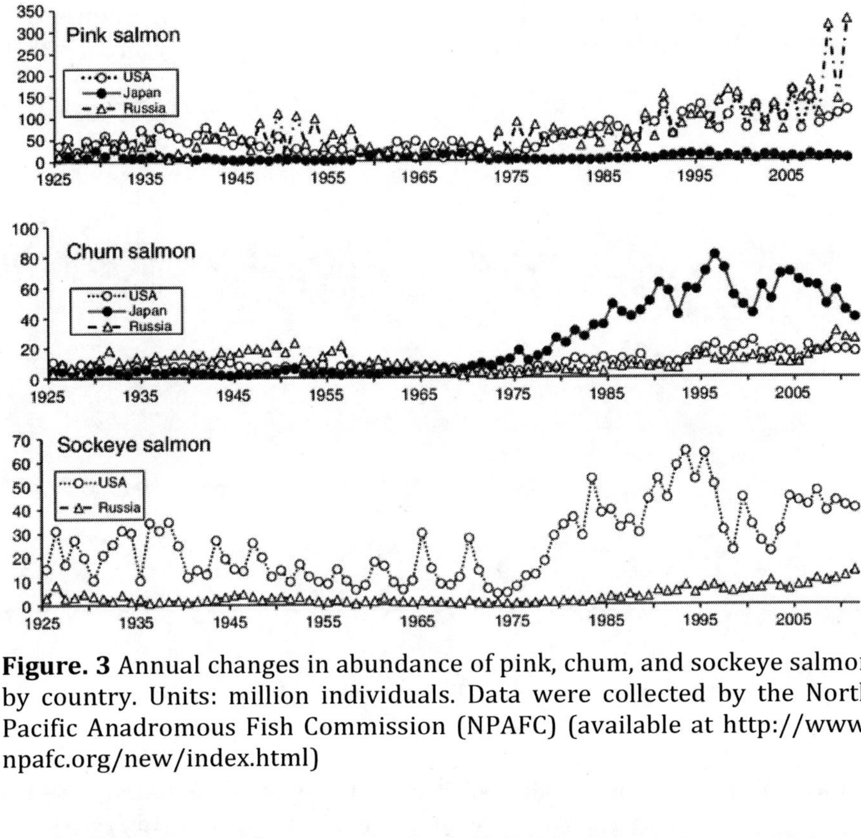

Figure. 3 Annual changes in abundance of pink, chum, and sockeye salmon by country. Units: million individuals. Data were collected by the North Pacific Anadromous Fish Commission (NPAFC) (available at http://www.npafc.org/new/index.html)

3. GLOBAL WARMING AND BIOMASS DYNAMICS OF SALMON

Explanations for the mechanism and causes of the recent trend in global warming are largely divided into the rebound theory from the Little Ice Age [18] and the theory of greenhouse gas effects stemming from anthropologic activities [Fourth Assessment Report of the Intergovernmental Panel on Climate Change (IPCC); available at: http://www.ipcc.ch/publications_and_data/ar4/syr/en/contents.html]. A discussion of these proposed mechanisms are beyond the scope of our study, but global warming as currently observed is a fact. The SST of the ocean increased from 1.2 to 1.7 °C/100 years in the Sea of Japan, and from 0.6 to 1.3 °C/100 years in the Pacific Ocean around Japan between 1900 and 2011 (Japan Meteorological Agency; available at: http://www.data.kishou.go.jp/kaiyou/shindan/a_1/japan_warm/japan_warm.html). Observations of temporal changes in the SST in the Sea of Japan per season since 1980 have revealed that the greatest impact of global warming occurs in the autumn (rate of increase in temperature: 0.066 °C/year, R^2 = 0.551) and the least in the spring (0.029 °C/year, R^2 = 0.181; Fig. 4; Table 1).

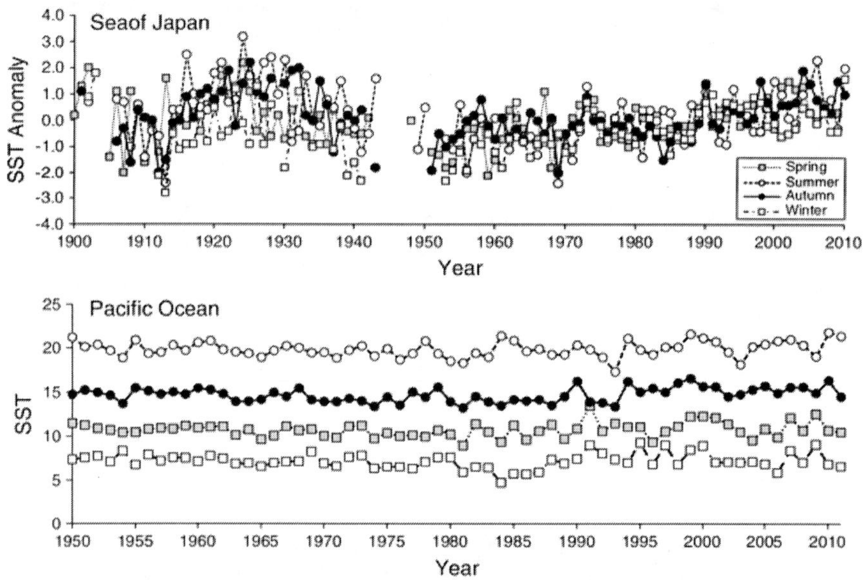

Figure. 4 Temporal changes in seasonal sea surface temperature (SST) in the Sea of Japan since 1900 and Pacific Ocean since 1950. Data are from the Japan Meteorological Agency (available at: http://www.data.kishou.go.jp/kaiyou/shindan/a_1/japan_warm/japan_warm.html)

For better or for worse, global warming is affecting the numbers of salmon. Chum salmon populations in Hokkaido, which migrated en masse in the 1990s, are suspected to have been positively affected by this trend. Salmon populations are considered to suffer from severe mortality immediately after their seaward migration and during the first winter in the ocean. The mortality of Japanese chum salmon released from the hatchery, however, is significantly less during their seaward migration than that of wild fish because they are reared and released from the hatchery and migrate to the sea with a larger body size. After living along the coast for a few months, the Japanese chum salmon spends its time and grows in the Okhotsk Sea during the summer and autumn, then survives the winter in the Western Subarctic Gyre [19]. Consequently, the survival of Japanese chum salmon during their overwintering is significantly affected by its growth up to the autumn in the Okhotsk Sea.

Table 1 Relationship between year and sea surface temperature anomaly in the Sea of Japan and the Pacific Ocean around Japan since 1980[a]

Parameters describing relationship between year and SST anomaly	Winter	Spring	Summer	Autumn
Sea of Japan				
R^2	0.079	0.181	0.282	0.551
F	2.473	6.402	11.400	35.566
P	NS	0.017	0.002	<0.001
a	0.020	0.029	0.051	0.066
b	−40	−59	−102	−130
$(T = aY + b)$				
Pacific Ocean around Japan				
R^2	0.098	0.066	0.166	0.361
F	3.273	2.113	5.968	16.946
P	NS	NS	0.021	<0.001
a	0.038	0.028	0.047	0.062
b	−69	−45	−74	−108
$(T = aY + b)$				

SST, Sea surface temperature; NS, not significant

[a] Data were obtained from the Japan Meteorological Agency at: http://www.data.kishou.go.jp/kaiyou/shindan/a1/japan_warm/japan_warm.html

We have analyzed the scales of adult chum salmon returning to the Ishikari River between the 1940s and early 2000 and estimated first-year growth using the back-calculation method (Fig. 5) [20]. The results show that the first-year growth of salmon was significantly favorable from the 1990s to the early 2000s. A significant correlation was also observed between growth and survival rate. Taken together, these results suggest that the survival rate of chum salmon is higher when the growth rate during the period in coastal Japan and in the Okhotsk Sea is greater. On the other hand, first-year growth showed a positive correlation with the SST during the summer and autumn in the Okhotsk Sea and a negative correlation with the sea ice area in the winter [21]. A path analysis using the various climate change indices was used to analyze the mechanism behind the favorable growth for the Hokkaido chum salmon in the Okhotsk Sea between the 1990s and the early 2000s. A path model analysis is used to describe the directed dependencies among a set of variables, and this analysis includes models equivalent to any form of multiple regression analysis, factor analysis, canonical correlation analysis, and discriminant analysis [Wikipedia: the free encyclopedia; see 'Path analysis' (available at: http://en.wikipedia.org/wiki/Main_Page; accessed 9 Sept 2013)]. The results show that temperature anomalies at the earth's surface [surface air temperature (SAT)] directly affected SSTs during the summer and autumn. Furthermore, the SSTs directly affected the growth of juvenile salmon, increasing the survival rate and population size of Hokkaido chum salmon (Fig. 6). This leads to the conclusion that recent global warming has imparted a positive effect on the Hokkaido chum salmon. The results of the path model revealed that SAT also directly affected the Aleutian Low Pressure Index (ALPI) in the winter season, which in turn affected the PDO.

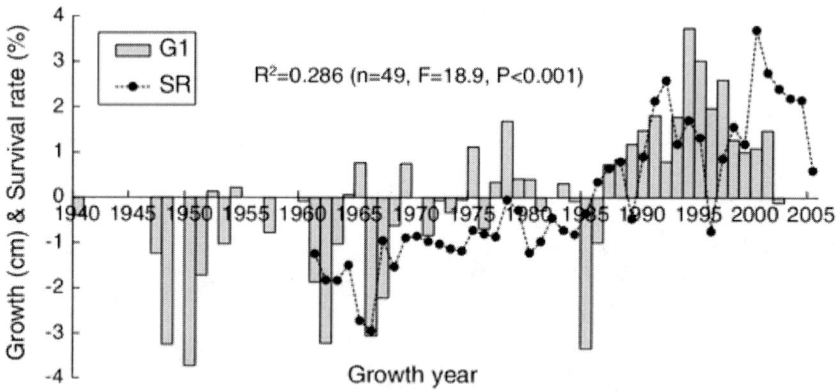

Figure. 5 Temporal changes in anomalies of growth at age 1 year (*G1*) and survival rate (*SR*) of Hokkaido chum salmon. Growth anomaly was calculated for adult chum salmon returning to the Ishikari River using scale analysis and back-calculation. Modified from Seo et al. [20]

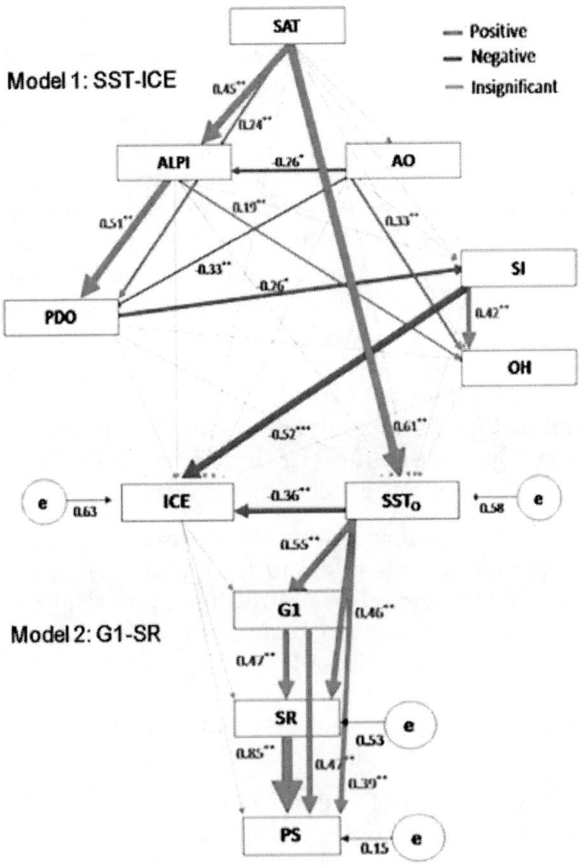

Figure. 6 Path model analyses relationships among global surface air temperature (*SAT*), Aleutian Low Pressure Index (*ALPI*), PDO, Arctic oscillation (*AO*), Siberian high (*SI*), Okhotsk high (*OH*), ice cover area (*ICE*), and summer SST (*SST*$_O$) in the Sea of Okhotsk (*Model 1*), and relationships among ICE, SST$_O$, growth during the first year of life history (*G1*), survival rate (*SR*), and population size (*PS*) of Hokkaido chum salmon (*Model 2*) between 1961 and 2002. The tested path model is indicated by *arrows*; significantly positive and negative direct paths are shown by *unbrokenred* and *blue arrows*, respectively. Significantly positive and negative indirect paths are shown by *broken red* and *blue arrows*, respectively. Non-significant paths are shows by *gray arrows*. To aid in the interpretation, the *width of lines* is approximately proportional to the strength of the relationships. *P < 0.05, **P < 0.01, ***P < 0.001. *Numeral values* and *asterisks* show correlation coefficients and probabilities, respectively. The residual variables are shown for ICE, SST$_O$, SR, and PS (*thin black arrows*). Modified from Seo et al. [20]

In contrast, in the Sea of Japan, the Tsushima Warm Current flows northward. Since the early 1990s, the current has been influenced by the warming climate (Fig. 7a) and, consequently, the abundance of early-run chum salmon returning to the coast of the Sea of Japan decreased during the years when the current was strong (Fig. 7b). The early-run abundance of chum returning by the end of September was found to differ between years when the Tsushima Warm Current was weak (2,407 ± 1,028 thousand individuals, n = 14) and when it was strong (1,446 ± 745 thousand individuals, n = 11) [analysis of variance (ANOVA) $F = 6.795$, $P = 0.016$], although analysis of the middle-run abundance returning in October and November indicated no difference between weak years (1,525 ± 802 thousand individuals, n = 14) and strong years (1,490 ± 754 thousand individuals, n = 11) (ANOVA $F = 0.012$, $P > 0.05$). One explanation may be that it is difficult for adult chum salmon to migrate for spawning in the sea area at high temperatures or those above 20 °C [22]. This trend can be interpreted as a negative effect of global warming.

By examining projected temperature isotherms and applying an optimum (8–12 °C; optimum growth rate and feeding habit), adaptable (5–13 °C; available habitat for swimming and feeding), and wintering temperature range (4–6 °C) of chum salmon based on the SRES-A1B scenario in the Fourth Assessment Report of the IPCC [11], we are capable of predicting the potential marine distribution of chum salmon populations in the North Pacific during the next years and possibly decades (Fig. 8). Based on the simulation, the following predictions can be made:

1) The distribution area of chum salmon has decreased in the Okhotsk Sea, the Gulf of Alaska, and even in the Bering Sea, thereby indicating a decrease in carrying capacity which may possibly have a density-dependent effect.

2) On the other hand, the distribution area is likely to expand to the Arctic Sea such as the Chukchi Sea.

3) The migration area in the winter may also shift from the Gulf of Alaska to the Western Subarctic Gyre.

4) In southern distribution areas, such as Japan, there is a risk that the early run of chum salmon lose their route of spawning migration into the coastal sea.

5) Juveniles will be not able to spend adequate time in the coastal sea and be forced to migrate offshore at an earlier developmental stage without attaining sufficient growth.

6) The salmon populations are at risk of losing their migration route to the Okhotsk Sea, and their survival greatly depends on whether they are able to find a new migration route along the Chishima Islands [11].

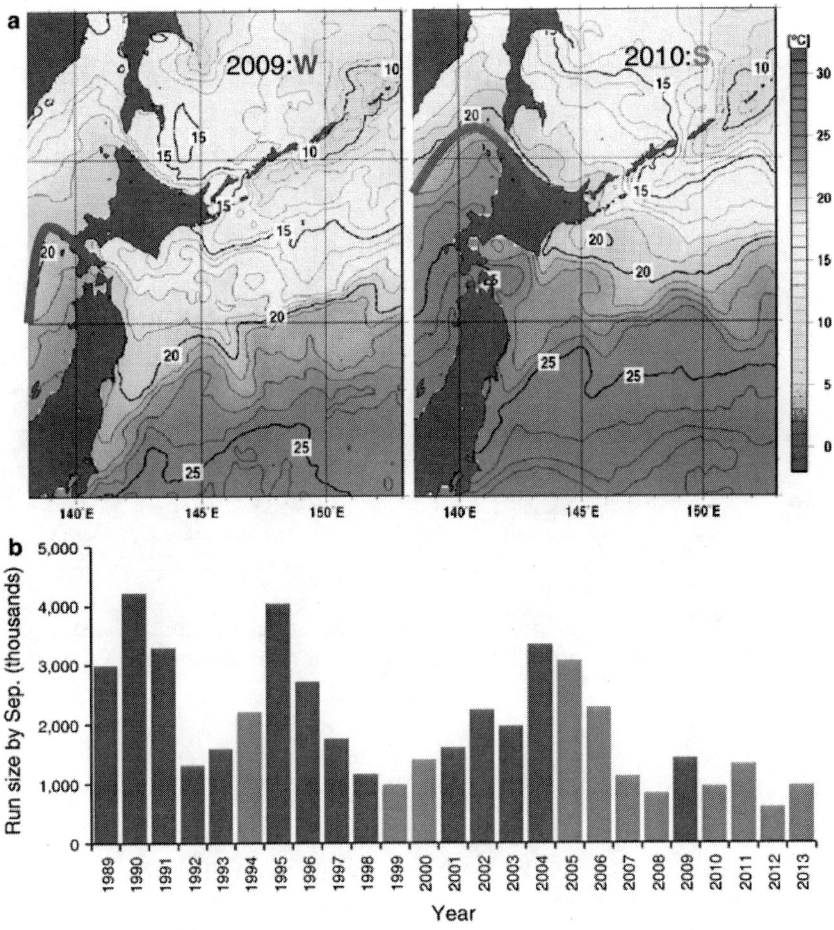

Figure. 7 Fluctuation in the strength of Tsushima warm current (TWC) compared to the run size of early-population chum salmon returning to the Sea of Japan coast of Hokkaido. **a** Mean SST isothermal diagrams around Japan in September 2009 (typical of a weak TWC) and 2010 (typical of a strong TWC). **b** Annual change in the run size of early-population chum salmon returning to the Sea of Japan coast of Hokkaido in years of a weak TWC (*blue bars*) and strong TWC (*red bars*), respectively. Mean SST isothermal diagrams are from the Japan Meteorological Agency (http://www.data.kishou.go.jp/kaiyou/db/hakodate/Monthly/sst_h.html). Modified from Qin and Kaeriyama [22]

Our predictions also correspond to the results generated by Mantua et al. [23] on the effects of the PDO.

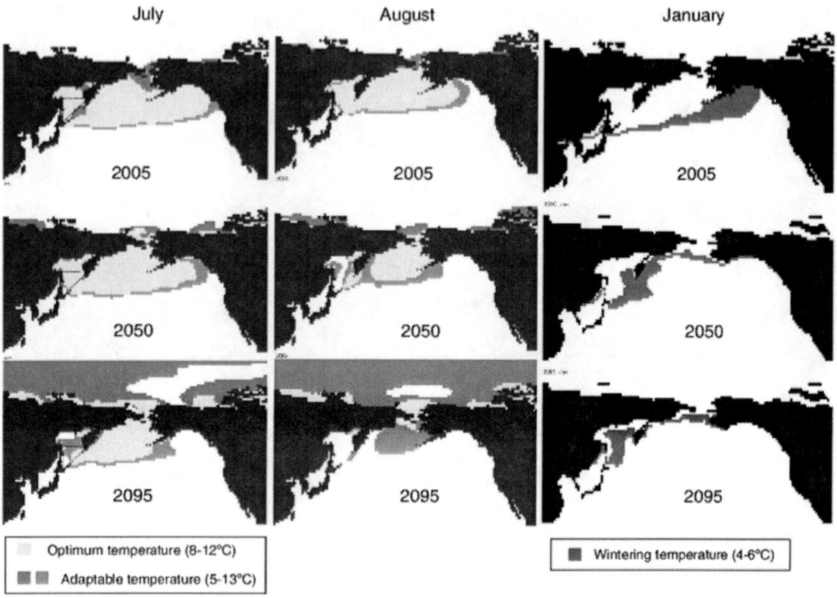

Figure. 8 Prediction of the global warming effect for chum salmon in the North Pacific based on the SRES-A1B scenario. Modified from Kaeriyama et al. [11]

4. SALMON AND ECOSYSTEM SERVICES

An ecosystem is a system of great uncertainty and reflects the results of an interaction between an inorganic system and an aggregate of organisms that have evolved over a long period. Its function has been maintained by biodiversity in terms of genes, species, and ecosystems. This biodiversity is a product of inter-organismal interactions, including prey–predator relationships, symbiosis and parasitic relationships, and competition over resources, such as food and living space. When any one component of a network disappears, biodiversity declines. Ecosystem services are defined as the benefits that humans have obtained from functions of ecosystems and organisms in the ecosystem. For example, the salmon, which is a keystone species of the North Pacific ecosystem, contributes to ecosystem services by transporting marine-derived nutrients (supporting service), maintaining biodiversity (regulating service), serving as a food source (provisioning service), and acting as an environmental and emotional educational factor, as well as a comfort (cultural service) factor, to the terrestrial ecosystems by returning to the birth river for reproduction.

Hatchery-produced chum salmon (hatchery salmon) exceeds 50 % of the total catch of chum salmon in the North Pacific [8]. The interaction between wild-origin (wild salmon) and hatchery salmon is a serious issue in conservation programs of Pacific salmon in the North Pacific ecosystems. Wild and hatchery chum salmon have been found to exhibit similar reproductive success, with both wild and hatchery males obtaining similar access to nesting females, and females of both types exhibiting similar breeding behaviors and duration of breeding periods [24]. In the Yurappu River, southern Hokkaido, the wild population of chum salmon reproduces naturally during the winter (from December to January) at the lower reach of the river [25], while many hatchery-derived salmon have been observed to reproduce naturally in the fall (from October to middle November) at the upper reach near by the hatchery. The wild population has a higher genetic endemism than the hatchery populations [26]. The genetic differentiation among hatchery populations is very weak despite high genetic diversity in Hokkaido [26, 27] because of excessive egg-transplantation effects [28]. In one study [29], carbon ($\delta^{13}C$) and nitrogen stable isotope ($\delta^{15}N$) analyses of adult chum salmon returning to the Yurappu River showed that wild salmon had significantly higher values than hatchery fishes. The $\delta^{13}C$ analysis showed that hatchery salmon varied more widely in $\delta^{13}C$ levels than wild salmon [29]. Chum salmon mainly feed in coastal habitats and tend to consume prey enriched in $\delta^{13}C$ relative to those found in offshore regions [11, 30, 31, 32]. Wild steelhead trout also foraged on more coastally derived sources, suggesting that they did not travel as far offshore as the hatchery fish [33]. Japanese chum salmon distribute widely throughout the Bering Sea [19]. They have been found to feed on diverse prey in the offshore Bering Sean are low in nutrients and to predominantly consume nekton in the coastal waters of the Aleutian Islands [11]. These observations suggest that wild chum salmon have a higher trophic level than hatchery salmon when at sea. Wild salmon will distribute throughout coastal waters and do not migrate as far offshore as hatchery fish in the Bering Sea, suggesting that wild chum salmon have a higher ecological niche and adaptability than hatchery salmon in the sea environment despite there being no significant difference in the reproductive success of wild and hatchery salmon in the river environment.

Kawasaki [34] pointed out that the greatest external causes of disturbance of the atmosphere–marine system and fishery resource system are global warming and overfishing, thereby raising the question of how these issues should be managed. This researcher continued the discussion by emphasizing that resource variation is a balance between an increase in the amount of available resources and the intensity of fishery. As such, he stated, mankind needs to escape from the concept of the maximum sustainable yield in which 'fishing' is an internal force—rather, mankind should treat the 'environment' as a noise and an external force in order to sustainably manage fishery resources. In this sense, there is a limit to business-oriented aquatic resources management schemes (e.g., the total allowable catch) at a population or species level. At the very least, fisheries resource management should be operated based on the marine ecosystem approach.

The marine ecosystem is a complex and dynamic system with a high degree of uncertainty. It is constantly disturbed by natural factors, such as climate change and the El Niño Southern oscillation, and by anthropogenic factors, such as global warming and overfishing. Artificial control of a marine ecosystem is unrealistic. The structure and functions of the ecosystems depend on biodiversity—specifically, the diversity of species and genes—that has been nurtured by the long history of the Earth and on the process of evolution. The decline in biodiversity observed today means that (some of) the components of a biological interaction are missing, causing a decline in the ecosystem services and simplification of the system. A classic example is the 'global warming' phenomenon.

5. FUTURE CHALLENGES

Intrinsically, marine organisms are sustainable resources for human and fishery industries. According to the statistics generated by the Food and Agriculture Organization of the United Nations [35], approximately 154 million tons of marine organisms are caught globally per year, of which 131 million tons (85 %) are consumed as food by humans. Of these 131 million tons of food, 63.6 million tons (41 %) are produced by aquafarming/aquaculture. The amount of harvested marine organisms that is naturally reproduced has declined since 2000 and is currently estimated to be 90 million tons. Consequently, fishery can no longer be considered as a sustainable resource management, and some marine organisms, such as eels (*Anguilla* spp.) and tuna (*Thunnus* spp.), are now classified as endangered due to overfishing [36, 37]. On the other hand, the amount produced by aquafarming around the world has exponentially increased despite various problems. For example, shrimp farms in Southeast Asia destroy coastal ecosystems, such as mangrove forests, and water pollution caused by large amounts of organic matters and antibiotics used in the farms is becoming a serious problem [38]. Compared to wild fish, high concentration of dioxins and PCBs have been found to accumulate in farmed Atlantic salmon *Salmo salar* [39].

These problems highlight the importance of implementing an ecosystem approach-based risk management scheme compounded by precautionary principles and adaptive management principles to sustainably conserve the aquatic ecosystems and organisms. Therefore, management should shift to the back-cast method that defines future visions and goals and constantly monitors the current situation. As such, management should reject the forecast method that accepts current conditions without defining a long-term vision and instead merely predicts the future [11].

Ecosystems in Japanese rivers have extensively deviated from their natural state and have become artificial. Conservation of biodiversity in ecosystems is an important and basic task. While the number of artificially hatched and stocked chum salmons has increased, that of wild salmon from natural reproduction has significantly decreased due to the degradation of river ecosystems [40]. Despite the small population size, wild chum salmon are reproducing naturally in about 60 rivers across Hokkaido [41]. As previously noted, these wild salmon have higher trophic levels and genetic-endemism than hatchery fish, as well as an extensive adaptability to environmental change.

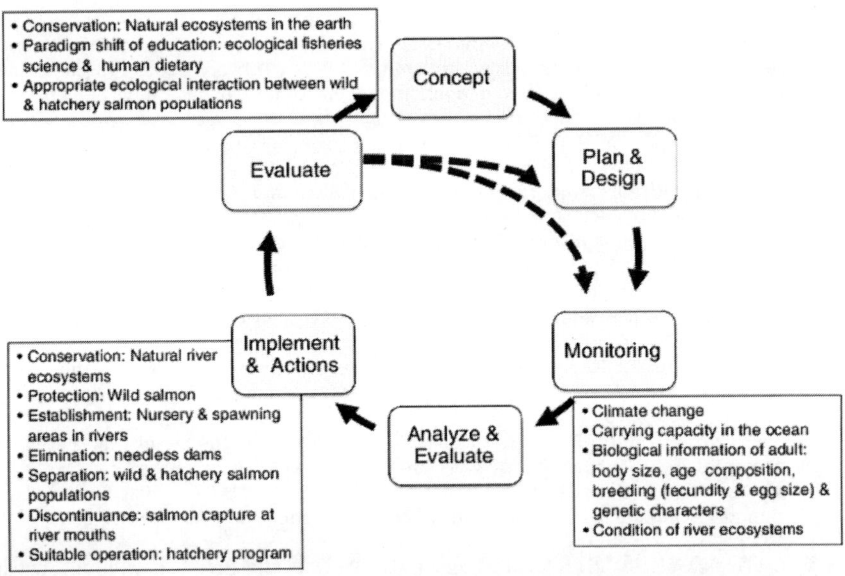

Figure. 9 Conceptual diagram showing the sustainable adaptive management of Pacific salmon. Modified from Kaeriyama et al. [11]

To address the threat of impending global warming, efforts should focus on rehabilitating and strengthening the resilience of wild fish to generate species that have naturally evolved by natural selection and which are highly adaptable to environmental changes. It is well understood that the restoration of river ecosystems in which fish can live is a basic requirement. Hence, the ecosystem-based approach would be best method for conserving not only the remaining wild salmon populations but also the biodiversity in the freshwater ecosystem in Japan. This action plan for sustainable conservation of Pacific salmon should be carried on by employing a sustainable adaptive management approach based on feedback control between monitoring (such as climate change, carrying capacity, breeding and genetic characters of salmon, and condition of river ecosystem) and actions (e.g., conservation and rehabilitation of natural riparian ecosystems, protection of wild salmon populations, and zoning between hatchery and wild salmon) under the conditions of a changing climate (Fig. 9) [11].

REFERENCES

1. Gross MR (1987) Evolution of diadromy in fishes. Am Fish Soc Symp 1:14–25
2. McDowall RM (1988) Diadromy in fishes. Croom Helm, London
3. Kaeriyama M (1994) Life history strategy of sockeye salmons-diversity and identity of life history pattern. In: Goto A et al (eds) Freshwater fish migrating between rivers and seas-life history and evolution. Tokai University Press, Tokyo, pp 101–113 (in Japanese)
4. Kaeriyama M (1996) Effects of population density and habitat environment on life history strategy and migration of juvenile sockeye (*Oncorhynchus nerka*) and chum salmon (*O. Keta*). Sci Rep Hokkaido Salmon Hatchery 50:101–111
5. Takasaki N, Park L, Kaeriyama M, Gharrett AJ, Okada N (1996) Characterization of species-specifically amplified SINEs in three salmonid species—chum salmon, pink salmon, and kokanee: the local environment of genome may be important for the generation a dominant source gene at a newly retroposed locus. J Mol Evol 42:103–116
6. Quinn TP (2005) The behaviour and ecology of Pacific salmon and trout. University of Washington Press, Seattle
7. Kruse GH (1998) Salmon run failures in 1997-1998: a link to anomalous ocean conditions? Alaska Fish Res Bull 5:55–63
8. Stabeno PJ, Bond NA, Kachel NB, Salo SA, Schumacher JD (2001) On the temporal variability of the physical environment over the south-eastern Bering Sea. Fish Oceanogr 10:81–98
9. Kaeriyama M (2003) Does regime shift have an impact on the life history and biomass dynamics of the Pacific salmon? Kaiyo Mon 392:127–132 (in Japanese)
10. Mantua NJ, Hare SR, Zhang Y, Wallace JM, Francis RC (1997) A Pacific interdecadal climate oscillation with impacts on salmon production. Bull Am Meteor Soc 78:1069–1079
11. Kaeriyama M, Seo H, Kudo H, Nagata M (2012) Perspectives on wild and hatchery salmon interactions at sea, potential climate effects on Japanese chum salmon, and the need for sustainable salmon fishery management reform in Japan. Environ Biol Fish 94:165–177
12. Hunt GL, Stabeno P, Walter G, Sinclair E, Brodeur RD, Napp JM, Bond NA (2002) Climate change and control of the southeastern Bering Sea pelagic ecosystem. Deep-Sea Res II 49:5821–5853
13. Hunt GL, McKinnell S (2006) Interplay between top-down, bottom-up, and wasp-waist control in marine ecosystems. Prog. Oceanogr 68:115–124
14. Drinkwater KF, Beaugrand G, Kaeriyama M, Kim S, Ottersen G, Perry RI, Pörtner HO, Polovina JJ, Takasuka A (2010) On the processes linking climate to ecosystem changes. J Mar Syst 79:374–388

15. Kawasaki T (1983) Why do some pelagic fishes have wide fluctuation in their numbers? FAO Fish Rep 291:1065–1080
16. Peterson WT, Schwing FB (2003) A new climate regime in northeast Pacific ecosystem. Geophys Res Lett 30:1896. doi:10.1029/2003GL017528
17. Irvine JR, Tompkins A, Saito T, Seong KB, Kim JK, Klovach N, Bartlett H, Volk E (2012) Pacific salmon status and abundance trends—2012 update. NPAFC Doc 1422:1–89
18. Akasofu S (2008) Know your global warming: to not be confused with false theories. Seibundo-Shinkosha, Tokyo (in Japanese)
19. Urawa S (2000) Migration route of Japanese chum salmon and future research topics. Natl Salmon Resour Cent News 5:3–9 (in Japanese)
20. Seo H, Kudo H, Kaeriyama M (2011) Long-term climate-related changes in somatic growth and population dynamics of Hokkaido chum salmon. Environ Biol Fish 90:131–142
21. Kaeriyama M, Yatsu A, Noto M, Saitoh S (2007) Spatial and temporal changes in the growth patterns and survival of Hokkaido chum salmon populations in 1970–2001. N Pac Anadr Fish Comm Bull 4:251–256
22. Qin Y, Kaeriyama M (2011) Recent production trends of chum salmon *Oncorhynchus keta* under conditions of warming climate. NPAFC Tech Rep 8:113–116
23. Abdul-Aziz OI, Mantua NJ, Myers KW (2011) Potential climate change impacts on thermal habitats of Pacific salmon (*Oncorhynchus* spp.) in the North Pacific Ocean and adjacent seas. Can J Fish Aquat Sci 68:1660–1680
24. Berejikian BA, Van Doornik DM, Scheurer JA, Bush R (2009) Reproductive behaviour and relative reproductive success of natural- and hatchery-origin Hood Canal summer chum salmon (*Oncorhynchus keta*). Can J Fish Aquat Sci 66:781–789
25. Kobayashi T, Abe S (1977) Studies on the Pacific salmon in the Yurappu River and Volcano Bay 2. On the migration and the growth of the fry during seaward migration and the return of marked adults. Sci Rep Hokkaido Salmon Htachery 31:1–11 (in Japanese with English abstract)
26. Yokotani R, Azuma N, Kudo H, Abe S, Kaeriyama M (2009) Genetic differentiation early- and late-run populations of chum salmon (*Oncorhynchus keta*) naturally spawned in the Yurappu River inferred from mitochondrial DNA analysis. Fish Genet Breed Sci 39:9–16
27. Sato S, Ando J, Ando H, Urawa S, Urano A, Abe S (2001) Genetic variation among Japanese populations of chum salmon inferred from the nucleotide sequences of the mitochondrial DNA control region. Zool Sci 18:99–106
28. Nagai A, Yamada A, Qin Y, Edpalina RR (2012) Genetic influence of hatchery program on chum salmon (*Oncorhynchus keta*) populations in the Japan Sea Coast, Honshu Island of Japan, inferred from the mitochondrial DNA sequence variation. Fish Genet Breed Sci 42:33–40 (in Japanese with English abstract)
29. Qin Y, Nagai A, Kudo H, Kaeriyama M (2013) Stable isotope comparison between presumably wild and hatchery chum salmon *Oncorhynchus keta* in the Yurappu River, Hokkaido, Japan. Nippon Suisan Gakkaishi 79:872–874 (in Japanese with English tables and figures)
30. McConnaughey T, McRoy CP (1979) Food-web structure and the fractionation of carbon isotopes in the Bering Sea. Mar Biol 53:257–262
31. Fry B, Sherr EB (1984) $\delta^{13}C$ measurements as indicators of carbon flow in marine and fresh-water ecosystems. Contrib Mar Sci 27:13–47
32. Hobson KA, Piatt JF, Pitocchelli J (1994) Using stable isotopes to determine seabird trophic relationships. J Anim Ecol 63:786–798
33. Quinn TP, Seamons TR, Johnson SP (2012) Stable isotopes of carbon and nitrogen indicate differences in marine ecology between wild and hatchery-produced steelhead. Trans Am Fish Soc 141:526–532
34. Kawasaki K (2007) Overview of regime shift—a new point of view on the earth system management. In: Kawasaki K et al (eds) Regime shift—climate change and biological resource management. Seizando, Tokyo, pp 1–20
35. Food and Agriculture Organization of the United Nations (FAO) (2012) The state of world fisheries and aquaculture 2012. FAO, Geneva

36. Myers RA, Worm B (2003) Rapid worldwide depletion of predatory fish communities. Nature 423:280–283
37. Pauly D, Christensen V, Dalsgaard J, Froese R, Torres F Jr (2005) Fishing down marine food webs. Science 279:860–863
38. Primavera JH (2005) Mangroves, fishponds, and the quest for sustainability. Science 310:57–59
39. Hites RA, Foran J, Carpenter DO, Hamilton MC, Knuth BA, Scwager SJ (2004) Global assessment of organic contaminants in farmed salmon. Science 303:226–229
40. Kaeriyama M, Edpalina RR (2004) Evaluation of the biological interaction between wild and hatchery population for sustainable fisheries management of Pacific salmon. In: Leber KM et al (eds) Stock enhancement and sea ranching, 2nd edn. Blackwell Publishing, Oxford, pp 247–259
41. Miyakoshi Y, Urabe H, Saneyoshi H, Aoyama T, Sakamoto H, Ando D, Kasugai K, Mishima Y, Takada M, Nagata M (2012) The occurrence and run timing of naturally spawning chum salmon populations in northern Japan. Environ Biol Fish 94:197–206

INDEX